BISHOP BURTON LRC
WITHDRAWN

ACCESSION No. T057250

CLASS No. 333.72

FARMING, FORESTRY AND THE NATURAL HERITAGE
Towards a more integrated future

THE NATURAL HERITAGE OF SCOTLAND

Each year since it was founded in 1992, Scottish Natural Heritage has organised or jointly organised a conference which has focused on a particular aspect of Scotland's natural heritage. The papers read at the conferences, after a process of refereeing and editing, have been brought together as a book. The thirteen titles already published in this series are listed below (No. 6 was not based on a conference).

1. *The Islands of Scotland: a Living Marine Heritage*
   Edited by J.M. Baxter and M.B. Usher (1994), 286pp.

2. *Heaths and Moorlands: a Cultural Landscape*
   Edited by D.B.A. Thompson, A.J. Hester and M.B. Usher (1995), 400pp.

3. *Soils, Sustainability and the Natural Heritage*
   Edited by A.G. Taylor, J.E. Gordon and M.B. Usher (1996), 316pp.

4. *Freshwater Quality: Defining the Indefinable?*
   Edited by P.J. Boon and D.L. Howell (1997), 552pp.

5. *Biodiversity in Scotland: Status, Trends and Initiatives*
   Edited by L.V. Fleming, A.C. Newton, J.A. Vickery and M.B. Usher (1997), 309pp.

6. *Land Cover Change: Scotland from the 1940s to the 1980s*
   By E.C. Mackey, M.C. Shewry and G.J. Tudor (1998), 263pp.

7. *Scotland's Living Coastline*
   Edited by J.M. Baxter, K. Duncan, S.M. Atkins and G. Lees (1999), 209pp.

8. *Landscape Character: Perspectives on Management and Change*
   Edited by M.B. Usher (1999), 213pp.

9. *Earth Science and the Natural Heritage: Interactions and Integrated Management*
   Edited by J.E. Gordon and K.F. Leys (2000), 344pp.

10. *Enjoyment and Understanding of the Natural Heritage*
    Edited by Michael B. Usher (2001), 224pp.

11. *The State of Scotland's Environment and Natural Heritage*
    Edited by Michael B. Usher, Edward C. Mackey and James C. Curran (2002), 354pp.

12. *Birds of Prey in a Changing Environment*
    Edited by D.B.A Thompson, S.M. Redpath, A. Fielding, M. Marquiss and C.A.Galbraith (2003), 570pp.

13. *Mountains of Northern Europe: Conservation, Management, People and Nature*
    Edited by D.B.A Thompson, M.F. Price and C.A.Galbraith (2005), 416pp.

This is the fourteenth book in the series.

FARMING, FORESTRY AND THE NATURAL HERITAGE
Towards a more integrated future

Edited by Richard Davison & Colin A. Galbraith

EDINBURGH: TSO SCOTLAND

© Scottish Natural Heritage 2006

First published in 2006 by The Stationery Office Limited
71 Lothian Road, Edinburgh, EH3 9AZ

Applications for reproduction should be made to Scottish Natural Heritage,
2 Anderson Place, Edinburgh EH6 5NP

British Library Cataloguing in Publication Data
A catalogue record for this book is available from the British Library

ISBN 0 11 497324 5

Cover photography: Front cover: Eildon Hills from Scott's View, near Melrose
© Laurie Campbell, Scottish Natural Heritage.

PREFACE

Farming and forestry are going through a time of significant change, with increasing economic pressures, trade liberalisation, policy evolution and growing public interest in the wildlife, landscape and recreational value of the countryside. This brings uncertainty as well as opportunities. A key opportunity is to move decisively towards integrating landscape, biodiversity and recreational objectives in land management. Looking at progress towards this goal and at ways of helping further progress was the aim of Scottish Natural Heritage's 2003 conference.

By 2003, a number of significant land management initiatives had been launched in Scotland, notably the *Scottish Forestry Strategy* in 2000 (Forestry Commission, 2000) and the *Forward Strategy for Scottish Agriculture* in 2001 (Scottish Executive, 2001). Following on from the *Forward Strategy*, the report, *Custodians of Change*, was published in 2002 (Scottish Executive, 2002) and this highlighted biodiversity, landscape and the effects of diffuse pollution as the three key environmental priorities in agriculture. In 2003, major progress was also being made towards a *Scottish Biodiversity Strategy*, and this was published in 2004 (Scottish Executive, 2004). In 2002, Scottish Natural Heritage launched *Natural Heritage Futures* (Scottish Natural Heritage, 2002) which set out natural heritage objectives for farmland and for forests and woodlands.

The Land Reform (Scotland) Act 2003 established statutory access rights to most land and inland water in Scotland, and placed responsibilities on both the public and land managers. Since then, a new *Scottish Outdoor Access Code* has been approved by the Scottish Parliament and the new access rights and responsibilities came into effect in early 2005 (Scottish Natural Heritage, 2005). Delivering practical advice on the management of access and recreation is an early priority in helping to make the new arrangements work on the ground. The last few years have also seen further changes to funding support for agriculture and rural development. Early steps towards a system of land management contracts – the best opportunity in many years to achieve real integration on the ground – have also been taken.

The conference, therefore, was a timely one with major new initiatives being announced and work beginning to focus on how to put these initiatives into practice. There are 31 chapters in this volume which indicates the scale of the subject area covered by the conference. Some chapters review current knowledge of the effects of farming and forestry on biodiversity, landscape and recreation, whilst others look at the ways in which farmers and other land managers can be supported and what the advisory and research needs will be in the future. In putting together the conference, Scottish Natural Heritage was keen to look at a number of case studies that helped to indicate what progress was being made towards integration and what the challenges were to making better progress. This book looks at 11 case studies, some written by academic researchers and others who are working on the ground to achieve integration. All are relevant in getting us to think about the challenges ahead.

As indicated above, change continued apace since the conference was held in November 2003 and, consequently, some chapters may feel slightly dated already. However, the book provides a good deal of information on where we had got to by the end of 2003 and illustrates the steps being taken to move towards better integration of natural heritage objectives in land management. The challenges of integration remain the same and so the chapters remain relevant.

This book brings together the majority of the contributions at the conference. A conference and a publication of this nature inevitably involves a tremendous amount of work by many people. In planning for the conference, we should like to give particular thanks to Daniel Gotts, who chaired the conference organising group and provided valuable support, and to Helen Forster, who worked tirelessly on the administration of the conference. Many other staff, including Jenny Bryce, Ralph Blaney, Claudia Rowse, Barbara Bremner, Duncan Stone, Nigel Buchan, Alan Macpherson and Sylvia Conway helped to make the conference a success. Three site visits were held at the end of the conference and special thanks go to Charlie Taylor (Forest Enterprise), Andrew Barbour and Polly Freeman (Atholl Estates), and Ian Crawford for hosting these. The conference was held at the Pitlochry Theatre, which is a wonderful venue, and we thank all of the staff who work there.

The speakers at the conference, and the people who stimulated the conference discussions, have contributed a great deal to this book. However, we owe a special 'thank you' to the unseen and generally anonymous referees. As with previous volumes in the *Natural Heritage of Scotland* series, the chapters have generally been sent to two referees – some to more than two, some to fewer. It is the critical comments of the referees that assist in achieving a high quality publication, and although the referees are unnamed it is their hard work that, as editors, we really appreciate.

Particular thanks also need to be given to Tim Reed and Lissie Wright of Ecotext Limited. They provided valuable editorial support in overseeing comments from referees and then securing the necessary changes to the chapters. We also owe a tremendous thanks to Jo Newman who has worked extremely hard behind the scenes in organising referees, chasing reports and papers, and finalising manuscripts. Without this support, this book would not have been published.

Richard Davison and Colin Galbraith
Scottish Natural Heritage

June 2006

References

Forestry Commission (2000). *Forests for Scotland: The Scottish Forestry Strategy*. Forestry Commission, Edinburgh.
Scottish Executive (2001). *A Forward Strategy for Scottish Agriculture*. The Stationery Office, Edinburgh.
Scottish Executive (2002). *Custodians of Change*. Agriculture and Environment Working Group, Scottish Executive, Edinburgh.
Scottish Executive (2004). *Scotland's Biodiversity: It's In Your Hands. A Strategy for the Conservation and Enhancement of Biodiversity in Scotland*. Scottish Biodiversity Forum, Scottish Executive, Edinburgh.
Scottish Natural Heritage (2002). *Natural Heritage Futures*. Scottish Natural Heritage, Perth.
Scottish Natural Heritage (2005). *Scottish Outdoor Access Code*. Scottish Natural Heritage, Perth.

CONTENTS

Preface		v
Contents		vii
Foreword		x
Contributors		xi
PART ONE: SETTING THE SCENE		1
1.	The contribution of farming and forestry to our natural heritage Margaret Gill, Bob McIntosh, Deborah Slater & Iain Wright	5
2.	Key changes in forestry: current situation and future prospects Martin Gale	17
3.	What does the public want from the countryside? Steve Sankey	25
PART TWO: FARMING, FORESTRY AND BIODIVERSITY		35
4.	How do farming practices influence biodiversity? Nigel Boatman	39
5.	The influence of forest and woodland management on biodiversity in Scotland: recent findings and future prospects Jonathan Humphrey, Chris Quine & Kevin Watts	59
6.	The Scottish Biodiversity Strategy: developing the Rural Land Use Implementation Plan John Henderson	77
7.	Advisory and planning tools to inform natural heritage management Daniel Gotts	85
PART THREE: FARMING, FORESTRY AND ACCESS		97
8.	The Land Reform (Scotland) Act 2003 and the draft Scottish Outdoor Access Code: what does it mean for land managers? Marian Silvester	101

9. Agri-environment schemes, forestry schemes and access – are these policies working together?
   Steve Hunt — 109

10. Helping farmers and foresters to manage public access
    Ken Taylor, Peter Scott & Vyv Wood-Gee — 117

PART FOUR: FARMING, FORESTRY AND LANDSCAPE — 129

11. The role of Landscape Character Assessment in farming, forestry and the natural heritage
    Carys Swanwick & Julie Martin — 133

12. Supporting landscape design through guidance and management strategies
    Sue Bennett — 147

13. From past to present: understanding and managing the historic environment
    Lesley Macinnes — 157

14. Scotland's future lowland landscapes: encouraging a wider debate
    Simon Brooks — 171

15. Towards integrated management: future landscape research and advisory needs
    Alison Farmer — 179

PART FIVE: MOVING TOWARDS INTEGRATION: SOME CASE STUDIES — 189

16. From forest patches to functional habitat networks: the need for holistic understanding of ecological systems at the landscape scale
    P. Angelstam, J. Törnblom, E. Degerman, L. Henrikson, L. Jougda, M. Lazdinis, J.C. Malmgren & L. Myhrman — 193

17. Perceptions and realities: the motivations and practices of farmers within Nitrate Vulnerable Zones
    Colin J. Macgregor & Charles R. Warren — 211

18. The 4 Point Plan: helping livestock farmers to reduce diffuse pollution risk
    Rebecca Audsley — 217

19. The Royal Highland Education Trust
    Lindsey Gibb — 221

20.	Integrating access and forestry: the 7stanes case study Karl Bartlett	225
21.	Scottish Borders LBAP project: farming for partridges and sparrows Andy Tharme	233
22.	Determining the effects of grazing on moorland birds: a summary of work underway at the RSPB Murray Grant, James Pearce-Higgins, Graeme Buchanan & Mark O'Brien	239
23.	The Breadalbane Initiative for Farm Forestry Ruth Anderson	245
24.	Hill farming and environmental objectives: conflict or consensus? C. Morgan-Davies, A. Waterhouse, K. Smyth & M.L. Pollock	249
25.	The Pontbren Farmers' Group David Jenkins	255
26.	A brief review of Land Management Contracts and their possible use in the Cairngorms National Park Fiona Newcombe	261
27.	European rural development policies and the Land Use Policy Group Ralph Blaney & Maria de la Torre	269

PART SIX: LOOKING TO THE FUTURE: ACHIEVING BETTER INTEGRATION — 273

28.	The European policy and funding horizon and its implications for integration David Baldock	277
29.	A countryside for everyone: towards multi-benefit land use John Thomson	285
30.	Achieving integration on the ground: a view from the land Robert Balfour	291
31.	The main priorities for research and advice on farming, forestry and the natural heritage in the next 10 years Colin A. Galbraith & Richard Davison	297

Index — 305

FOREWORD

The Scottish Natural Heritage Conference 2003 was the twelfth in our annual series. For this conference, the theme was very wide: faming, forestry and the natural heritage. At its centre was the desire to look at how the delivery of new and challenging public policies on farming, forestry, biodiversity, landscape and access could be done in a more integrated way.

Many of the building blocks for achieving better integration of farming, forestry and the natural heritage already exist, most notably in terms of our knowledge of the impacts of farming and forestry on the natural heritage and the publication of new strategies for agriculture, forestry and biodiversity. There remain two particular challenges for us to overcome. First, our vision for the future remains incomplete until it fully encompasses landscape and access. Second, securing delivery of integration on the ground requires the human dimension – people's perceptions and needs – to be addressed. It also requires the focus to be at a landscape scale rather than just at an individual farm level. So much more could be achieved by working at a larger scale, including better prospects for maintaining viable wildlife populations, for co-ordinating access provision and management, and for conserving and improving the distinctive landscapes of an area.

This conference was an ambitious one in looking at farming, forestry, biodiversity, landscape and access, but it did help us to review progress towards meeting these challenges. Indeed, many people at the conference were already addressing them. I hope that the publication of this conference volume in the *Natural Heritage of Scotland* series will help everyone to take stock of the good progress made so far and to push towards better integration on the ground.

John Markland
Chairman, Scottish Natural Heritage

June 2006

CONTRIBUTORS

Ruth Anderson, Breadalbane Initiative for Farm Forestry, Dundavie, Glen Fincastle, Pitlochry, PH16 5RN.

Per K. Angelstam, School for Forest Engineers, Faculty of Forest Sciences, Swedish University of Agricultural Sciences, SE-739 21 Skinnskatteberg, Sweden.

Rebecca Audsley, Scottish Agricultural College, Farm Business Services, Auchincruive, KA6 5HW.

David Baldock, Institute for European Environmental Policy, Fifth Floor, Dean Bradley House, 52 Horseferry Road, London, SW1P 2AG.

Robert Balfour, Pitillock Farm, Freuchie, Fife, KY15 7JQ.

Karl Bartlett, 7stanes Project, 55 Moffat Road, Dumfries, DG1 1NP.

Sue Bennett, Planning and Environment, Dumfries and Galloway Council, Newall Terrace, Dumfries, DG1 1LW.

Ralph Blaney, Scottish Natural Heritage, Great Glen House, Leachkin Road, Inverness, IV3 8NW.

Nigel D. Boatman, Central Science Laboratory, Sand Hutton, York, YO41 1LZ.

Simon Brooks, Scottish Natural Heritage, Great Glen House, Leachkin Road, Inverness, IV3 8NW.

Graeme Buchanan, RSPB Scotland, Dunedin House, 25 Ravelston Terrace, Edinburgh, EH4 3TP.

Richard Davison, Scottish Natural Heritage, Battleby, Redgorton, Perth, PH1 3EW.

Erik Degerman, National Board of Fisheries Regional Office, Pappersbruksallén 22, SE-702 15 Örebro, Sweden.

Alison Farmer, Alison Farmer Associates, 50 Cambridge Place, Cambridge, CB2 1NS.

Colin A. Galbraith, Scottish Natural Heritage, 2 Anderson Place, Edinburgh, EH6 5NP.

Martin Gale, c/o Director General's Office, Forestry Commission Scotland, Silvan House, 231 Corstorphine Road, Edinburgh, EH12 7AT.

Lindsey Gibb, Scottish Natural Heritage, Battleby, Redgorton, Perth, PH1 3EW.

Margaret Gill, The Macaulay Institute, Craigiebuckler, Aberdeen, AB15 8QH (current address: Scottish Executive, Environment & Rural Affairs Department, Pentland House, 47 Robb's Loan, Edinburgh, EH14 1TY).

Daniel Gotts, Scottish Natural Heritage, 2 Anderson Place, Edinburgh, EH6 5NP (current address: Scottish Executive, Pentland House, 47 Robb's Loan, Edinburgh, EH14 1TY).

Murray Grant, RSPB Scotland, Dunedin House, 25 Ravelston Terrace, Edinburgh, EH4 3TP.

John Henderson, Scottish Executive Environment & Rural Affairs Department, Pentland House, 47 Robb's Loan, Edinburgh, EH14 1TY.

Lennart Henrikson, WWF-Sweden, Ulriksdals slott, SE-170 81 Solna, Sweden.

Jonathan Humphrey, Forest Research, Northern Research Station, Roslin, Midlothian, EH25 9SY.

Steve Hunt, Farming and Wildlife Advisory Group Scotland, Algo Business Centre, Glenearn Road, Perth, PH2 0NJ.

David Jenkins, Coed Cymru, The Old Saw Mill, Tregynon, Newton Powys, SY16 3PL.

Leif Jougda, Swedish Forest Agency, Volgsjövägen 27, SE-912 32 Vilhelmina, Sweden.

Marius Lazdinis, Vilnius Law University, Ateities str. 20, LT-2057 Vilnius, Lithuania.

Colin J. Macgregor, School of Geography and Geosciences, University of St Andrews, St Andrews, Fife, KY16 9AL.

Lesley Macinnes, Historic Scotland, Longmore House, Salisbury Place, Edinburgh, EH9 1SH.

Jan C Malmgren, JM Natur – Ecological Restoration and Conservation, Baldergatan 48, SE-692 32 Kumla, Sweden.

Julie Martin, Julie Martin Associates, The Round House, Swale Cottage, Station Road, Richmond, North Yorkshire, DL10 4LU.

Bob McIntosh, Forestry Commission Scotland, Silvan House, 231 Corstorphine Road, Edinburgh, EH12 7AT.

C. Morgan-Davies, Scottish Agricultural College, Sustainable Livestock Systems Group, Hill & Mountain Research Centre, Kirkton Farm, Crianlarich, FK20 8RU.

Lennart Myhrman, Ludvika Municipality, SE-771 82 Ludvika, Sweden.

Fiona Newcombe, Cairngorms National Park Authority, 14 The Square, Grantown-on-Spey, Moray, PH26 3HG.

Mark O'Brien, RSPB Scotland, Dunedin House, 25 Ravelston Terrace, Edinburgh, EH4 3TP.

James Pearce-Higgins, RSPB Scotland, Dunedin House, 25 Ravelston Terrace, Edinburgh, EH4 3TP.

M.L. Pollock, Scottish Agricultural College, Sustainable Livestock Systems Group, Hill & Mountain Research Centre, Kirkton Farm, Crianlarich, FK20 8RU.

List of Contributors

Chris Quine, Forest Research, Northern Research Station, Roslin, Midlothian, EH25 9SY.

Steve Sankey, Gerraquoy, St. Margaret's Hope, South Ronaldsay Orkney, KW17 2TH.

Peter Scott, Peter Scott Planning Services Ltd., 86 Cammo Grove, Edinburgh, EH4 8HD.

Marian Silvester, Scottish Rural Property & Business Association, Stuart House, Eskmills Business Park, Musselburgh, EH21 7PB.

Deborah Slater, The Macaulay Institute, Craigiebuckler, Aberdeen, AB15 8QH.

K. Smyth, Scottish Agricultural College, Land Economy Group, West Mains Road, Edinburgh, EH9 3JG.

Carys Swanwick, Department of Landscape, University of Sheffield, Third Floor, Arts Tower, Western Bank, Sheffield, S10 2TN.

Ken Taylor, Asken Ltd., 17 Hayfell Rise, Kendal, Cumbria, LA9 7JP.

Andy Tharme, Scottish Borders Council, Scottish Borders Council Headquarters, Newtown St Boswells, TD6 0SA.

John Thomson, Scottish Natural Heritage, Caspian House, 2 Mariner Court, 8 South Avenue, Clydebank Business Park, Clydebank, G81 2NR.

Johan Törnblom, School for Forest Engineers, Faculty of Forest Sciences, Swedish University of Agricultural Sciences, SE-739 21 Skinnskatteberg, Sweden.

Maria de la Torre, Scottish Natural Heritage, Great Glen House, Leachkin Road, Inverness, IV3 8NW.

Charles R. Warren, School of Geography and Geosciences, University of St Andrews, St Andrews, Fife, KY16 9AL.

A. Waterhouse, Scottish Agricultural College, Sustainable Livestock Systems Group, Hill & Mountain Research Centre, Kirkton Farm, Crianlarich, FK20 8RU.

Kevin Watts, Forest Research, Alice Holt Research Station, Wrecclesham, Farnham, Surrey, GU10 4LH.

Vyv Wood-Gee, Shortrigg, Hoddom, Lockerbie, DG11 1AW.

Iain Wright, The Macaulay Institute, Craigiebuckler, Aberdeen, AB15 8QH.

Farming, Forestry and the Natural Heritage: Towards a More Integrated Future

PART 1:
Setting the Scene

Farmland near Bankfoot, Perthshire © Lorne Gill, Scottish Natural Heritage

Farming, Forestry and the Natural Heritage: Towards a More Integrated Future

PART 1:
Setting the Scene

Farming and forestry have seen major changes over the last few decades and, more recently, there has been growing public interest in the wildlife, recreational and landscape value of the countryside. Given the wide-ranging nature of the subject, the first three chapters play an important role in setting the scene by summarising the key changes to the countryside, and the views of the public.

In the opening chapter, Gill *et al.* summarise the main drivers of land use change in recent decades and look at the evidence of impacts on the natural heritage. We are reminded that much of the natural heritage that we see around us today is the result of how the land is used for farming and forestry, and that although farming, forestry and the natural heritage are increasingly being managed in an integrated way, this has not always been the case. The chapter looks at how we might forecast future impacts and what challenges might be faced in moving towards a more integrated approach. Crucially, Gill *et al.* stress that how people manage, interact with, and appreciate the countryside will be increasingly important in delivering this approach.

In Chapter 2, Gale provides a wide-ranging view of the changes to forestry. With forestry, these changes can take place over a long time so adopting a long-term approach to planning is essential. The chapter looks at changes and trends at the global, European and Scottish scales and concludes by setting out a vision for forestry in Scotland in 2050. This vision sees forestry meeting three key needs: the continuing supply of wood for increasingly sophisticated processing industries; supporting the natural heritage through continued restructuring, through fitting forests into the landscape and through more native woodland conservation and expansion; and the provision of more opportunities for recreation.

The needs of the public are the focus of Chapter 3 by Sankey. This chapter summarises key changes in agriculture, forestry, biodiversity and access over the last few decades and interprets them from the perspective of what the public wants from the countryside. After reviewing surveys of the general public, Sankey argues that people want environmentally responsible production in return for the significant levels of financial support provided. He concludes that land management contracts can be used to show how each farmer will deliver this. In short, multi-functional forestry is now settled policy and practice, and what we now need to secure is multi-functional agriculture.

Farming, Forestry and the Natural Heritage: Towards a More Integrated Future

1 The contribution of farming and forestry to our natural heritage

Margaret Gill, Bob McIntosh, Deborah Slater & Iain Wright

Summary

1. Farming and forestry are an integral part of our natural heritage, having exerted impacts over millennia.
2. Single objectives for farming (food production) and forestry (development of a strategic timber reserve) in the last century led to conflict with environmentalists, but the current multiple objectives for both farming and forestry overlap with the aims of environmentalists and, if sufficient knowledge of cause and effect is available, then it should be possible to minimise such conflict in future.
3. Increasingly, it is recognised that we need to know not just about the natural environment but also to understand the human dimension of land use change – to integrate the social and the natural sciences.
4. We also need to know not just about the visible impacts, such as those on the landscape, but also the less visible impacts such as on water quality and soil microbes.
5. Key challenges in the context of the Forward Strategy for Scottish Agriculture and the Scottish Forestry Strategy include understanding the decision making processes of those turning policy into practice and integrating the local effects of practice at the field/tree scale to forecast the impact on the wider environment.

1.1 Introduction

The title of this chapter encompasses the breadth of contributions in this book! However, the approach adopted here has the subtitle of 'the part played by farming and forestry in shaping Scotland's natural heritage as it is now'. Up until recently, single objectives for farming and forestry led to conflict with environmentalists. Today, much time and effort is being invested in agreeing multiple objectives for farming, forestry and the natural heritage.

This chapter considers some of the main drivers of land use change in recent decades and looks at the evidence of impacts on the natural heritage. It draws attention to some of the potential constraints to forecasting future impacts and concludes by exploring some of the key challenges facing different communities in adopting a more integrated approach.

The document *Sustainable Development and the Natural Heritage: The SNH Approach* (Scottish Natural Heritage, 1993) gave the definition of the 'natural heritage of Scotland' from the Natural Heritage (Scotland) Act 1991 as including 'the flora and fauna of Scotland, its geological and physiographical features, its natural beauty and amenity ...'. 'Natural heritage' therefore embraces the combination and interrelationships of landform, habitat, wildlife and landscape and their potential for enjoyment'. However, it went on to recognise that 'The natural heritage has cultural and historic dimensions which must also be considered, because there is not much which is pristine about the natural heritage in Scotland' (Scottish Natural Heritage, 1993, p. 1).

1.2 Drivers of land use change
1.2.1 Demand for food and timber

Farming started in the UK many thousands of years ago when hunter-gatherers realised the benefits of cultivating crops. In the last two centuries, a growing population drove a demand for food, which was not comfortably met in the UK until the latter half of the 20th century (Blaxter & Robertson, 1995). Food surpluses started to appear in the 1970s in Western Europe, although on a global scale over 800 million people remain malnourished. Despite this overproduction, subsidies were still paid to UK agriculture and it took some time for the emphasis of both policy and practice on food production to change. This undoubtedly fuelled the controversy with environmentalists, since the subsidies supported farming systems which produced unnecessary quantities of food, and had potentially negative impacts on the environment, which might have been avoided by using lower input systems. The demand for specific types of foods is known to change with income. For example, in India and China, in recent years, per capita income has been a major determinant of per capita demand for meat (Delgado et al., 1999), presumably reflecting trends which happened in Europe in the past. In recent times, demand for organic produce has increased (Kristensen & Thansborg, 2002) and there is renewed interest in some parts of Scotland for locally produced food sold in farm shops or at farmers' markets (Scottish Executive, 2001). However, such niche markets are unlikely to dominate food production systems in the way that demand for food did in Europe in the past.

In terms of forestry, the natural forests of Great Britain were destroyed well before 1900, while the early part of the 20th century saw a Government-inspired afforestation programme, largely involving fast growing exotic conifer species on land that was not valuable for agricultural production, with the aim of creating a strategic reserve of timber. The successful, but single-minded, implementation of this programme brought forestry into increasing conflict with the natural heritage, not least because the newly created plantations were often created on upland habitats considered to have nature conservation value (Nature Conservancy Council, 1986). Little attention was paid to forest design principles and sometimes the newly created plantations had adverse landscape impact.

1.2.2 Demographics

On a global scale, the human population is still increasing and passed 6 billion during 1999 (FAO, 2002). However, on a more local scale, the population of Scotland is on a downward trend, from 5.23 million in 1971 to 5.1 million in 1991, (Copus et al., 1999) to 5.06 million in 2001, with predictions of a continued downwards trend to 4.96 million by 2021 (General Register Office for Scotland, undated). The trends in the rural population, on the other hand, are the reverse. A previous downwards trend in rural population was reversed during the 1980s (Gelan, 2003).

We have examined data for two rural case study areas: the Cairngorms, in which population increased from 6,488 in 1971 to 7,850 in 1991 (Copus et al., 1999) and the Formartine parish in the Ythan catchment, where population increased from 20,890 to 34,570 over the same period. However, in order to understand the impact of even local population numbers on land use, it is necessary to disaggregate the changes to identify the degree of involvement of people living in rural communities with agriculture. Figure 1.1 illustrates the decreasing trend of people in direct employment in agriculture, and a slight decrease in full-time occupiers, but a marked increase in the last decade in part-time occupiers according to June Agricultural Census data for 1982, 1992 and 2002 (Scottish Executive, 2003). The amount of time people living on farms are able to spend on land management will have a major impact on how they manage the land.

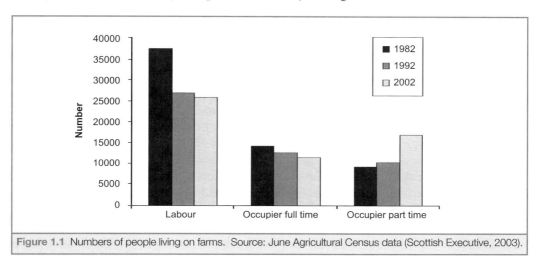

Figure 1.1 Numbers of people living on farms. Source: June Agricultural Census data (Scottish Executive, 2003).

1.2.3 Impact of policies

In researching this chapter, it was interesting to look back at the range of policies that have been put in place to ensure that people were fed and timber was available. As far back as the reign of King James I, there was an Act of Parliament in 1426 relating to the 'Compulsory sowing of wheat, peas and beans' (Symon, 1959). If farmers now feel that they do not have the opportunity to do what they want with their land, it would appear that they did not in centuries past either! Other examples are highlighted in Table 1.1. However, the key policies impacting on present day farming and forestry were introduced in the last century (Blaxter & Robertson, 1995).

Table 1.1 Examples of historical policies relating to farming and forestry. Source: Symon (1959).

Date	Policy
1424	Destruction of trees forbidden
1426	Compulsory sowing of wheat, peas and beans
1551	Killing of young lambs forbidden
1587	Destroyers of corn to be treated as thieves
1641	Punishment for destroyers of trees and enclosures
1669	Fencing of arable land alongside roads
1703	Butchers forbidden to be graziers
1703	Irish imports of grain forbidden

1.2.4 Agricultural policies

Blaxter & Robertson (1995) drew attention to the Wheat Act of 1932, and the Cattle Industry Act of 1934, as being the pivotal policies in stimulating the modern revolution in food production, policies which arose as a response to the difficulties in which the farming industry found itself at that time. However, the Agriculture Act of 1947 was the policy which started economic support, through 'promoting and maintaining … a stable and efficient agricultural industry capable of producing such part of the nation's food and other agricultural produce as in the national interest it is desirable to produce in the United Kingdom.' This Act allowed for fixed prices and assured markets of 11 commodities, making up 80% of the industry. This was modified by the Agriculture Act of 1957, so that guarantees only applied to the amount of produce which the Government thought fit to support.

The principles underlying these Acts were sustained when the UK entered the European Community, and the concept of subsidizing food production has continued until now. The emphasis on food production changed during this time in response to oversupply, with the introduction of, for example, payments for the set-aside of land. But it is only now, with the reform of the Common Agriculture Policy (CAP), in response to enlargement of the European Union, the rising cost of the CAP and concerns about the environmental impact of agriculture, that there is a real opportunity for bringing the objectives of the farming industry closer to those of environmental agencies. In particular, the shift in funding from so-called Pillar 1 support (market support and direct payments) to Pillar 2 support for environmental maintenance and enhancement and wider rural development initiatives, is the main mechanism by which the European Union hopes to achieve this (European Commission, 1999, 2004).

1.2.5 Forestry policies

By the beginning of the 20th century, woodland cover had been progressively reduced to an all-time low of 5%. The difficulty in sourcing timber during the First World War led to a decision by Government to establish the Forestry Commission by means of the Forestry Act 1919 (Pringle, 1994). The Forestry Commission was given the remit of establishing a strategic reserve of timber as a matter of national security. The aim was to achieve this both by direct action (such as the acquisition and planting of trees) by the Forestry Commission itself and by providing incentives to private owners to manage existing woodlands and create new ones. This was very much a pioneering stage with new techniques being developed. The objective was fairly focussed on getting as much timber growing as soon as possible. It was recognised that this would bring into production land that was hitherto considered unproductive, and that jobs in rural areas would be created. There was little interest in, or concern about, the landscape and nature conservation implications of the programme (Nature Conservancy Council, 1986). The policy was successful, in that in just 80 years more than 1.5 million ha of new woodlands were created, doubling the area of woodland in Great Britain from 5% to 10% of land area (Forestry Commission, 2003).

Concern about the preponderance of exotic conifers was being expressed as early as the 1930s (Pringle, 1994), but the Second World War again drew attention to the requirement for home supplies of timber to be available. Formal recognition of the need for forestry policy to give due regard to the effects of forestry on the wildlife and beauty of the countryside therefore had to wait until 1957 when, through the Zuckerman report on Forestry, Agriculture and Marginal Land (Zuckerman, 1957), the Government acknowledged that creating a strategic reserve of timber was no longer the central plank of forestry policy. In deciding where trees should be planted, the Government determined that special attention should be paid to upland areas, particularly in Scotland and Wales, where expansion of forestry would provide much needed diversification of employment and important social benefit. This signal, coupled with the advent of private forestry companies and allied to the benefits available to anyone investing in forestry, led to significant afforestation programmes throughout the 1960s, 1970s and 1980s in upland areas. This led to many conflicts between forestry interests and those interested in landscapes and nature conservation (Nature Conservancy Council, 1986; Gee & Stoner, 1988). One of the most high profile issues was the so-called 'Flow Country debate' where an increase in afforestation activity by private forestry companies in the peatlands of Caithness and Sutherland brought into sharp focus the potential detrimental effects of afforestation on upland habitats (Stroud *et al.*, 1987). This issue raged for several years and was instrumental in changes to the tax regimes in the 1988 budget, which effectively brought to an end large-scale planting programmes in the uplands.

Current forest policy is rooted in principles of sustainable forest management, with forests expected to deliver a mixture of social, economic and environmental outputs. Since devolution in the late 1990s, the development of forest policy has been devolved and the administrations in Scotland, England and Wales all have a Forestry Strategy. These set out the key policies which will ensure that forestry remains politically, and socially, relevant, and that it delivers outputs and outcomes of relevance to the wider rural development agenda and to the delivery of other Government agendas. All three strategies envisage an increase in the area of forest, and increasing diversification of existing forests, to achieve the delivery of an appropriate balance of social, economic and environmental outcomes (Scottish Executive, 2000).

1.3 The impacts of land use on the natural heritage
1.3.1 Farming practice

Agriculture contributed approximately £2 billion to the Scottish economy in 2003, employing 68,000 people (Scottish Executive, 2004). A recent study looking at agriculture's contribution to Scottish society, economy and environment (Slee *et al.*, 2001) found quantitative measures of the costs but not of the benefits of agriculture. Identified benefits included landscape and aesthetic value, water accumulation and supply, nutrient recycling and fixation, soil formation, wildlife, storm protection and flood control, and carbon sequestration. The costs were not available at a Scottish level, but in terms of the UK as a whole the negative impacts on natural capital (i.e. water, air, soil, landscape and biodiversity) were estimated by Pretty *et al.* (2000) to be £1.7 billion per year, with £1.1 billion being attributed to emissions to the atmosphere, £450.0 million for water pollution, £96.0 million for erosion and organic matter loss from soil, and about £125.0 million for loss of biodiversity and landscape features. However, these estimates are based on gross assumptions, many of which are open to challenge, as are discussions on who should pay. As many of the environmental consequences of farming can be traced back to public demand for cheap food, the question needs to be asked as to where responsibility for environmental impact associated with the production of food stops – is it with farmers, policy-makers or consumers?

One practice which has made a major contribution to increased food production is the use of chemicals. Prior to the Second World War, agriculture in lowland areas was dominated by mixed cropping and livestock systems, with rotations of crops with grassland used for livestock production. The months under grass allowed the accumulation of soil organic matter and nitrogen. The ploughing of the ley led to mineralization of the soil organic matter, providing nutrients for the subsequent crop. Crop rotation also contributed to the prevention of the build up of weeds, pests and diseases. The development of artificial fertilizers, along with herbicides and pesticides, made intensification of cereal farming possible, as well as the more intensive use of grassland (Wilkins, 2000). Until the 1940s, the use of fertilizers in the UK remained low, with phosphatic fertilizers being the main fertilizers used, and nitrogen accounting for only 15% of total nutrients supplied in the first quarter of the 20th century (Wilkins, 2000). In the 1940s, the average input of nitrogen fertilizer to grassland in the UK was less than 5 kg ha^{-1} yr^{-1}, but increased steadily from then until a peak in the mid-1980s. Since then, fertilizer use has decreased as awareness of the extent of diffuse pollution has increased and this has been followed by legislation. However, recent research such as Edwards *et al.* (2003) suggests that there is still much to learn about the relationship between nitrogen cycles and ecological changes in catchments. As more information becomes available, policies may need to be adjusted to have the desired effect.

In the past 50 years or so, the intensification of agriculture has had a significant effect through two major changes. Firstly, there has been a reduction in the area of various semi-natural habitats associated with mixed farming. Secondly, there has been an increase in the intensity of management of those areas under cropping or grassland (Birnie *et al.*, 2002). A trend towards more continuous cropping, and a shift from spring to autumn cereals, along with herbicide and pesticide application, has decreased overall biodiversity and reduced the numbers of farmland birds (Wilson *et al.*, 1999), often seen as an indicator of wider biodiversity.

In the hills of Scotland, changes in livestock numbers and breeds are the main factors for change. Recent decades have seen a decline in the number of cattle grazing hill land, with some concerns being expressed about the environmental consequences. Research at the Macaulay Institute and elsewhere has shown that grazing by cattle, as opposed to sheep, can lead to a reduction in the cover of, for example, mat grass (*Nardus stricta*) with a concomitant increase in the cover of a variety of broad- and fine-leaved grasses and dicotyledons (Grant *et al.*, 1985, 1996). Such changes in the floristic composition, and more specifically the structure of vegetation, also influence the faunal diversity, and the invertebrate populations in particular (Dennis *et al.*, 2001), which, in turn, can influence the number and species of birds. However, it may not only be species but also breeds that have a differential impact on vegetation. There has been a trend over the past few decades towards much greater uniformity in the breeds of sheep, and especially cattle, in Scottish agriculture, with the importation of non-native breeds into the UK, particularly from continental Europe, replacing traditional breeds. A handful of breeds now accounts for the majority of agricultural production. The numbers of individual animals of some traditional Scottish breeds have declined to such an extent that they are now classified by the Rare Breeds Survival Trust as 'minority' or even 'rare' (Rare Breeds Survival Trust, 1994).

Although there is considerable anecdotal evidence that breeds may differ in their foraging behaviour, little research has been conducted on this topic.

1.3.2 Forestry practice

Forestry practice has changed markedly over the last 20 to 30 years in response to the changing drivers of forest policy. Two of the key changes have been a virtual cessation of afforestation on upland habitats, and the recognition of the nature conservation value of these habitats. There has thus been a move towards new forests and woodlands being created on surplus agricultural land, such as marginal arable land, where conflicts with nature conservation are minimal or non-existent and where the amenity, landscape and biodiversity impacts from the planting of mixed woodlands are generally positive.

There has also been much interest and action in the restoration and expansion of native woodland remnants and the creation of new native woodlands. In Scotland, the proportion of native species in the total woodland resource, by area, is now 21%, but the proportion of native species used in grant-aided afforestation schemes between 1999 and 2003 was 60%, indicating a significant trend towards re-introduction of native species. In each country the Forestry Commission administers a grants scheme which provides support for woodland creation and management activities, which help to deliver the Forestry Strategies of England, Scotland and Wales.

There has also been a major programme of restructuring of the forests created in the early part of the 20th century through a sophisticated process of forest design planning. Associated with this has been public consultation to ensure a gradual conversion from even-aged plantations to forests, with structural and species diversity, which are carefully landscaped and which take account of opportunities to conserve and enhance key species and habitats, as well as increasing the general level of biodiversity (Forestry Commission, 1998).

1.4 Case Studies

To investigate the links between policy, practice and impact, annual AgCensus (available on the Scottish Executive's Agriculture and Fisheries Statistics page at www.scotland.gov.uk/Topics/Statistics/15631/8884) data for two case study areas were compared over a 20 year period.

In the Ythan catchment, the area planted as oilseed rape increased from zero in 1982 to a maximum of 8% of the farmed area in 1992 in response to Arable Aid payments, decreasing again to 5% in 2002, when the payments for cereals and oilseed rape were the same. Over the same time period, the area of grass decreased steadily, corresponding to marked decreases in total cattle numbers (Table 1.2). Total sheep numbers increased between 1982 and 1992 and decreased thereafter. In the Cairngorms, the total area of grass increased, and this was reflected by an increase in grass for mowing whilst rough grazing decreased. In this catchment, where the number of cattle is much less than in the Ythan, the changes were less marked, though following the same downwards trend. Sheep numbers showed the same response to policy changes as in the Ythan data.

Table 1.2 Time series data on land use and livestock numbers in the two case study areas: Ythan and Cairngorms. Source: Scottish Executive (2003).

Year	Total grass (ha)		Rough grazing (ha)		Sheep		Cattle	
	Ythan	C'gorms	Ythan	C'gorms	Ythan	C'gorms	Ythan	C'gorms
1982	45,540	24,369	3,372	372,290	108,656	235,760	130,000	42,744
1992	40,771	26,976	3,462	372,002	166,094	268,132	100,507	36,062
2002	35,723	41,445	3,038	335,506	128,714	230,562	81,970	33,363

In terms of forestry, the impact of the policy of increasing planting in the uplands can be seen in Table 1.3, where there was considerably more planting in the uplands (Cairngorms) than in the lowlands (Ythan). Comparing case study areas highlights the risks of drawing conclusions from statistics at too aggregated a level, but these brief illustrations show the potential for using the census data to learn lessons from the impacts of past policies.

Table 1.3 Time series of the area planted to woodland in the two case study areas: Ythan and Cairngorms. Source: Scottish Executive (2003).

Year	Woodland (ha)	
	Ythan	Cairngorms
1982	863	4,343
1992	1,529	5,281
2002	2,874	16,244

1.5 Forecasting future impacts – scaling up from the microbe to the mountain

The case studies indicate links between policies and land use, but there is still much to learn about the links between farming and forestry practice and environmental impact. For example, while cutting down a plantation has an obvious visual impact, there are secondary impacts resulting from opening up the understorey vegetation to light, which may set ecological processes in motion which continue for years. Forecasting the future impacts of farming and forestry practice, and hence the development of policies to facilitate the adoption of practices which will achieve positive environmental outcomes, is thus a key to future success. Three examples of research projects involving the Macaulay Institute illustrate the complexities involved.

The first, entitled 'Processes and biodiversity at different ecological levels in woodland ecosystems', is testing the hypotheses that i) diversity within one functional group is coupled to the diversity of other groups, and ii) chemical composition of a dominant ecosystem component influences diversity and function of associated communities. This research is being undertaken in a Scots pine (*Pinus sylvestris*) forest and is looking at how differences in the chemical composition of the pine needles (which appear to be genetically controlled) can have an impact on their decomposition in the soil, and in turn on the soil microbes and the surrounding flora and fauna (Iason *et al.*, 2005).

The second study (Dennis *et al.*, 2003) is a collaboration between the Macaulay Institute, the Centre for Ecology and Hydrology, Scottish Agricultural College (SAC), Biomathematics and Statistics Scotland, Royal Society for the Protection of Birds (RSPB) and the University of Aberdeen. The study is exploring how grazing management can affect the structure of grass swards, which in turn can affect the invertebrate population which lives in the sward, which in turn will affect the number and species of upland birds which feed on the invertebrates. A better understanding of these interactions will help to forecast how changes in agricultural policy can affect upland birds. However, understanding the natural science linkages alone is insufficient because the final impacts depend on the practical responses to changes in policy, which have a human dimension.

The third example, a recent study by Burton (2005) in the Huntly area of Aberdeenshire, has shown that a major factor influencing the decision on whether or not to change from the current farming system was the availability of labour provided by family members who have full time jobs elsewhere, but return to help with the peak labour requirements, such as at harvest and calving. Not all of the relevant data are captured in the statistics collected in the June AgCensus forms, indicating the need to understand fully the household characteristics which affect decision making.

1.6 Key challenges

This chapter has highlighted how policy drivers have changed over the years and how practice has responded to these drivers. In terms of current policy drivers, both *A Forward Strategy for Scottish Agriculture* (Scottish Executive, 2001), and the *Scottish Forestry Strategy* (Scottish Executive, 2000) refer to the importance of sustainability. Sustainability implies meeting multiple objectives: economic, environmental and social. Forecasting how practice will respond to multiple objectives presents particular challenges. Two challenges are highlighted below – the need to understand the factors which affect the decisions which farmers and foresters take in response to policy, and the links between the effects of turning policy into practice at the field or tree level, to the impact at the landscape scale.

1.6.1 Understanding decision-making

A Forward Strategy for Scottish Agriculture sets out how 'a more joined up approach to policy' can bring together the expertise of different interest groups to help to meet the multiple objectives simultaneously. As part of this process, there is a need for a better understanding of how people take decisions. How do landowners decide how to use their land and what information do they need in making these decisions? How do farmers and foresters decide what practices to adopt and what information do they need to do this? How do policy-makers decide how to turn European Directives into 'fit for purpose' policies for Scotland and what information do they need? The challenge will be to share that understanding, matching language to audience and being willing to listen.

1.6.2 Scaling up knowledge from the field/tree level to the wider environment

When the objectives of farming and forestry were principally about producing food and timber, the impacts of inputs such as fertiliser were measured solely in terms of yields. One consequence of addressing multiple objectives is the recognition of the need to be aware of the impact of farming practice on diffuse pollution, biodiversity and landscape value (Scottish Executive, 2002). The challenge will be to understand the interactions between farming and forestry practice, not only on yields, but also on soil microbial communities or the level of pollutants in water, to enable the forecasting of the likely consequences for other components of the environment. This is no easy task. It requires long-term commitment from individual researchers and funders, patience from agencies requiring immediate answers and trust from the public who want to know why we do not have the answers already.

Acknowledgments

We are very grateful to many other colleagues who willingly contributed data or their time to help in the preparation of this chapter. In particular, Ann Malcolm, Dick Birnie, Tony Edwards, Jim Gauld and Colin Campbell made major contributions and their help is very much appreciated. We are also grateful to the Census Branch, SEERAD-Agricultural Statistics Division for allowing access to the June Census Data, and also Aberdeenshire Council for the provision of statistics.

References

Birnie, R., Dennis, P., Dunn, S., Edwards, A., Horne, P., Hill, G., Hulme, P., Paterson, E., Langan, S. & Wynn, G. (2002). *Review of Recent UK and European Research Regarding Reduction, Regulation and Control of the Environmental Impacts of Agriculture.* Report prepared on behalf of the Agriculture and Environment Working Group. Macaulay Institute, Aberdeen.

Blaxter, K. & Robertson, N. (1995). *From Dearth to Plenty: The Modern Revolution in Food Production.* Cambridge University Press, Cambridge.

Burton, R. (2005). The drivers of agricultural land use change in Scotland over the past decade: preliminary results from Huntly, Aberdeenshire. In *The Countryside in the 21st Century – Anglo-German perspectives,* ed. by D. Schmeid & O. Wilson. *Bayreuther Geographische Arbeiten,* **26**, 121-133.

Copus, A.K., Gourlay, D., Petrie, S., Cook, P., Palmer, H. & Waterhouse, T. (1999). Land use and economic activity in possible National Park areas in Scotland. *Scottish Natural Heritage Review No. 115.*

Delgado, C., Rosegrant, M., Steinfeld, H., Ehui, S. & Courbois, C. (1999). *Livestock to 2020: The Next Food Revolution. 2020 Vision for Food, Agriculture, and Environment Discussion Paper 28.* International Food Policy Research Institute, Washington, DC. Available from www.ifpri.org/2020/dp/dp28.pdf.

Dennis, P., Young, M.R. & Bently, C. (2001). The effects of varied grazing management on epigeal spiders, harvestmen and pseudoscorpions of *Nardus stricta* grassland in upland Scotland. *Agriculture, Ecosystems and Environment,* **86**, 39-57.

Dennis, P., Evans, D., Skartveit, J., Stephen, L., Pakeman, R., Pearce-Higgins, J., Redpath, S., Mayes, R., McCracken, D., Gordon, I., Grant, M., Elston, D., Kunaver, A., Quinn, S., Benton, T., Bryant, D. & Marquiss, M. (2003). Effects of grazing management on upland bird populations: disentangling habitat structure and arthropod food supply. *Proceedings of the British Ecological Society Annual Meeting.* Manchester Metropolitan University, 9-11 September 2003. p. 52.

Edwards, A.C., Sinclair, A.H. & Domburg, P. (2003). Identification, designation and formulation of an action plan for a nitrate vulnerable zone: a case study of the Ythan catchment, NE Scotland. *European Journal of Agronomy,* **20**, 165-172.

European Commission (1999). Council Regulation (EC) No 1257/1999 of 17 May 1999 on support for rural development from the European Agricultural Guidance and Guarantee Fund (EAGGF) and amending and repealing certain Regulations.

European Commission (2004). Proposal for a Council Regulation on support for rural development by the European Agricultural Fund for Rural Development (EAFRD). Commission communication entitled 'Building our common future – Policy challenges and budgetary means of the enlarged Union 2007-13': COM(2004) 101; Bulletin of the European Union 1/2-2004, point 1.7.2.

FAO (2002). *World Agriculture: Towards 2015/2030 – Summary Report.* Food and Agriculture Organization of the United Nations, Rome.

Forestry Commission (1998). *Forest Design Planning: A Guide to Good Practice.* Forestry Commission, Edinburgh.

Forestry Commission (2003). *Annual Report and Accounts Great Britain and England.* Forestry Commission, Edinburgh.

Gee, A.S. & Stoner, J.H. (1988). The effects of afforestation and acid deposition on the water quality and ecology of upland Wales. In *Ecological Change in the Uplands*, ed. by M.B. Usher & D.B.A. Thompson. Blackwell Scientific Publications, Oxford. pp. 273-288.

Gelan, A. (2003). Commuting, migration, and rural development. Integrative modelling of biophysical, social, and economic systems for resource management solutions. In *Proceedings of MODSIM 2003: International Congress on Modelling and Simulation*, ed. by D.A. Post. MODSIM, Canberra. pp. 1493-1498.

General Register Office for Scotland (undated). *Population projections and estimates, 2001- and 2003-based.* Available from www.gro-scotland.gov.uk/statistics/library/index.html.

Grant, S.A., Suckling, D.E., Smith, H.K., Torvell, L. & Forbes, T.D.A. (1985). Comparative studies of diet selection by sheep and cattle: the hill grasslands. *Journal of Ecology*, **73**, 987-1004.

Grant, S.A., Torvell, L., Sim, E.M., Small, J.E. & Armstrong, R.H. (1996). Controlled grazing studies on *Nardus* grassland: effects of between-tussock sward height and species of grazer on *Nardus* utilization and floristic composition in two fields in Scotland. *Journal of Applied Ecology*, **33**, 1053-1064.

Iason, G.R., Lennon, J.R., Pakeman, R.J., Thoss, V., Beaton, J.K., Sim, D.A. & Elston, D.A. (2005). Does chemical composition of individual Scots pine trees determine the biodiversity of their associated ground vegetation? *Ecology Letters*, **8**, 364-369.

Kristensen, E.S. & Thansborg, S.M. (2002). Future European market for organic products. In *Organic Meat and Milk from Ruminants*, ed. by I. Kyriazakis & G. Zervas. European Association for Animal Production No. 106. pp. 5-13. Wageningen Academic Publishers, Wageningen.

Nature Conservancy Council (1986). *Nature Conservation and Afforestation in Britain.* Nature Conservancy Council, Peterborough.

Pretty, J.N., Brett, C., Gee, D., Hine, R.E., Mason, C.F., Morrison, J.I.L., Raven, H., Rayment, M.D. & van der Bijl, G. (2000). An assessment of the total external costs of UK agriculture. *Agricultural Systems*, **65**, 113-136.

Pringle, D. (1994). *The Forestry Commission – The First 75 Years.* Forestry Commission, Edinburgh.

Rare Breeds Survival Trust (1994). *Rare Breeds Facts and Figures.* Rare Breeds Survival Trust, Stoneleigh.

Scottish Executive (2000). *Forests for Scotland: The Scottish Forestry Strategy.* Scottish Executive, Edinburgh.

Scottish Executive (2001). *A Forward Strategy for Scottish Agriculture.* Scottish Executive, Edinburgh.

Scottish Executive (2002). *Custodians of Change.* Agriculture and Environment Working Group, Scottish Executive, Edinburgh.

Scottish Executive (2003). *Abstract of Scottish Agricultural Statistics 1982 to 2002.* Environment and Rural Affairs Department, Scottish Executive, Edinburgh. More information relating to the June Agricultural

Census Data can also be found at www.scotland.gov.uk/Topics/Statistics/15631/8884.

Scottish Executive (2004). *Scottish Economic Statistics, 2004.* Scottish Executive, Edinburgh.

Slee, W., Barnes, A., Thomson, K., Roberts, D.J. & Wright, I.A. (2001). *Agriculture's Contribution to Scottish Society, Economy and Environment. A Literature Review for the Scottish Executive Rural Affairs Department and CRU.* Department of Agriculture and Forestry, University of Aberdeen and Macaulay Land Use Research Institute, Aberdeen.

Scottish Natural Heritage (1993). *Sustainable Development and the Natural Heritage – The SNH Approach.* Scottish Natural Heritage, Perth.

Stroud, D.A., Reed, T.M., Pienkowski, M.W. & Lindsay, R.A. (1987). *Birds, Bogs and Forestry.* Nature Conservancy Council, Peterborough.

Symon, J.A. (1959). *Scottish Farming Past and Present.* Oliver and Boyd Ltd., Edinburgh.

Wilkins, R.J. (2000). Grassland in the Twentieth Century. *IGER Innovations*, **4**, 26-33.

Wilson, J.D., Morris, A.J., Arroyo, B.E., Clark, S.C. & Bradbury, R.B. (1999). A review of the abundance and diversity of invertebrates and plant foods of granivorous birds in northern Europe in relation to agricultural change. *Agriculture, Ecosystems and Environment*, **75**, 13-30.

Zuckerman, S. (1957). *Forestry, Agriculture and Marginal Land.* HMSO, London.

2 Key changes in forestry: current situation and future prospects

Martin Gale

Summary

1. Forestry is a long-term business, the effects of our tree planting and felling will be inherited by generations to come. It is particularly important, therefore, to look forward over extended timescales. This chapter looks ahead to Scotland's contribution to global forestry in 2050 and the wider global environmental and economic issues that might affect this, including deforestation, climate change, biodiversity and sustainability, demand for fuel wood, illegal logging and globalization. The impact of EU enlargement and CAP reform on forestry is also explored.
2. Scotland can offer unique contributions in terms of expertise, proximity to markets and special habitats. In 2050, wood production will still play a serious role in Scotland, supplying increasingly sophisticated processing industries that will require consistent, high quality timber supplies. The importance of environmental considerations including restructuring, landscaping and native woodland conservation and expansion will continue with landscape-scale ecological restoration becoming much more significant. The extent to which forests can serve the needs of the Scottish people in terms of recreation and health will perhaps be of over-riding importance.

2.1 Introduction

Crystal ball gazing is always difficult, but in a long-term business like forestry it cannot be avoided. This is because, in a very real sense, we are prisoners of the past, inheriting the trees, woods and forests that were planted by our predecessors. So we, too, owe it to our successors to think carefully about what we should bequeath them. What kind of trees, woods and forests will future Scottish generations be grateful for?

This chapter will, therefore, conclude with an attempt to look ahead, with a vision for 2050. Before doing so, the chapter examines the global context – both environmental concerns and economic realities. Then, focusing in on this part of the globe, the impact of European Union (EU) enlargement and Common Agricultural Policy (CAP) reform will be explored. Focusing in even more closely on Scotland, this chapter will then look at what Scotland's contribution can be to global forestry in the middle of the century. A vision for 2050 can then be developed.

2.2 Global context: environmental concerns

Forests cover about 30% of the earth's land area, but each year we are still losing over 9 million ha of forest (FAO, 1999), an area considerably bigger than the size of Scotland. Why does this matter? The answer is simple – we cannot isolate what we do here in Scotland from the rest of the world.

Amidst all of the press reports of post-Kyoto negotiations it is sometimes easy to lose sight of the central role of forests in the global carbon cycle equation. Evidence suggests that deforestation in the 1980s may have accounted for one quarter of all man-made carbon emissions (Watson *et al.*, 2000). Thus, conservation of existing forest carbon stocks has very significant potential for the mitigation of climate change. Of course, forestry measures alone will not be enough to halt the increase in atmospheric carbon dioxide concentration, but they can complement efforts to reduce carbon emissions from the burning of fossil fuels.

The conservation of biological diversity is also recognised as an urgent and critical task. The world's forests contain a very significant proportion of global biodiversity. Although Scotland may feel a long way from these global problems and challenges, the UK does have a serious responsibility for its own contributions to climate change and mitigation and for its own biodiversity. In terms of the UK's responsibilities globally, it is also important to recognise that the UK is the world's third largest net importer of timber (Martin, 2003).

Following the World Summit on Sustainable Development which was held in Johannesburg in 2003, the UK Forest Partnership for Action was established to promote sustainable development in the forest sector, both at home and internationally. Comprising Government, industry and non-governmental organisations (NGOs), this Partnership is taking action in the areas of forest certification, forest restoration and protection, illegal logging and sustainable timber procurement.

2.3 The global context: economic realities

A key outcome from the Johannesburg Summit was international recognition that the eradication of poverty is a key to sustainable development. At present, hundreds of millions of people throughout the world depend on forests: living in them as forest dwellers; living near them and making use of forest products; or working in them on commercial activities. Half of the world's wood production is still used as fuel wood as a primary source of energy for households (FAO, 1999): reducing the distance that women and children have to go to gather firewood can have a major impact on their quality of life.

Illegal logging is another unfortunate economic reality. The fact that forests are often remote, that the value of an individual tree can represent many weeks' wages, and that it is difficult to keep track of log movements makes forestry particularly susceptible to illegal and corrupt practices. There is little doubt about the significance of illegal and corrupt activities and their threat to the world's forests. Importantly, this issue is now being discussed openly and addressed.

The third economic reality is industry globalization. Opinions may vary on the merits and demerits of globalization, but it is undoubtedly a fact of life and likely to remain so. With international trade in forests and forest products worth US$130 billion per year (FAO, 2004) this applies as much to the forestry sector as any other. The forces behind globalization are, if anything, likely to grow stronger. These include rapid communication and the internet revolution, cheaper transport systems, rapid flows of capital, the coming together of markets and human migration.

Of course, globalisation brings with it threats as well as opportunities. If forests are not managed sustainably, they can suffer from increased accessibility and exploitation. In tackling these threats, individual countries will undoubtedly benefit from concerted international action. Neither the UK nor Scotland can stand aside from what is happening in the rest of the world as they develop their own forestry and land use policies.

Globalization will also continue to bring economic pressures and technological change. Forest industries are subject to fierce competition from powerful and effective importers. They must keep abreast of technological and market changes to ensure that they can maintain their competitive edge.

2.4 The impact of the European Union: enlargement

Already the EU contains about 116 million ha of forest and produces around 85% of its wood raw material needs, from an annual harvest that represents only about half the gross annual increment. Following the accession of Cyprus, the Czech Republic, Estonia, Hungary, Latvia, Lithuania, Malta, Poland, Slovakia and Slovenia, the forest area will grow to 160 million ha, with production of over 350 million m^3 per annum (PricewaterhouseCoopers, 2004). In a European context, there will be no shortage of wood.

Joining the EU will create both opportunities and threats for the forest environments in the acceding countries. Whilst fears have been expressed that accession might lead to increased logging of ancient and conservation value forests, such as those in the Carpathian mountains, accession will also require compliance with EU environmental regulations. Increased trade with existing EU countries is also likely to lead to increased incentives, in terms of access to consumers requiring sustainable timber products, for producers to seek forest certification (PricewaterhouseCoopers,, 2004).

Given what has been said above about globalization it would, of course, be wrong to consider the enlarged EU in isolation from the rest of the world. Increasingly, throughout the world, wood will come from planted forests: at present about 35% of global roundwood originates in planted forests and this is expected to increase to 44% by 2020 (Dyck, 2003). Thus, planted forests have a very important role to play in meeting the world's future wood needs.

2.5 The impact of the European Union: Common Agricultural Policy reform

The crucial significance of the CAP for forestry and, in particular, opportunities for forestry expansion is now understood. For individual farmers considering a possible switch of land use to forestry, the financial consequences are bound up with agricultural support regimes. Issues such as stocking densities for livestock can have an important impact on the way in which semi-natural woodland is managed and, because afforestation of farmland is an 'accompanying measure' under CAP reform (Edwards, 2004), about 40% of the funding for the Scottish Forestry Grants Scheme comes from Europe.

As is so often the case, however, and particularly in relation to the impact of CAP reform on forestry, the 'devil is in the detail'. Seemingly minor details of CAP implementation can have a significant effect on forestry. So, at the moment we are left with a number of unanswered questions or issues.

- Until the details are fully worked out, there will be a period of uncertainty during which many farmers will, wisely, choose not to make major decisions about future land use on their farms – this will adversely affect the level of tree planting.
- Expenditure on rural development is sometimes referred to as 'Pillar 2' in contrast with 'Pillar 1' which is the mainstream agricultural support. Critical questions for forestry in Scotland are the future size of the Pillar 2 funding envelope in Scotland and, within that, the extent to which Pillar 2 funding is available for forestry as opposed to other measures.
- Another question for the forestry sector in Scotland is how Land Management Contracts will play out. Will they simply be 'farm management contracts' or will they embrace other land uses such as forestry? At present, the indications are that they will be 'farm management contracts', so perhaps we should be looking forward in the medium term to their evolution into more broadly based 'Land Management Contracts'. These Land Management Contracts would embrace integrated land use and could cover issues such as management of native woodland through controlled grazing.
- It will also take time to see how the new 'single farm payments' play out in terms of land prices and the medium- to long-term availability of land for woodland expansion.

2.6 What Scotland can offer
2.6.1 Expertise

Scotland has a wood processing industry of which it should be proud. It has made remarkable strides over the last 20 years, investing over £1 billion to ensure that it holds its own against international competition. Since 1980, the wood harvest has nearly quadrupled, and it will almost double again over the next 15-20 years (Scottish Executive, undated). One illustration of this increase in efficiency has been the development of the sawmilling industry. Just eight years ago there were only two mills producing over 50,000 m^3 per annum – now there are eight mills.

The recent development in Scotland of the Centre for Timber Engineering at Napier University in Edinburgh is another good example of what we can do. It will link engineering requirements with detailed analysis of the fundamental properties of wood. Wood is an immensely versatile material and the surface of its potential application has only just been 'scratched'.

Some Scottish companies are leading the way in terms of product development. For example, one company, James Jones & Sons Ltd. of Forres, is now manufacturing 'I-beams' which are engineered wood products making use of finger-jointed timber and oriented strand board to produce beams with totally consistent and predictable engineering qualities. In addition, the 'wood for good' campaign is aiming to grow the market for wood in the UK, justifiably claiming the environmental advantages of using wood rather than other raw materials.

A lot has been learned in Scotland about what sustainability means in practice and how it can be applied to forest management through the development of environmental guidelines (on issues such as forests and water, landscape and conservation) and their incorporation into the UK Forestry Standard and certification. Scotland has charted a course that others have followed.

These lessons in sustainability are of practical significance in Scotland, but they also mean that our expertise can be exported to the rest of the world. This expertise allows us to 'punch above our weight' in international discussions. Furthermore, we often forget the extent to which foresters who have trained and worked in Scotland continue to work overseas sharing their knowledge with others.

2.6.2 Proximity to markets

The size of the UK timber market – one of the biggest net importers in the world – has already been referred to. Scotland's share of this market is growing steadily. In 1980, for example, Scottish sawmills had just 1% of the UK market: now they have nearly 10%. Scotland has natural advantages in terms of accessibility and of being able to respond to the 'just in time' requirements of customers. However, the industry will continue to face serious competition from importers – the figures quoted earlier on European wood supplies and demand demonstrate that Scottish processors will need to continue to innovate, to improve efficiency and to add value in order to remain profitable.

2.6.3 Special habitats

Scotland has some unique natural habitats and we need to learn from past controversies. Examples such as the Flow Country offer important lessons. A key lesson for the forestry sector from this particular example is the need to look more broadly – beyond the forestry horizon – and to recognise and respect other interests and concerns.

The Flow Country contains an almost unique natural habitat that was worthy of conservation. Developing this point, we must recognise the economic value of our natural heritage. The contribution of the tourist industry to the Scottish economy is significantly greater than that of forestry, farming or fishing (Borders Forest Trust, 2003). Through imaginative design of woods and forests, and the provision of recreation facilities, much can be done to help make Scotland a more attractive tourist destination.

Native woodlands also represent something of unique value that needs to be conserved and capitalised on. Considered over a timescale of centuries and millennia, the story of native woodland loss is not a pretty one, but the opportunity exists to try to right past wrongs as the expertise is being developed (e.g. Forestry Commission, 2003). Dozens of projects around Scotland, from Glen Affric to Carrifran, demonstrate that we have the will and the vision (Caledonian Partnership, 1997). There are opportunities now for large-scale ecological restoration through developing forest habitat networks.

The timber production forests of tomorrow will be very different from the uniform monocultures of the past. As we continue to restructure our forests; diversifying age classes, diversifying species and leaving more open space, these plantations will evolve into attractive woodland that continues to grow timber but which is also rich in wildlife and attractive for recreation. Many of the forests in Perthshire prove this point admirably.

2.7 A vision for 2050

So what does all of this mean for the forests and woodlands of Scotland that we bequeath to our successors?

There will continue to be a serious role for wood production, supplying increasingly sophisticated processing industries. What the industry will continue to require is consistency of raw material and supply. The industry needs to be able to plan investments on the basis of reliable information about future wood supplies, and this information needs to capture critical quality parameters as well as volumes.

For the environment we need to continue the good work of restructuring, of fitting forestry into the landscape, and of native woodland conservation and expansion. We should also be undertaking more landscape-scale ecological restoration.

Above all, we need to serve the needs of the people of Scotland. We all know that forests and woods can provide opportunities for healthy exercise and spiritual refreshment. But, with a few notable exceptions, there is a serious mismatch between the location of many of our most attractive forests and where people live. How many woodland walks are there within 35 km of central Glasgow?

We need to think hard about the future role of Scotland's national forests – the forests managed by Forestry Commission Scotland. The Deputy Minister of the Environment announced a review in the autumn of 2003, and a consultation paper was published (Forestry Commission Scotland, 2003).

Sustainable forest management is very much about balance – the right trees in the right places. As we move forward, the view is that we are getting better at striking this balance. The Scottish Forestry Strategy's vision is that

"Scotland will be renowned as a land of fine trees, woods and forests which strengthen the economy, which enrich the natural environment and which people enjoy and value" (Scottish Executive, 2000, p. 3).

There is no reason to change that vision, nor the guiding principles that underpin it. The fifth of those guiding principles – that of diversity and local distinctiveness – deserves a little more attention. In central Scotland, the woodlands on former industrial land are highly valued by local people for their recreation and amenity value. In the Cairngorms, or Loch Lomond and the Trossachs, are some of Scotland's most magnificent forests, visited by hundreds of thousands of people each year, and now of course a key element in our new National Parks. In Galloway and Grampian are forests that are also attractive, but where there is a serious focus on wood production, with their proximity to many of our wood processing mills. And in the Northern and Western Isles we are encouraging small-scale planting to provide shelter, conservation and amenity.

To ensure that these local and regional differences are captured, the Scottish Forestry Forum is establishing regional fora throughout Scotland. These will help to ensure that local people can help to develop and realise their vision for the future of forestry in their area.

So, a wide range of complex forces will shape our forests of the future. Our role is to try to think imaginatively, and to listen intelligently to others, so that – whatever the future holds – future generations will thank us for the forests we leave them.

References

Borders Forest Trust (2003). *Restoring Borders Woodland: the Vision and the Task*. Unpublished report.

Caledonian Partnership (1997). *Scotland's Caledonian Forests: Resource Assessment and Implementation of a Restoration Programme for Glen Affric*. Final LIFE–Nature Progress Report Contract No. B4-3200/94/769. Highland Birchwoods, Munlochy.

Dyck, B. (2003). Benefits of planted forest; social, ecological and economic. Paper delivered at UNFF intersessional experts meeting on the role of planted forests in sustainable forest management, 2003, Wellington, New Zealand. www.maf.govt.nz/mafnet/unff-planted-forestry-meeting/conference-papers.

Edwards, T. (2004). *SPICe briefing: Support for rural development through the CAP.* www.scottish.parliament.uk/business/research/briefings-04/sb04-13.pdf.

FAO (1999). *State of the World's Forests 1999.* Food and Agriculture Organization of the United Nations, Rome.

FAO (2004). *FAOSTAT – online statistical service.* http://apps.fao.org. Food and Agriculture Organization of the United Nations, Rome.

Forestry Commission (2003). *Forestry Facts and Figures 2003.* Forestry Commission, Edinburgh.

Forestry Commission Scotland (2003). *Review of Land Managed by Forestry Commission Scotland.* Forestry Commission Scotland, Edinburgh.

Martin, P. (2003). Corporate Social Responsibility, DFID Illegal Logging and associated trade programme. www.ttf.co.uk/industry/csr/dfid.asp.

PricewaterhouseCoopers (2004). *Distinguishing the Wood from the Trees: the Impact of EU Enlargement on the Forestry, Forest Products and Paper Industries.* PricewaterhouseCoopers, London.

Scottish Executive (undated). Scottish Environment Statistics Online. www.scotland.gov.uk/stats/envonline.

Scottish Executive (2000). *Forests for Scotland: the Scottish Forestry Strategy.* Scottish Executive, Edinburgh.

Watson, R.T., Noble, I.R., Bolin, B., Ravindranath, N.H., Verado, D.J. & Dokken, D.J. (2000). *Land Use, Land Use Change and Forestry. A Special Report of the Intergovernmental Panel on Climate Change.* Cambridge University Press, Cambridge

2 Farming, Forestry and the Natural Heritage: Towards a More Integrated Future

3 What does the public want from the countryside?

Steve Sankey

Summary

1. This chapter summarises key changes in agriculture, forestry, biodiversity and access over the last few decades and interprets them from the perspective of what the public wants from the countryside. The chapter draws on government research and public policy objectives and tests progress against stated public desires for Scotland's countryside.
2. A model for future development is proposed, which accepts that the general public wants to support farmers, but not at the expense of damaging the environment.
3. Visitors to Scotland want to experience the cultural and natural heritage and to go walking, while visitors to forests want access, scenery and exercise.

3.1 Introduction

The aim of this chapter is to summarise what the public wants from farming and forestry. The chapter will primarily focus on farming, but will also consider forestry.

It is a commonly held belief – especially by many farmers – that not much has changed in the last generation, other than an exponential rise in paperwork. However, a study of some agricultural statistics over the last 20 years (Scottish Executive, 2003a) shows that:

- the total land used for agriculture has decreased by 4%, mainly due to the expansion of woodland;
- average fertiliser applications to crops have increased by 6% to 217 kg/ha per annum;
- some land transfers have occurred;
- the area of grass harvested for animal feed has increased by 13%;
- the cropped areas have remained stable, but grazed areas have decreased by 13%; and
- there have also been profound stock changes, with more sheep and fewer cattle.

The farming industry and the general public can also be forgiven for thinking that current farming policy is all about maximising food production. This is not actually the case, however, as the current European Commission (EC) objectives for farming policy include, *inter alia,* objectives to protect farm incomes, protect the environment, and other socio-economic factors such as employment (European Commission, 1999). The extent to which these Common Agricultural Policy (CAP) objectives are matched by what the public wants from farmers and crofters is an important question.

With respect to food quality, a plethora of scares concerning eggs, chicken, pork, beef and lamb in recent years – from Edwina Currie MP and her eggs, to Bovine Spongiform Encephalopathy (BSE) and foot and mouth disease (FMD) – has left the public low in confidence about the production of its food. This does not need to be over-emphasised, but equally these perceptions must not be ignored, as organisations such as Quality Meat Scotland and others that are concerned with quality food production recognise in their remits. It is much harder to regain lost consumers than to retain them.

With respect to the environment, research clearly shows losses of biodiversity in the countryside. This is extremely serious as over 70% of Scotland is farmed. In the last half century, 50% of lowland raised mires have been lost, 78% of grey partridges (*Perdix perdix*) have gone, and the total length of Scottish hedgerows has been reduced by 50% (Mackey *et al.*, 2002). Forestry is implicated in the 23% loss of upland heath, with most plantings up to 1988 taking place on uplands or peatlands. Furthermore, the uplands suffered greatly from a 20% increase in sheep numbers in the 1980s and a 10% loss in cattle numbers as farmers in north-east Scotland and elsewhere switched from beef to grain. In the uplands, the rise in sheep numbers, coupled with a switch from grazing pasture to forestry, has had a serious negative impact on the biodiversity of the remaining grasslands. Yet the perceptions of the farming community remain somewhat defensive in view of the extent of these changes. These perceptions need to change, and change quickly, because we need to alter the way that public support underpins land management, and to recover these biodiversity losses. Government and industry champions, principally NFU Scotland and sector bodies, need to lead this change.

Last, but by no means least, income declines have been dramatic. UK farm income dropped by as much as 75% in the late 1990s, driving more than 20,000 farmers from the land (Gorelick, 2000). Although there has been some recovery of income since 2000, the loss of labour from our countryside is also something that our politicians in particular lament. However, there are interesting examples from France and elsewhere of how Rural Development Plans, funded by modulation under the CAP, have prioritised socio-economic criteria such as employment creation or retention (BIMA, 2000).

A Forward Strategy for Scottish Agriculture (Scottish Executive, 2001) recognises that greater integration, 'multi-functionality' of objectives and new ways of working are required in future. The main task is to put the case for the public's aspirations for farming and forestry, once established, and ask the question – in some innocence – "How well have the policy makers taken on board the public's wishes?".

3.2 What does the public want from its countryside?

The most recent research, based on a sample size of just over 2,000, was published by the Scottish Executive's Environment and Rural Affairs Department in October 2003, in *Public Perceptions of Food and Farming* (Scottish Executive, 2003b). Its main findings were:

- the top three priorities for improvement in the farming sector were animal health, animal welfare and the protection of the countryside and wildlife;
- the top two areas where the public thought future Government financial support should be invested were animal health (disease control) and reduction of pollution/use of pesticides and fertilisers, whilst the lowest ranked priorities were for

maintaining current production levels and supporting other businesses in the rural economy;
- older respondents (aged over 55 years), those living in remote rural areas and those with a link to the farming industry, all tended to see farming issues as more important than younger people and those living in urban areas;
- when prompted as to the potential impacts if farming were to cease to exist in Scotland, the key concerns were on the impact on the economy in terms of job losses. However, there were also fears about the lack of Scottish products being available and land being lost to property development;
- with prompting, respondents listed health and the environmental effects of pesticides and fertilisers, as well as the use of drugs and hormones to promote animal growth, to be the areas of most concern about food production standards;
- running throughout the research there were a number of threads suggesting that support for farmers was strong and that mistrust of the Government was high. It is known from other public opinion polls that the public is more likely to believe a representative of an environmental non-governmental organisation (NGO) than civil servants or politicians (Worcester, 1999);
- awareness and understanding of assurance schemes was poor on the whole. Those schemes for which levels of understanding were greater were the more straightforward ones and those with strongest publicity;
- when purchasing fresh meat, the origin of the meat appeared to be of most importance and some certainly felt there was room for improvement in labelling; and
- when asked to trade off a range of attributes relating to price and quality however, price was the most important.

In summary, the public supports farmers but is concerned about health, environment and quality, and mistrusts government. The public also wishes to see its Scottish food produced in an environmentally responsible way. Conventional production scored the lowest on this issue, whilst organic production was middle ranking, and environmentally responsible production had the highest scores. These wishes need to be built into policy.

3.3 Tourism

Another important sector of public opinion in Scotland is the tourist. Tourism is Scotland's biggest industry – bigger than finance, farming and forestry. In 2002, over 20 million tourists took overnight trips and spent almost £4.5 billion, supporting around 9% of all Scottish employment (VisitScotland, 2003). What, then, do tourists want? Why do visitors – both domestic and international – come to Scotland? The activities undertaken by tourists outdoors are shown in Table 3.1.

An earlier survey in 1999 (System Three, 1999) found that 69% of tourists went on a walk of under two miles, whilst 40% had a walk of more than 2 miles. Some 39% watched wildlife. These figures compare quite favourably with shopping (71%) and cultural visits (69%). So the countryside and its wildlife are why four million tourists visit Scotland each year, and these tourists support the Scottish economy to the tune of more than £2 billion.

Table 3.1 Activities enjoyed by tourists. Source: VisitScotland (2003).

	UK Holiday Trips 2002 (%)	Overseas Holiday Trips 1996 (%)
Visiting castles, monuments, churches etc.	39	83
Hiking/hillwalking/rambling/other walking	30	39
Field/nature study	21	9
Golf	8	2
Visiting theme parks/activity parks	9	6
Fishing	8	3

3.4 Environment

The Scottish Executive asks the public regular, and varied, questions on the environment. The last time they did so in 2002 (Scottish Executive, 2002a), the following key findings emerged (see Table 3.2).

- Respondents were particularly (very) worried about two environmental issues – raw sewage put into the sea and nuclear waste.
- Levels of concern had increased for issues such as new development in the countryside, protection of areas of conservation interest, lack of access to parks and fish farming.
- Levels of concern about issues of more direct relevance to farming and forestry included protection of wildlife (73% were worried about this) and the use of pesticides, fertilisers and chemical sprays (68%).

3.5 Forestry

For forests, similar reasons for countryside visits were noted from Forestry Commission (FC) research at UK, regional and local levels. Visitor surveys in the Tweed Valley Forest Park at Glen Tress in the Borders (Forestry Commission, 2002), Inverness (Forestry Commission, 2000a) and Lochaber (Forestry Commission, 2000b) all point towards walking, relaxing or cycling as being the principal reasons for visits (Table 3.3).

Research on public opinion in Scotland in 2003 (based on a sample size of 1,000) (Forestry Commission, 2003a) found that:

- 91% of adults in Scotland selected at least one public benefit as a good reason to support forestry with public money;
- the top reasons to support forestry were to provide places for wildlife to live, to provide places to visit and walk in, and to help prevent the greenhouse effect and global warming;
- 64% of adults have visited a wood in Scotland in the last few years for walks, picnics or other recreation; and
- 58% of adults would like more woodland in Scotland.

Table 3.2 Level of concern about a range of environmental issues. Figures are percentages based on a sample size of 4,119 (note that rows may not sum to 100% due to rounding). The issues of most relevance to this chapter are shown in bold. Source: Scottish Executive (2002a).

Issue	Very worried (%)	Quite worried (%)	Not Very worried (%)	Not worried at all (%)	Don't know (%)
Raw sewage put into sea	49	35	9	4	2
Nuclear waste	47	33	13	5	2
Damage to the ozone layer	34	42	15	6	2
Pollution of rivers, lochs and seas	**30**	**45**	**18**	**6**	**1**
Protection of wildlife	**28**	**45**	**18**	**7**	**2**
Road traffic	27	38	25	9	1
Quality of drinking water	27	30	26	17	1
Pesticides, fertilisers and chemical sprays	**26**	**42**	**21**	**8**	**3**
Waste disposal	25	42	22	9	2
Global warming by greenhouse effect	25	42	21	8	4
Genetically modified crops	**24**	**32**	**25**	**14**	**6**
Using up non-renewable resources	21	38	26	11	6
Fumes and smoke from factories	20	36	27	15	2
Acid rain	**20**	**36**	**26**	**11**	**7**
Over fishing	19	34	28	15	4
Generation of electricity by nuclear power	19	34	27	15	5
Protection of areas of conservation interest	**16**	**41**	**28**	**12**	**4**
New development in the countryside	15	33	33	16	2
Derelict land in towns and cities	13	34	34	17	2
Forestry	**11**	**32**	**36**	**18**	**4**
Farming methods	**11**	**31**	**35**	**18**	**5**
Lack of access to parks	**10**	**26**	**38**	**23**	**2**
Fish farming	7	21	39	27	5

Table 3.3 Selected forestry site surveys undertaken in Scotland during the period 2000-2002 Source: Forestry Commission (2000a,b, 2002).

Site	Date	Reasons for visit
Scottish Borders	2002	82% repeat visits; 62% scenic beauty; 70% cycling, 53% walking, 50% scenery
Inverness	2000	85% holidaymakers; 35% picnicking/relaxing; 28% walking
Lochaber	2000	64% holidaymakers; 77% walking; 23% cycling

On a UK basis (based on a sample size of 4,120), over 80% of adults have visited woodlands in the countryside and 50% have visited woodlands in and around towns (Forestry Commission, 2003b). Over 60% had visited at least once a month, with the reasons for visiting being given as:

- wildlife (65%);
- peace and quiet (65%);
- attractive scenery (62%); and
- a safe environment (57%).

Given that Forestry Commission Scotland has around 227 sites in Scotland with parking facilities and 21 visitor centres, it is obvious that these facilities are important, valued and used by the visiting public, whether domestic or otherwise (Forestry Commission, 2003c). Indeed, on a UK basis, 80% of adults have visited a forest in the countryside, 50% of adults have visited a woodland in or around a town, and 60% of adults undertake this visit once a month (Forestry Commission, 2003b).

3.6 Developing policy to meet public needs

The mismatch between the CAP's principal objectives and the Scottish public's wishes has been summarised. The CAP has not helped the environment, the farmer, or the consumer. The public has expressed a clear wish for access, for safe food, for quality, for environmental protection, and yet, paradoxically, at a low price. This tension will not easily be removed, so the obvious solution is to focus on value-for-money and use taxpayers' money (through agricultural subsidies) to reward farmers and foresters for delivering what the public wants.

A number of key questions arise. Can we design farming and forestry systems for the future that are predicated upon satisfying several objectives, and starting from the perspective of the consumer? How can we assure the Scottish and visiting public that the £660 million invested every year by the taxpayer in Scottish farming is good value-for-money? Are we satisfied that the 10,000 ha of woodland under annual management grants is giving good value-for-money?

The general public wants to support farmers, but not at the expense of damaging the environment. Visitors to Scotland want to experience the cultural and natural heritage, and to go walking. Users of forests want access, scenery and exercise, either walking or cycling. We need to wrap all these wishes into policy.

3.6.1 Biodiversity

Replacing much of our lost biodiversity ought to be relatively easy. Recent successes for farmland birds and arable weeds that are a by-product of set aside give cause for real optimism. Farmland bird populations can respond quickly and positively to the right prescriptions being introduced, for example, through cross-compliance. Field margins, hedgerow management and replacement, and beetle bank creation across all farms could re-build biodiversity quickly. Keeping forestry away from sensitive habitats has already largely been resolved, but the continued removal of conifers from the sensitive sites planted in the past – notably the Flow Country - is essential.

Moreover, farmers and crofters know how to manage land and many know how to look after the habitats and species dependent on such management. The most important priorities need to be identified through the Scottish Biodiversity Strategy (Scottish Executive, 2004) and incorporated into Scotland's interpretation of the CAP mid-term review, through cross-compliance on Pillar 1 payments and agri-environment programmes in Pillar 2. In theory this should be easy, but there seem to be many cultural barriers to overcome at both governmental and producer level.

3.6.2 Access

The Land Reform (Scotland) Act 2003 gives statutory rights of responsible access to land over farms and forests, so why not reward land managers for managing some of this? This should be straightforward, and structured around easier access from our towns and cities to the countryside. The concept of core path networks is now understood and the first signs of implementation can now be seen. Why not reward land managers who create new routes or wish to manage these facilities? This is already done in the public sector through the Forestry Commission Scotland's (FCS) provision of over 200 sites, and by other agencies such as Scottish Natural Heritage and Local Authorities. Environmental NGOs, such as the Scottish Wildlife Trust, but also the John Muir Trust, National Trust for Scotland, Royal Society for the Protection of Birds and the Woodland Trust, welcome visitors freely onto their several hundred sites in Scotland. Several private estates may also be regarded as exemplars in this regard – Buccleuch and Rothiemurchus for example – but the routine management of access onto private land, farms and forests, in return for public subsidies routed through CAP and other payments, should be supported.

3.6.3 Other needs of consumers

There is a whole host of these which may be gleaned from the surveys quoted above, and from many other sources. At root, though, consumers – be they tourists or residents – want our Scottish landscape to be well looked after by land managers. They value our scenery and accept that farmers and foresters created it, but they are also very worried about pollution, animal welfare and farming methods. There should be no difficulty in paying land managers on behalf of society to continue to manage the land responsibly, but it must be payment for what the public wants, and that is the challenge ahead. So how might we accommodate all of these wishes into policy?

3.6.4 A model for the future

The FCS has already embraced multi-functionality both in terms of how it works and in the delivery of its products. Our forests are well used for recreation, as landscape assets and as homes for biodiversity, in addition to being wood factories. The FCS should be encouraged to supply more of the same on behalf of the consumer – things are going well. It is pleasing to see well-used FCS sites being re-structured to accommodate multiple objectives. With respect to policy, the Scottish Forestry Strategy (Scottish Executive, 2000) recognises that targeting should happen, through locational supplements and other techniques, to ensure that biodiversity priorities such as capercaillie (*Tetrao urogallus*) recovery or the Central Scotland Forest actually materialise on the ground.

With respect to agriculture, change is slower but we are all aware of the enormous potential for change and improvement on the immediate horizon in the form of the CAP mid-term review. It is our best chance since subsidies were invented to deliver multi-functionality in Scotland. Importantly, after devolution Scotland can make many of its own decisions on these issues for the first time. To meet the needs of the public, I propose that what should be developed in future, delivered through Land Management Contracts, agreed with every single Scottish farmer and crofter, are:

- a fully de-coupled payment funded by CAP mainstream subsidies with public goods delivered through cross-compliance rules relating to access, landscape management and biodiversity. It is important that the definition of 'Good Agricultural and Environmental Condition' under this regulation does not become diluted to the minimum legal requirement on farmers. I suggest that such a weak interpretation of practice would be unacceptable to the public. We need quality food production in Scotland to counter the constraints of our climate and of our geographic peripherality in Europe. That quality means taking a pride in what we manage and deliver. Around £550 million of public subsidies is available under this option to buy public goods, including a world class environment and quality food;
- Pillar 2 (Rural Development Plan) initiatives funded by modulation and the UK Treasury. In this way we can genuinely add resources to Scottish farmers and buy specialist management (and therefore provide financial reward) of public goods and production methods such as organic farming;
- the Scottish Executive's extensive consultation on CAP is to be welcomed, and the Scottish Wildlife Trust is pleased to see environmental representatives on the stakeholder group. We look forward to the implementation of the many good ideas generated by *Custodians of Change* (Scottish Executive, 2002b), and we would encourage the Executive to embrace multi-functionality in agriculture, to follow the lead set by the Forestry Commission, and to deliver what it is that the taxpayer – the provider *and* consumer of public goods – actually wants.

References

BIMA (2000). *Land Management Contracts*. French Government Ministry of Agriculture and Fisheries, Paris.

European Commission (1999). *Europe's Agenda 2000: Strengthening and Widening the European Union*. European Commission, Brussels.

Forestry Commission (2000a). *Inverness Visitor Survey 2000*. Forestry Commission, Edinburgh.

Forestry Commission (2000b). *Lochaber Visitor Survey 2000*. Forestry Commission, Edinburgh.

Forestry Commission (2002). *Scottish Borders Visitor Survey 2002*. Forestry Commission, Edinburgh.

Forestry Commission (2003a). *Public Opinion of Forestry 2003: Scotland*. Forestry Commission, Edinburgh.

Forestry Commission (2003b). *UK Public Opinion of Forestry 2003*. Forestry Commission, Edinburgh.

Forestry Commission (2003c). *Forestry Facts and Figures 2003: A Summary of Statistics about Woodland and Forestry in Great Britain*. Forestry Commission, Edinburgh.

Gorelick, S. (2000). *Facing the Farm Crisis: Farming at Peril*. The Ecologist, **30**, 22, May 2000.

Land Reform (Scotland) Act 2003. www.opsi.gov.uk/legislation/scotland/acts2003/20030002.htm.

Mackey, E.C., Shaw, P., Holbrook. J., Shewry, M.C., Saunders, G., Hall, J. & Ellis, N.E. (2002). *Natural Heritage Trends: Scotland 2001*. Scottish Natural Heritage, Perth.

Scottish Executive (2000). *Forests for Scotland: The Scottish Forestry Strategy.* Scottish Executive, Edinburgh.

Scottish Executive (2001). *A Forward Strategy for Scottish Agriculture.* Scottish Executive, Edinburgh.

Scottish Executive (2002a). *Public Attitudes to the Environment in Scotland.* Scottish Executive Environment and Rural Affairs Department, Edinburgh.

Scottish Executive (2002b). *Custodians of Change.* Report of the Agriculture and Environment Working Group, Scottish Executive, Edinburgh.

Scottish Executive (2003a). *Key Scottish Environment Statistics.* Scottish Executive Environment and Rural Affairs Department, Edinburgh.

Scottish Executive (2003b). *Public Perceptions of Food and Farming.* Scottish Executive Environment and Rural Affairs Department, Edinburgh.

Scottish Executive (2004). *Scotlands Biodiversity: It's In Your Hands. A Strategy for the Conservation and Enhancement of Biodiversity in Scotland.* Scottish Executive, Edinburgh.

System Three (1999). Tourism Attitudes Survey 1999. Unpublished Commissioned Report F99NC07. Scottish Natural Heritage, Perth.

VisitScotland (2003). *Scottish Tourism Survey 2002.* VisitScotland, Edinburgh.

Worcester, R.M. (1999). *Science and Democracy: Public Attitudes to Science and Scientists.* MORI, London. Available at www.mori.com/pubinfo/rmw/budapestpaper.pdf.

3 Farming, Forestry and the Natural Heritage: Towards a More Integrated Future

PART 2:
Farming, Forestry and Biodiversity

Farmer near Old Meldrum, Aberdeenshire © Lorne Gill, Scottish Natural Heritage

Farming, Forestry and the Natural Heritage: Towards a More Integrated Future

PART 2:
Farming, Forestry and Biodiversity

This part focuses on biodiversity issues, particularly on the extent of our current knowledge of the impacts of farming and forestry on biodiversity and on what steps are being taken to ensure that biodiversity improves.

In Chapter 4, Boatman provides a detailed review of how farming practices influence biodiversity. He notes that major changes have occurred in farming practices and that these have affected biodiversity in many ways. Boatman then shows how this knowledge is being used to look at ways of conserving and restoring biodiversity on farmland: the likely outcomes for biodiversity from further changes in farming practice and from policy developments can then be predicted with greater confidence.

A similarly detailed review of the influence of forest and woodland management on biodiversity is provided by Humphrey *et al*. (Chapter 5). This chapter looks at recent developments in woodland management and reviews a wide range of research into the effects on biodiversity of these developments. The need to manage forestry and biodiversity at the landscape scale is emphasised. The chapter looks at the role of Forest Habitat Networks as a tool for guiding woodland restoration, at a landscape scale, to enhance biodiversity. A key conclusion is the need for a much wider range of stakeholders to be involved in the decision-making process.

One of the major policy developments in recent years has been the launch of the Scottish Biodiversity Strategy by the Scottish Executive. Although the Strategy and its implementation plans were published in the months following the conference, Henderson (Chapter 6) provides a helpful insight into the early development of the Strategy. This work included wide consultation and the involvement of key partners through various working groups.

In Chapter 7, Gotts looks at the ways in which farmers and other land managers can be supported. He provides a good reminder that it is not just advisory publications that are useful but a wide range of other tools, such as demonstration projects and research, too. Thinking about the needs of land managers is also very important and Gotts emphasises the need to look at their motivations and at how best to influence and build on these.

Farming, Forestry and the Natural Heritage: Towards a More Integrated Future

4 How do farming practices influence biodiversity?

Nigel Boatman

Summary

1. Agriculture is the dominant land use in the UK, and farmed habitats are key to the survival of much of our wildlife. In recent decades, intensification of agriculture has led to declines in many species characteristic of farmland.
2. Sources of evidence for changes in the status of various taxonomic groups are reviewed, along with the use of indicators to monitor population and habitat changes for policy purposes.
3. Changes have occurred across a wide range of farming practices, and these have affected biodiversity in a variety of ways. Important changes include specialisation and geographical polarisation, larger fields and farms, greater agrochemical use, simpler rotations, more autumn sowing and more efficient harvesting of crops, more non-inversion tillage, drainage and reseeding of grassland, a switch from hay to silage, increased stocking rates and the use of avermectin wormers. Some of these changes have been less pronounced in Scotland than elsewhere in the UK.
4. Evidence for the effects of these changes on biodiversity is reviewed, and some key examples highlighted.
5. Various mechanisms are now available to conserve biodiversity on farmland and reverse observed declines. These include less intensive farming systems, regulation and measures provided for under CAP reform, such as cross-compliance and agri-environment schemes. Although each has a role to play, agri-environment schemes probably offer the greatest potential for the restoration of farmland biodiversity. Outcome-orientated and geographical targeting at a landscape scale will increase the chances of achieving objectives.

4.1 Introduction

There is a growing realisation of the importance of agricultural land and farming practices as drivers of change in the status of much of our biodiversity, and its potential contribution to the achievement of conservation objectives. This is now recognised in the formulation of policy on agriculture and the countryside, which is likely to have profound effects on the way farming is carried out in future.

This chapter seeks to provide an overview of the relationship between farming practices and biodiversity. Policy indicators used to measure changes in biodiversity and habitats are reviewed briefly, along with supporting evidence of changes on farmed land. Changes in farming practice over the last few decades are then considered, along with examples to illustrate how these have affected biodiversity. Finally, potential approaches to conserving and restoring biodiversity on farmland are discussed.

4.2 The importance of agriculture for biodiversity

Agriculture is the dominant form of land use in the UK, covering around 75% of the land area, and has been significant for several millennia. Much of our wildlife has become adapted to living in farmland habitats, so much so that many plants and animals are now dependent on some form of agricultural management for their continued survival. For example, the arable flora is dependent on continued cultivation, whilst the flora of species-rich grasslands, such as the machair of north-western Scotland, the Hebrides and Orkney, and its associated fauna, are dependent on the maintenance of appropriate grazing regimes (Birnie *et al.*, 2002). Restricting wildlife conservation to specific areas such as nature reserves and Sites of Special Scientific Interest (SSSIs) is not a viable option, because many animal species characteristic of farmland range over large areas and can only be conserved effectively by appropriate measures implemented across the wider countryside. For example, only 28% of Biodiversity Action Plan (BAP) species are found mainly in protected areas, compared to 33% found mainly in the wider countryside, the remainder occurring in both. For BAP bird species, an even greater proportion (58%) is dependent on the wider countryside, compared to only 8% found mainly in reserves and SSSIs (Birnie *et al.*, 2002).

4.3 Changes in agricultural policy and practice

Farming practices have changed through the centuries and the fortunes of our fauna and flora have fluctuated at the same time, reflecting the changes in habitats and food resources provided (Stoate, 1995). The second half of the 20th century, however, saw an unprecedented level of change and intensification, in response to the policy of maximising food production during and after the Second World War, accompanied by subsidies which ensured a prolonged period of price stability (see Section 4.7). This policy was highly successful in terms of increasing agricultural productivity but, at the same time, a major decline in biodiversity occurred, and agricultural intensification has been held responsible for much of this observed decline, both in the UK (e.g. Chamberlain *et al.*, 2000; Benton *et al.*, 2002; Robinson & Sutherland, 2002) and Europe (Donald *et al.*, 2001).

Concern over farmland biodiversity has increased in recent decades, partly because the greater availability of survey data has allowed declines in biodiversity to be quantified. In particular, the Common Birds Census, which began in the late 1960s, indicated substantial declines in some farmland bird species by the 1980s (Gregory *et al.*, 2002). At the same time, surplus production of farm crops and livestock within the European Union (EU), combined with pressure from the World Trade Organisation, has led to successive modifications of the Common Agricultural Policy (CAP) to reduce the cost of subsidies and decouple them from productivity. As part of this process, the principle of providing support for maintenance and enhancement of the rural environment has gained increasing acceptance. This has arisen from the recognition that agricultural policy, as practised pre-CAP reform, is unsustainable, both economically and environmentally, and that future policy development needs to take into account not only productivity, but also environmental and socio-economic sustainability (Tilzey, 1998).

4.4 Evidence of changes in biodiversity

Changes in biodiversity on arable farmland were reviewed by Robinson & Sutherland (2002). They concluded that around half of plant species, a third of insect species and four-fifths of bird species characteristic of farmland have declined. Declines were more marked for specialised species than for generalist species, which have remained stable or increased.

The most detailed datasets available which record recent changes in the status of wildlife are the annual breeding bird surveys carried out by the British Trust for Ornithology (BTO). For farmland, the most important is the Common Bird Census (CBC), now replaced by the Breeding Bird Survey (BBS). The CBC has shown substantial declines in many farmland bird species since recording began in the late 1960s, though a few species have increased. Overall, farmland specialists declined by 46% between the mid-1970s and 2000 (Gregory *et al.*, 2002). CBC coverage was not adequate to measure trends for Scotland only, but recent trends in Scotland, as estimated from the BBS, have generally mirrored those in the rest of the UK. However, two farmland species, house sparrow (*Passer domesticus*) and starling (*Sturnus vulgaris*), have increased in Scotland, but decreased in the UK as a whole (Raven *et al.*, 2003), though these species appear to have suffered range contractions between the late 1960s/early 1970s and the late 1980s in Northern Scotland, which were not evident elsewhere in the UK during this period (Gibbons *et al.*, 1993).

The *New Atlas of Breeding Birds in Britain and Ireland, 1988-1991* (Gibbons *et al.*, 1993) records changes in the distribution of breeding birds in the UK since the earlier *Atlas of Breeding Birds, 1968-1972* (Sharrock, 1976). For non-farmland habitats, there were similar numbers of species showing positive and negative changes in distribution. For farmland species, however, 24 out of 28 species had declined in range. Pain *et al.* (1997) found that there were fewer birds characteristic of extensive pastoral systems in south-western and central Scotland in 1988-1991 than in 1968-1972, and there was a retraction in range away from the north-east. The Western Isles, where extensive farming is still widely practised, remain the most important areas for these species.

Changes in plant and arthropod biodiversity on lowland farmland were reviewed by Sotherton & Self (2000). They note that all of the vascular plant species, and most of the bryophytes, for which farmland is important have suffered range contractions in the latter part of the 20th century. Further information is provided by the *New Atlas of the British and Irish Flora*: of the 100 species which have shown the greatest relative decrease between 1930-1969 and 1987-1999, 37 occur on arable land, including five species that are now extinct in the wild in the UK (Preston *et al.*, 2002). There is less information for arthropods. However, data from the Rothamsted Insect Survey showed a decline in moth diversity and abundance between 1933-1959 and 1960-1989, though no change was recorded in woodland. Annual monitoring of over 100 cereal fields carried out by the Game Conservancy Trust in Sussex showed no change in abundance of ground beetles (Coleoptera: Carabidae), aphid-specific predators such as ladybirds (Coleoptera: Coccinellidae) and predatory flies (Diptera), but decreases in cereal aphids (Homoptera), their parasitoid wasps (Hymenoptera: Parasitica), leaf beetles (Coleoptera: Curculionidae), rove beetles (Coleoptera: Staphylinidae), spiders (Araneae) and sawfly larvae (Hymenoptera: Symphyta) (Sotherton & Self, 2000). Many of these groups are important as food items in the diet of farmland bird chicks. Recently, Benton *et al.* (2002) linked changes in invertebrate catches from a Rothamsted suction trap in Scotland to changes in agricultural practice and farmland birds. Using Principal Components Analysis, they showed that average daily catches per year for 12 arthropod groups were significantly correlated over 27 years with axes representing a range of variables indicating changes in farming practice.

In addition to the changes recorded by Sotherton & Self (2000), Robinson & Sutherland (2002) note that bumblebees (Bombidae) have also declined, particularly in eastern and central England. Data for other groups such as mammals, reptiles and amphibians are even more sparse, but it is likely that reptiles and amphibians, and at least some species of mammal, have declined on farmland (Robinson & Sutherland, 2002).

The status of the rural landscape and natural environment have been monitored by the Countryside Survey (CS) 2000 survey and its predecessors, carried out by the Centre for Ecology and Hydrology. These provide information on land cover, landscape features, freshwaters, habitats and vegetation (Firbank *et al.*, 2003a). Losses of hedges and ponds in the 1980s were halted or reversed during the 1990s. However, plant species richness declined in 34 out of 37 tests of change in broad habitats between 1990 and 1998 (Smart *et al.*, 2003).

4.5 Policy indicators of biodiversity change

The UK Government has adopted 147 core 'Quality of Life' indicators, and 15 headline indicators, to measure the success of its sustainable development strategy (Department of the Environment, Transport and the Regions, 1999). One of the headline indicators is an index of breeding bird populations, based on the results of the CBC and BBS for 105 species. It is considered to be "a good indicator of biodiversity generally, due to the wide ranging habitat distribution of wild bird species" (Department for Environment, Food and Rural Affairs, 2002). Birds also tend to be at or near the top of the food chain, which also serves to make them good indicators of biodiversity generally. In addition to the main indicator for all species, subsidiary indicators for farmland birds and woodland birds have also been developed. The farmland bird indicator, based on 19 species, has declined by over 40% since 1970. Part of the Government's Public Service Agreement target is to reverse this decline by 2020.

In addition to the headline indicator, core Quality of Life indicators of relevance for biodiversity include:

- trends in plant diversity;
- Biodiversity Action Plans;
- landscape features;
- extent and management of SSSIs; and
- native species at risk.

In addition to the UK-wide indicators, Scotland has its own set of sustainable development indicators. The indicator relevant to biodiversity is 'Percentages of Biodiversity Action Plan species and habitats which are identified as stable or increasing' (Scottish Executive, 2002a). Although Scottish policy regarding biodiversity conservation is linked to BAP species and habitats, it is axiomatic that achievement of policy objectives for target species is dependent upon the maintenance of a healthy ecosystem which provides the resources required to maintain populations of target species.

4.6 Changes in farming practices and their influence on biodiversity

The major habitat-related determinants of animal abundance and diversity on farmland are food availability and structure, which influence habitat use for shelter, breeding and foraging. Because many species require a range of habitat types and conditions to complete their life cycle, habitat diversity needs to be considered, as well as the condition of individual habitats.

The major changes in farming practices which have occurred since the 1940s include:

- a polarisation of farming towards arable in the east, and grassland/livestock in the west, with a concomitant decline in mixed farming and the use of grass leys;
- a trend towards larger, more specialised farms with larger fields and fewer hedgerows (though hedgerows are less common in Scotland than in other parts of the UK); and
- increased use of fertiliser and pesticides.

On arable land, the changes include:

- simpler crop rotations and block cropping;
- a switch from spring to autumn sowing (less pronounced in Scotland);
- more efficient harvesting and sealed grain storage; and
- recently, more non-inversion tillage.

On grassland the changes include:

- more drainage and reseeding;
- a switch from hay to silage;
- a switch from dicotyledonous fodder crops to grass;
- increased stocking rates; and
- introduction of avermectins.

Space does not permit a detailed description of all of these changes here, but further details may be found in Stoate (1996), Chamberlain *et al.* (2000), Vickery *et al.* (2001) and Robinson & Sutherland (2002). A further potential change which may affect biodiversity in arable areas is the introduction of genetically modified crops, though at least in the near future such effects are likely to be mainly a result of modification of pesticide use rather than genetic modification *per se* (Firbank *et al.*, 2003b).

Changes in farming practices in Scotland have not always been as great as in other parts of the UK. For example, spring barley and fodder crops are still widely grown, and in the north and west, the crofting system has maintained many elements of traditional management. Nowhere is this more evident than in the machair of the Hebrides, Orkney and Shetland, where the persistence of traditional techniques has permitted the survival of a diversity of wildlife which has been lost from most of the British Isles (Taylor, 2000).

Because the changes listed above have generally occurred more or less concurrently, it is not always easy to infer the causes of associated wildlife population changes from survey data unless supporting experimental data are available. Fortunately, there has been a number of such experimental studies in recent years which have greatly increased our understanding of the relationship between farming practices and biodiversity, and ways in which they can be better integrated. There follows a review of the effects of different farming practices and changes in these practices on biodiversity, using selected examples from the literature.

4.6.1 Mixed farming

The trend towards grassland-dominated farming in the west and arable farming in the east has reduced the diversity of habitats and resources associated with mixed farming. Chamberlain & Fuller (2000) found that the probability of local extinction of bird species was related to this polarisation of farming, with the rate of local extinction being greater in pastoral areas. Robinson *et al.* (2001) modelled the effect of the amount of arable land present in grassland landscapes on the abundance of 15 bird species. They found a positive correlation between numbers of some species and arable land, which was strongest when arable land was rare in the area. For the 11 seed-eating species which they studied, a statistically significant positive effect of arable land was recorded for six of the eight declining species studied, whilst for the three stable or increasing species, non-significant or negative effects were recorded. They concluded that the loss of arable habitat may be responsible for the declines of some granivorous species in pastoral areas, and that the retention of pockets of arable management may be important in allowing populations of granivorous species to persist in these areas.

In a survey of habitat associations of seed-eating birds in winter in Scotland, Hancock & Wilson (2003) found weedy fodder crops and stubble fields to be important habitats. Where fodder crops are still grown, they perform a similar function to arable crops in providing seed food for birds, but in this respect they may be more valuable because higher levels of weeds are often tolerated than in cash crops.

Similar effects to those detected by Robinson *et al.* (2001) may occur in relation to levels of grassland in predominantly arable areas, but this has not yet been tested in a similar manner. For example, lapwings (*Vanellus vanellus*) nesting in arable land prefer fields which are close to short pasture, which they use as feeding habitat (Galbraith, 1988). Traditional mixed rotations incorporated leys, which in the past were often established by undersowing a previous spring cereal, a practice which encourages sawflies (Hymenoptera: Symphyta) (Aebischer, 1990). These insects are among the most favoured food items of the rapidly declining grey partridge (*Perdix perdix*), and are also eaten by chicks of other declining species such as skylark (*Alauda arvensis*) and corn bunting (*Miliaria calandra*).

Brown hares (*Lepus capensis*) have also decreased in the pastoral west, probably due to a loss in habitat diversity; in surveys they were positively associated with arable land (Hutchings & Harris, 1996; Vaughan *et al.*, 2003).

4.6.2 Field size and hedgerows

Increased mechanization and the development of larger machinery have resulted in a trend towards increasing field size in order to achieve economies of scale. This has resulted in a reduction of habitat diversity, as well as large-scale losses of hedgerows during the 1960s, 1970s and 1980s, particularly in arable areas. Barr & Parr (1994) estimated net losses in Scotland of around 3,300 km between 1978 and 1984, and 13,300 km between 1984 and

1990. However, data from the most recent Countryside Survey indicate that the net loss in hedgerows has been halted (Petit *et al.*, 2003). Hedgerows and field margins are important habitats for a range of wildlife (Boatman *et al.*, 1999; Barr & Petit, 2001), and it is probable that removal has resulted in reductions in abundance and diversity of a number of species in areas where it has been widespread. However, species inhabiting such habitats are generally also widespread in other habitats. Gillings & Fuller (1998) found no significant effect of habitat loss on population trends for 38 bird species, and concluded that loss of hedgerows and other habitats was not the main factor responsible for bird declines. They suggested that habitat degradation was more important.

Although in Scotland there is a Biodiversity Action Plan for Ancient and/or Species-rich Hedgerows, hedges were far less common than in England and Wales before enclosure began around 1750, with most boundaries being marked by stone walls or turf banks. Survey work during 1998 and 1999 covering much of lowland Scotland found only 169 species-rich hedges (Hepburn, 2000).

4.6.3 Fertiliser

Fertiliser, particularly nitrogen, is one of the main yield determinants for grassland and most arable crops. Levels of nitrogen fertiliser use in Britain increased from around 500,000 tonnes in 1960 to around 1,600,000 tonnes in the 1980s, though they have since decreased to around 1,400,000 tonnes. Over the same period, average annual application rates to all crops increased from around 35 kg ha^{-1} to around 130 kg ha^{-1} (Robinson & Sutherland, 2002). There is no doubt that this increased fertiliser use has had a major influence on the flora and associated fauna of agricultural land. Increased nutrient availability was identified as a major driver of vegetation change in the analysis of data from the most recent Countryside Surveys (Smart *et al.*, 2003). At very low levels of fertility, addition of fertiliser can increase species diversity, but above a standing crop of around 7,000 kg ha^{-1}, diversity falls rapidly with increasing fertility (Marrs, 1993). Slower growing plant species are suppressed by competition from high yielding species which are able to respond to added fertiliser (Mountford *et al.*, 1993; Kirkham *et al.*, 1996). Increased fertiliser use, often in conjunction with reseeding of grassland to ryegrass (*Lolium perenne*), which is more responsive to fertiliser than other species, has resulted in major losses of botanical diversity in the majority of UK grasslands in the lowlands and the more accessible uplands. Nitrogen fertiliser is also detrimental to the survival of a number of species of rare arable flora (Wilson, 1999).

Whilst this loss of botanical diversity may have reduced the diversity of phytophagous invertebrates such as plant hoppers (Auchenorrhyncha), effects on invertebrates are complex, and some species may increase in numbers (Vickery *et al.*, 2001). Earthworms benefit from moderate fertiliser use, but decrease at high levels. Fertiliser also affects the structure of vegetation, which may change its suitability as nesting and foraging habitat for birds (Vickery *et al.*, 2001). A number of bird species, including lapwing, skylark, meadow pipit (*Anthus pratensis*), yellow wagtail (*Motacilla flava*), and starling prefer to forage in short, open vegetation, and the dense, rapidly growing swards resulting from high fertiliser use are likely to be less suitable as foraging habitat. Some species, such as lapwing, require short vegetation for nesting, and even species such as snipe (*Gallinago gallinago*) and redshank (*Tringa totanus*) which nest in long swards for concealment, use tussocky grass rather than uniformly dense swards (Vickery *et al.*, 2001)

4.6.4 Pesticides

Declines in some bird and mammal species due to the toxic effects of organochlorine pesticides during the 1960s and 1970s have been well documented (e.g. Newton, 1995). With the withdrawal of these pesticides, the affected populations have largely recovered and such direct effects are not considered to be significant at the current time. There is concern, however, over the indirect impact of pesticides through the food chain. This concern was stimulated by the work of the Game Conservancy Trust (Potts, 1986) which showed that reduced availability of invertebrate food for the chicks due to pesticide use was implicated in the decline of the grey partridge. A review by Campbell *et al.* (1997) concluded that such effects had not only been conclusively demonstrated for this species, but that they were possible for a further 11 species. Recent work has shown that pesticide use can affect the foraging behaviour and breeding performance of yellowhammer (*Emberiza citrinella*) (Morris *et al.*, 2002; Boatman *et al.*, 2004), and corn bunting (Brickle *et al.*, 2000).

Numbers of chick food insects are affected not only by insecticide use, but also by the use of herbicides which reduce the availability of food plants for phytophagous species. Furthermore, herbicides reduce weed seed production, thus reducing the food supply for seed-eating species of birds during the winter. Poor overwinter survival as a result of low food availability is thought to be a major factor in the recent declines of seed-eating birds (Robinson & Sutherland, 2002). Herbicides are also likely to have been at least partly responsible for the declines observed in much of our arable flora (Wilson, 1992). The recently published results of the Farm Scale Evaluations of genetically modified herbicide tolerant crops have greatly increased the available information on the impact of broad-spectrum herbicide use on weeds, weed seeds and invertebrates (Brooks *et al.*, 2003; Haughton *et al.*, 2003; Hawes *et al.*, 2003; Heard *et al.*, 2003a,b).

4.6.5 Arable cropping practices

The availability of inorganic fertilisers and pesticides has reduced the need for diverse crop rotations to maintain fertility and control weed, pest and disease populations. A few restrictions remain which cannot be alleviated by agrochemical use, such as the cereal disease 'take-all' caused by the fungus *Geaumannomyces graminis* var. *tritici*, which builds up in successive wheat crops and makes the growing of continuous wheat uneconomic, but simple stockless arable rotations restricted to a few of the most profitable crops are now the norm over large areas, particularly in the east. Furthermore, economies of scale often dictate that not only are fields larger, but blocks of fields are sown to the same crop. This is detrimental to species such as the brown hare, whose numbers are positively associated with landscape diversity (Tapper & Barnes, 1986). There is a need for more research on the implications of the scale of landscape variation for biodiversity. Organisms utilise the rural landscape at a variety of scales, at both individual and population level, and these do not necessarily coincide with the human concept of 'field' or 'farm'. Many organisms require more than one type of habitat to complete their life cycle, and the ability of such organisms to obtain the range of resources they need, and hence survive, will be determined by the relationship between effective home range size and the scale at which habitat diversity occurs.

Formerly, most crops were sown in the spring. The availability of higher-yielding varieties capable of overwintering has resulted in a change to autumn sowing for most major crops, aided by larger, more powerful and efficient tractors and drills capable of sowing large areas in a relatively short time. Thus, over 90% of wheat and 60% of barley are now sown in the autumn. Most of the remaining spring-sown barley is in Scotland, partly due to the importance of this crop for malting. This change is likely to have contributed to the decline of spring-germinating weeds such as corn marigold (*Chrysanthemum segetum*), though there are also a number of autumn-germinating rare arable plant species (Wilson, 2000).

Numbers of skylarks, as measured by the Common Bird Census (CBC), were correlated with the UK area of spring-sown cereals between 1968 and 1996, and spring cereals supported higher breeding territory densities, and a higher number of breeding attempts per pair per season (Donald & Vickery, 2000). Skylarks prefer to nest in shorter crops, and it seems likely that the later development of spring crops means that they provide suitable nesting habitat for longer, resulting in higher numbers of successful nesting attempts. Lapwings also prefer to nest in spring-sown cereals rather than autumn-sown cereals. The CBC index for lapwings is also correlated with the proportion of cereals sown in the spring, and it has been suggested that the reduction in spring sowing has been a causal factor in the decline of this species (Shrubb, 1990).

Changes in cropping patterns have not always been detrimental. For example, the increase in the cultivation of oilseed rape (*Brassica napus*) has been beneficial for woodpigeons (*Columba palumbus*), which rely on it as a source of food in winter (Inglis *et al.*, 1997), and linnets (*Carduelis cannabina*), which feed unripe rape seeds to their chicks, instead of the formerly more abundant seeds of weeds such as charlock (*Sinapis arvensis*) (Moorcroft *et al.*, 1997). Oilseed rape is also commonly used for nesting by reed buntings (*Emberiza schoeniclus*), corn buntings and sedge warblers (*Acrocephalus schoenobaenus*), among other species (Watson & Rae, 1998).

4.6.6 Harvesting and grain storage

Modern combine harvesters are highly efficient, producing less wastage than formerly, and resulting in low densities of grain remaining in stubbles after harvest. This exacerbates the impact of efficient weed control practices on the production of weed seeds, further reducing the availability of food for seed-eating birds and other seed predators over the autumn and winter period. Species such as corn bunting, for which cereal grain forms a major part of the winter diet (Brickle & Harper, 2000), are likely to have been particularly affected. Furthermore, farms entering the Assured Combinable Crops Scheme now have to demonstrate that their grain storage facilities are bird-proof (www.assuredcrops.co.uk), and the increased emphasis on tidiness and hygiene around farm yards may have contributed to the recent decline in rural house sparrow populations (Hole *et al.*, 2002).

4.6.7 Cultivation practices

In recent years there has been a revival of interest in non-inversion tillage, because of the substantial potential savings in labour, fuel and time compared to ploughing as a primary cultivation method. Reduced cultivation systems increase earthworm abundance (Edwards & Lofty, 1982; El Titi, 1995), but effects on beetles (Coleoptera) and linyphiid spiders (Araneae: Linyphiidae) varied between species and seasons (Kendall *et al.*, 1995). In recent studies, greater productivity was recorded for some farmland birds in minimum tilled, compared with conventionally tilled, fields in Canada (Martin & Forsyth, 2003), and skylarks, granivorous passerines and gamebirds (Phasianidae) occupied a significantly greater proportion of fields established by non-inversion tillage than conventional tillage in the English Midlands (Cunningham *et al.*, 2003). However, whilst grass weeds tend to increase under reduced tillage, dicotyledonous weeds tend to decline (Froud-Williams, 1981). As dicotyledonous weed seeds are most abundant in the diet of granivorous birds, the long term impact of reduced tillage on these species is equivocal, and further studies are needed.

4.6.8 Drainage and reseeding of grassland

During the 20th century, most lowland grassland, and much upland grassland, was subjected to agricultural improvement so that, by 1984, only 11% of the grassland area of lowland England and Wales was composed of semi-natural and rough grasslands, and only 4% were semi-natural (Fuller, 1987). Drainage and reseeding are often the first steps in the process of improvement, followed by increased fertiliser use, though some grasslands have been improved by fertiliser use alone. The drainage and improvement of wet grassland has led to the widespread decline of a number bird species, including wildfowl (Anatidae) and waders (Charadriiformes) (Beintema *et al.*, 1997). This has occurred both in the lowlands (Green & Robins, 1993) and uplands (Baines, 1989).

The effects of reseeding have not generally been studied in isolation from those of drainage and increased fertiliser use, but clearly the replacement of a diverse sward with a monoculture has major effects on botanical diversity and the dependent invertebrate community. Reseeded grasslands tend to have a more uniform sward structure and surface topography than unimproved grasslands. This may lead to increased nest predation of ground-nesting species such as lapwing (Baines, 1989).

4.6.9 Cutting management

Since the 1970s, silage has replaced hay making as the dominant form of grass conservation for winter feed. This has a number of potential impacts for biodiversity. Silage is made at an earlier stage of development, when nutrient content, rather than dry matter, is at an optimum. This means that there is time for a second, and sometimes a third, cut. The earlier, and more frequent, cutting means that seed set is largely prevented, which removes a food source for birds and other animals both in the grass crop itself and in dung or manure from animals fed on silage, when it is spread on the fields. Swards are also rolled in spring to prevent contamination of the silage by soil, which can lead to spoilage. This practice removes variation in microtopography, thus potentially reducing the habitat value of the sward (Vickery *et al.*, 2001).

Management by cutting produces swards that are generally uniform in structure and species composition. This generally reduces the abundance and diversity of invertebrates, in comparison with hay or grazing management, though different groups vary in their response (Morris, 2000). Timing of cutting also affects invertebrate communities, with summer cutting usually being more detrimental than late spring or autumn (Duffey *et al.*, 1974; Baines *et al.*, 1998; Morris, 2000). Relaxation of cutting regimes at field edges has been proposed as a possible conservation measure. Haysom *et al.* (2004) found that leaving headland plots in grass fields uncut increased the diversity of carabid beetles.

Cutting can increase accessibility of food items for foraging birds such as wagtails (*Motacilla* spp.), but reduces the breeding success of ground-nesting birds, the effects being increased by earlier and more frequent mowing. The most significant example of this effect is the decline of the corncrake (*Crex crex*), from a once widespread, and common, bird to a relict population in northern and north western Scotland, as a result of chick mortality caused by mowing (Green & Stowe, 1993).

4.6.10 Grazing and stocking rates

The effects of grazing are complex, depending on the type of stock, the stocking rate, and timing of grazing. Space does not permit adequate description here, but Vickery *et al.* (2001) provide a summary of grazing effects on plants, invertebrates and birds. Sheep grazing produces shorter, more uniform swards than cattle grazing. Botanical diversity is greatest at intermediate grazing pressures and is reduced at very high or low pressures (Grime, 1979). Invertebrates vary in response to grazing, but in general high grazing pressure is detrimental because it reduces the structural complexity of the sward as well as the botanical diversity. For example, lepidopteran larvae, which are key food items for chicks of black grouse (*Tetrao tetrix*) are less abundant in heavily grazed rather than lightly grazed moorland (Baines, 1996). Seasonal grazing encourages sward diversity, and hence insect diversity, compared to continuous grazing, and autumn grazing is generally less deleterious than spring grazing (Vickery *et al.*, 2001). However, generalisations can be misleading; for example, ideal conditions for crane fly (*Tipula* spp.) oviposition and larval development are created by low grazing pressure in late summer and autumn, followed by heavy grazing in late winter and spring. Pastures managed in this way were selected by choughs (*Pyrrhocorax pyrrhocorax*) in the island of Islay when provisioning their young (Bignal & McCracken, 1993).

Short swards are often preferred by many bird species for feeding, because of greater access to prey, but small mammal populations are highest in long swards, which are therefore favoured by predatory birds such as barn owl (*Tyto alba*) and kestrel (*Falco tinnunculus*). Bird species vary in their sward height preferences for nesting, but high stocking rates can result in high rates of nest destruction by trampling and predation (Green, 1988; Shrubb, 1990).

The decline in ground-nesting birds in the Welsh uplands has been linked to the large increase in sheep numbers during the 1970s and 1980s (Fuller & Gough, 1999). The average number of bird species dependent on extensive pastoral systems declined as sheep livestock units per hectare increased between 1970 and 1990 (Pain *et al.*, 1997). Nevertheless, stocking rates remain much lower in Scotland than in England and Wales, and breeding wader populations are much higher (Shrubb, 2003). In Wales, where stocking rates are highest (7,399 sheep/1,000 ha), upland birds have declined more than in northern England and southern Scotland (3,350 sheep/1,000 ha), with the least decline in northern Scotland, where the average sheep density is only 1,152 sheep/1,000 ha (Shrubb, 2003).

4.6.11 Burning

Burning is used widely in the uplands as a method of promoting fresh growth. Where grouse (*Lagopus lagopus scoticus*) shooting is the major land use, burning is carried out carefully in small blocks on a long rotation to create a mosaic of heather (*Calluna vulgaris*) of different ages, and such burning helps to maintain the quality of the habitat. Where sheep (*Ovis aries*) dominate, however, burning is much larger scale and more frequent. Such large scale, frequent burning results in soil degradation and erosion, and reduces the value of upland vegetation as wildlife habitat (Shrubb, 2003; Tucker, 2004).

4.6.12 Introduction of avermectins

The introduction of avermectin anthelminthics (worming agents) in the 1980s to control livestock parasites gave rise to concern over the effects on dung-dwelling invertebrates, which break down the dung and also provide a food source for birds. Research has shown that insect activity in dung pats from treated animals is considerably reduced, and reproduction of some groups was affected. However, the consequences of these effects on populations of the species concerned, or their predators, are not yet clear (Vickery *et al.*, 2001).

4.7 Conserving and restoring biodiversity on farmland

Various mechanisms exist for encouraging the conservation of biodiversity on farmland, including regulation, cross-compliance, agri-environment schemes and the promotion of farming systems intended to improve environmental sustainability, such as integrated farm management and organic farming.

Cross-compliance refers to the application of conditions with which farmers must comply in return for the receipt of CAP payments. Farmers receiving subsidies will be obliged to maintain their land in 'Good Agricultural and Environmental Condition'. Regulation and proposed cross-compliance conditions are largely concerned with addressing issues of pollution, resource protection and conservation of historic and landscape features rather than with biodiversity. However, some proposed measures, such as restrictions on overgrazing, retention of landscape features, and the development of whole farm plans, have implications for biodiversity.

The impact of farming systems on biodiversity was recently reviewed by Leake (2002), who concluded that "It is not the farming system *per se* which exerts the greatest influence upon biodiversity but aspects associated with the system". The aspects to which he refers are the types and patterns of production and management practices which flow from the principles underlying the system, for example organic farms tend to be mixed and have diverse rotations; 'integrated' farming systems are more likely to practice non-inversion tillage.

Integrated farming systems seek to retain modern farming practices, such as the use of pesticides and inorganic fertilisers, but in a framework which minimises their use and promotes sustainability. The definition drawn up by the International Organisation for Biological Control (IOBC; www.iobc-wprs.org), which first promoted the concept, can be summarised as "a farming system that produces high quality food and other products by using natural resources and regulating mechanisms to replace polluting inputs and to secure sustainable farming". There are few studies of the effects of integrated farming systems on biodiversity, other than on invertebrate species important as natural enemies of crop pests, but those which have been carried out do not indicate major impacts. For example, Holland *et al.* (1998) found that, in a large split-field study on six sites over five years, numbers and diversity of non-target arthropods differed most between sites, years and crops and least between the two farming systems. Where evidence of differences has been found, this is generally in relation to differences in cultivation techniques (see above). There is a need for more research into the effects of integrated farming systems on biodiversity.

More studies have been carried out comparing biodiversity between organic and conventional farming systems. A useful summary is provided by the Soil Association (2000). Plant diversity tends to be greater in organic systems, but effects on invertebrates vary between groups (Moreby *et al.*, 1994). Studies on birds indicate that organic farms tend to support greater numbers of most species (e.g. Chamberlain *et al.*, 1999). Chamberlain & Wilson (2000) found that, of 19 species which were significantly more abundant on organic farms in at least one season of one year, the differences for eight species could be explained by hedge structure alone, emphasising the importance of specific management aspects within the system. Wickramasinghe *et al.* (2003) found that bat (Chiroptera) foraging activity was higher over organic farms than conventional farms, and concluded that bats benefit from less intensive farming, though their methods did not allow separation of effects due to differences in habitats from any crop management effects. However, they did note that key insect groups important to bats as food were more abundant on organic farms. Leake (2002) considered that diverse rotations incorporating livestock and both spring- and autumn-sown crops, and incomplete weed control, are the most beneficial features of organic farming. There is a need for more research on the effects of different components of organic farming and the interactions between them.

Although organic farming does appear to offer some benefits to biodiversity as outlined above, it is unlikely to occupy more than a small proportion of agricultural land in the foreseeable future. Agri-environment schemes appear to offer the greatest hope of achieving wide-scale biodiversity benefits across the farmed landscape, because they are targeted specifically at achieving biodiversity objectives and include a range of prescriptions which have been shown by research to be effective (Evans *et al.*, 2002). In Scotland, agri-environment schemes currently include the Rural Stewardship Scheme (RSS), Environmentally Sensitive Areas (ESAs) and the Organic Aid Scheme (OAS).

Some examples of outstanding successes for agri-environment schemes have been demonstrated (Aebischer *et al.*, 2000), though unfortunately, in many cases, the standard of monitoring has not been adequate to permit the rigorous evaluation of biodiversity benefits (Kleijn & Sutherland, 2003). Furthermore, the greatest successes to date have been for species with restricted distributions, such as corncrake, stone curlew (*Burhinus oedicnemus*) and cirl bunting (*Emberiza cirlus*), where specific and targeted research-based management programmes have been implemented over limited areas (Aebischer *et al.*, 2000). Achievement of BAP targets for more widely distributed species, and the Public Sector Agreement target for farmland birds, remains a significant challenge. Nevertheless, most commentators remain optimistic about the potential of agri-environment schemes to deliver biodiversity benefits (e.g. Evans *et al.*, 2002), and the current reform linked to the mid-term review of the Common Agricultural Policy offers opportunities to develop and improve agri-environment schemes and monitoring programmes, with encouraging prospects for the future of biodiversity on farmland.

4.8 The implementation of agri-environmental policy in Scotland

In 2001, the Scottish Executive Agriculture Strategy Steering Group published *A Forward Strategy for Scottish Agriculture* (Scottish Executive, 2001), which sets out a vision for the future development of farming in Scotland. The authors emphasize the need to tackle economic, social and environmental issues in a co-ordinated manner, and seek "solutions which are good for the environment *and* for business".

As part of the implementation of the strategy, an Agriculture and Environment Working Group was set up to "examine the environmental issues which will impact on farming … over the next 5-10 years, and to advise on how best to tackle them" (Scottish Executive,

2002b). In their report, biodiversity was identified as one of three priority environmental issues for Scotland over the next 5-10 years, the others being diffuse pollution to water, and landscape change. Recommendations relevant to this issue include revising set-aside prescriptions to maximise environmental benefit and link more appropriately with the RSS, the co-ordination of action at a bio-regional or catchment scale, the development of whole farm biodiversity management plans as a component of cross-compliance, the redesign of the RSS (incorporating ESAs), and the retention of the OAS as a discretionary scheme. Regional monitoring programmes for the RSS were also recommended. In crofting areas, measures to reduce overstocking with sheep and to encourage cattle rearing and cropping were proposed.

A further measure proposed in *A Forward Strategy for Scottish Agriculture* is the development of Land Management Contracts to encourage multi-functional farming, similar to those implemented in France. These would have a tiered structure, and include payments for agri-environment measures as well as forestry, socio-economic, quality and animal health measures. Farmers would choose from a menu, and the first two tiers would be generally available, though tier III would probably be discretionary and competitive.

4.9 Conclusion

Agricultural intensification has led to widespread declines in biodiversity through a variety of mechanisms. However, successive reforms of the CAP have resulted in the provision of a range of measures to promote the recovery and persistence of a diversity of wildlife on farmland. Agri-environment schemes offer the greatest potential to reverse past declines. Lessons learned from the implementation of such schemes so far indicate that, to increase the chances of success, options should be research based and targeted, both geographically but also in terms of desired outcomes. Many species require a range of habitats and resources to complete their life cycle, and the provision of suites of options may be necessary to provide maximum benefits. Wherever possible, implementation of options at a landscape scale (such as through co-operative action between farmers or crofters), rather than at the scale of individual farms, will increase the prospects of maintaining viable populations of target species. Schemes should be supported by appropriate monitoring and evaluation, with feedback into the further development of the scheme. Adherence to these principles in the future development of schemes under the Rural Development Regulation will ensure that the achievement of biodiversity conservation objectives within the context of a sustainable agriculture becomes a realistic objective.

References

Aebischer, N.J. (1990). Assessing pesticide effects on non-target invertebrates using long-term monitoring and time-series modelling. *Functional Ecology*, **4**, 369-373.

Aebischer, N.J., Green, R.E. & Evans, A.D. (2000). From science to recovery: four case studies of how research has been translated in to conservation action in the UK. In *Ecology and Conservation of Lowland Farmland Birds*, ed. by N.J. Aebischer, A.D. Evans, P.V. Grice & J.A. Vickery. British Ornithologists' Union, Tring. pp. 43-53.

Baines, D. (1989). The effects of improvement of upland, marginal grasslands on the breeding success of Lapwings *Vanellus vanellus* and other waders. *Ibis*, **131**, 497-506.

Baines, D. (1996). The implications of grazing and predator management on the habitats and breeding success of black grouse *Tetrao tetrix*. *Journal of Applied Ecology*, **33**, 54-62.

Baines, M., Hambler, C., Johnson, P.J., Macdonald, D.W. & Smith, H. (1998). The effects of arable field margin management on the abundance and species richness of Araneae (spiders). *Ecography*, **21**, 74-86.

Barr, C.J. & Parr, T.W. (1994). Hedgerows: linking ecological research and countryside policy. In *Hedgerow Management and Nature Conservation*, ed. by T.A. Watt & G.P. Buckley. Wye College Press, Wye. pp. 119-136.

Barr, C.J. & Petit, S. (Eds) (2001). *Hedgerows of the World: their Ecological Functions in Different Landscapes*. International Association for Landscape Ecology (UK), Cheshire.

Beintema, A.J., Dunn, E. & Stroud, D.A. (1997). Birds and wet grasslands. In *Farming and Birds in Europe: the Common Agricultural Policy and its Implications for Bird Conservation*, ed. by D.J. Pain & M.W. Pienkowski. Academic Press, London. pp. 269-296.

Benton, T.G., Bryant, D.M., Cole, L. & Crick, H.Q.P. (2002). Linking agricultural practice to insect and bird populations: a historical study over three decades. *Journal of Applied Ecology*, **39**, 673-687.

Bignal, E. & McCracken, D. (1993). Nature conservation and pastoral farming in the British Uplands. *British Wildlife*, **4**, 367-376.

Birnie, R.V., Curran, J., MacDonald, J.A., Mackey, E.C., Campbell, C.D., McGowan, G., Palmer, S.C.F., Paterson, E., Shaw, P. & Shewry, M.C. (2002). The land resources of Scotland: trends and prospects for the environment and natural heritage. In *The State of Scotland's Environment and Natural Heritage*, ed. by M.B. Usher, E.C. Mackey & J.C. Curran. The Stationery Office, Edinburgh. pp. 41-81.

Boatman, N.D., Davies, D.H.K., Chaney, K., Feber, R., de Snoo, G.R. & Sparks, T.H. (Eds) (1999). Field margins and buffer zones: ecology, management and policy. *Aspects of Applied Biology*, **54**. Association of Applied Biologists, Wellesbourne.

Boatman, N.D., Brickle, N.W., Hart, J.D., Holland, J.M., Milsom, T.P., Morris, A.J., Murray, A.W.A., Murray, K.A. & Robertson, P.A. (2004). Evidence for the indirect effects of pesticides on farmland birds. *Ibis*, **146** (supplement 2), 131-143.

Brickle, N.W. & Harper, D.G.C. (2000). Habitat use by corn buntings *Miliaria calandra* in winter and summer. In *Ecology and Management of Lowland Farmland Birds*, ed. by N.J. Aebischer, A.D. Evans, P.V. Grice & J.A. Vickery. British Ornithologists' Union, Tring. pp. 156-164.

Brickle, N.W., Harper, D.G.C., Aebischer, N.J. & Cockayne, S.H. (2000). Effects of agricultural intensification on the breeding success of corn buntings *Miliaria calandra*. *Journal of Applied Ecology*, **37**, 742-755.

Brooks, D.R., Bohan, D.A., Champion, G.T., Haughton, A.J., Hawes, C., Heard, M.S., Clark, S.J., Dewar, A.M., Firbank, L.G., Perry, J.N., Rothery, P., Scott, R.J., Woiwod, I.P., Birchall, C., Skellern, M.P., Walker, J.H., Baker, P., Bell, D., Browne, E.L., Dewar, A.J.G., Fairfax, C.M., Garner, B.H., Haylock, L.A., Horne, S.L., Hulmes, S.E., Mason, N.S., Norton, L.R., Nuttall, P., Randle, Z., Rossall, M.J., Sands, R.J.N., Singer, E.J. & Walker, M.J. (2003). Invertebrate responses to the management of genetically modified herbicide-tolerant and conventional spring crops. 1. Soil-active invertebrates. *Philosophical Transactions of the Royal Society London B*, **358**, 1847-1862.

Campbell, L.H., Avery, M.I., Donald, P., Evans, A.D., Green, R.E. & Wilson, J.D. (1997). *A Review of the Indirect Effects of Pesticides on Birds*. Report No. 227. Joint Nature Conservation Committee, Peterborough.

Chamberlain, D.E. & Fuller, R.J. (2000). Local extinctions and changes in species richness of lowland farmland birds in England and Wales in relation to recent changes in agricultural land-use. *Agriculture, Ecosystems and Environment*, **78**, 1-17.

Chamberlain, D.E. & Wilson, J.D. (2000). The contribution of hedgerow structure to the value of organic farms to birds. In *Ecology and Conservation of Lowland Farmland Birds*, ed. by N.J. Aebischer, A.D. Evans, P.V. Grice & J.A. Vickery. British Ornithologists Union, Tring. pp. 57-68.

Chamberlain, D.E., Wilson, J.D. & Fuller, R.J. (1999). A comparison of bird populations on organic and conventional farmland in southern Britain. *Biological Conservation*, **88**, 307-320.

Chamberlain, D.E., Fuller, R.J., Bunce, R.G.H., Duckworth, J.C. & Shrubb, M.J. (2000). Changes in the abundance of farmland birds in relation to the timing of agricultural intensification in England and Wales. *Journal of Applied Ecology*, **37**, 771-788.

Cunningham, H.M., Chaney, K., Wilcox, A. & Bradbury, R.B. (2003). Non-inversion tillage and farmland birds in winter. *BCPC International Congress – Crop Science and Technology 2003: Congress Proceedings*, **1**. British Crop Protection Council, Alton. pp. 533-536.

Department for Environment, Food and Rural Affairs (2002). *Foundations for our Future – Defra's Sustainable Development Strategy*. Department for Environment, Food and Rural Affairs, London.

Department of the Environment, Transport and the Regions (1999). *A Better Quality of Life: a Strategy for Sustainable Development in the UK* (Cm 4345). The Stationery Office, London.

Donald, P.F. & Vickery, J.A. (2000). The importance of cereal fields to breeding and wintering skylarks *Alauda arvensis* in the UK. In *Ecology and Conservation of Lowland Farmland Birds*, ed. by N.J. Aebischer, A.D. Evans, P.V. Grice & J.A. Vickery. British Ornithologists' Union, Tring. pp. 140-150.

Donald, P.F., Green, R.E. & Heath, M.F. (2001). Agricultural intensification and the collapse of Europe's farmland bird populations. *Proceedings of the Royal Society of London B*, **268**, 25-29.

Duffey, E., Morris, M.G., Sheail, J., Ward, L.K., Wells, D.A. & Wells, T.C.E. (1974). *Grassland Ecology and Wildlife Management*. Chapman & Hall, London.

Edwards, C.A. & Lofty, J.R. (1982). The effect of direct drilling and minimal cultivation on earthworm populations. *Journal of Applied Ecology*, **19**, 723-734.

El Titi, A. (1995). Ecological aspects of integrated farming. In *Ecology and Integrated Farming Systems*, ed. by D.M. Glen, M.P. Greaves & H.M. Anderson. John Wiley & Sons, Chichester. pp. 243-256.

Evans, A.D., Armstrong-Brown, S. & Grice, P.V. (2002). The role of research and development in the evolution of a 'smart' agri-environment scheme. In *Birds and Agriculture: Aspects of Applied Biology 67*, ed. by N.D. Boatman, D.H.K. Davies, K. Chaney, R. Feber, G.R. de Snoo & T.H. Sparks, Association of Applied Biologists, Wellesbourne. pp. 253-264.

Firbank, L.G., Barr, C.J., Bunce, R.G.H., Furse, M.T., Haines-Young, R., Hornung, M., Howard, D.C., Sheail, J., Sier, A. & Smart, S.M. (2003a). Assessing stock and change in land cover and biodiversity in GB: an introduction to Countryside Survey 2000. *Journal of Environmental Management*, **67**, 207-218.

Firbank, L.G., Heard, M.S., Woiwod, I.P., Hawes, C., Haughton, A.J., Champion, G.T., Scott, R.J., Hill, M.O., Dewar, A.M., Squire, G.R., May, M.J., Brooks, D.R., Bohan, D.A., Daniels, R.E., Osborne, J.L., Roy, D.B., Black, H.I.J., Rothery, P. & Perry, J.N. (2003b). An introduction to the Farm-Scale Evaluations of genetically modified herbicide tolerant crops. *Journal of Applied Ecology*, **40**, 2-16.

Froud-Williams, R.J. (1981). Potential changes in weed floras associated with reduced-cultivation systems for cereal production in temperate regions. *Weed Research*, **21**, 99-109.

Fuller, R.J. (1987). The changing extent and conservation interest of lowland grasslands in England and Wales: a review of grassland surveys 1930-94. *Biological Conservation*, **40**, 281-300.

Fuller, R.J. & Gough, S.J. (1999). Changes in sheep numbers in Britain: implications for bird populations. *Biological Conservation*, **91**, 73-89.

Galbraith, H. (1988). Effects of agriculture on the breeding ecology of lapwings *Vanellus vanellus*. *Journal of Applied Ecology*, **25**, 487-503.

Gibbons, D.W., Reid, J.B. & Chapman, R.A. (1993). *The New Atlas of Breeding Birds in Britain and Ireland*. T. & A.D. Poyser, London.

Gillings, S. & Fuller, R.J. (1998). Changes in bird populations on sample lowland English farms in relation to loss of hedgerows and other non-crop habitats. *Oecologia*, **116**, 120-127.

Green, R.E. (1988). Effects of environmental factors on the timing and success of breeding of common snipe *Gallinago gallinago* (Aves: Scolopacidae). *Journal of Applied Ecology*, **25**, 79-93.

Green, R.E. & Robins, M. (1993). The decline of the ornithological importance of the Somerset Levels and Moors, England and changes in the management of water levels. *Biological Conservation*, **66**, 95-106.

Green, R.E. & Stowe, T.J. (1993). The decline of the corncrake *Crex crex* in Britain and Ireland in relation to habitat change. *Journal of Applied Ecology*, **30**, 689-695.

Gregory, R.D., Noble, D.G., Robinson, J.A., Stroud, D.A., Campbell, L.H., Rehfish, M.M., Cranswick, P.A., Wilkinson, N.I., Crick, H.Q.P. & Green, R.E. (2002). *The State of the UK's Birds 2001*. Royal Society for the Protection of Birds, Sandy; British Trust for Ornithology, Thetford; Wildfowl & Wetlands Trust Slimbridge; and Joint Nature Conservation Committee, Peterborough.

Grime, J.P. (1979). *Plant Strategies and Vegetation Processes*. John Wiley & Sons, Chichester.

Hancock, M.H. & Wilson, J.D. (2003). Winter habitat associations of seed-eating passerines on Scottish farmland. *Bird Study*, **50**, 116-130.

Haughton, A.J., Champion, G.T., Hawes, C., Heard, M.S., Brooks, D.R., Bohan, D.A., Clark, S.J., Dewar, A.M., Firbank, L.G., Osborne, J.L., Perry, J.N., Rothery, P., Roy, D.B., Scott, R.J., Woiwod, I.P., Birchall, C., Skellern, M.P., Walker, J.H., Baker, P., Browne, E.L., Dewar, A.J.G., Garner, B.H., Haylock, L.A., Horne, S.L., Mason, N.S., Sands, R.J.N. & Walker, M.J. (2003). Invertebrate responses to the management of genetically modified herbicide-tolerant and conventional spring crops. II. Within-field epigeal and aerial arthropods. *Philosophical Transactions of the Royal Society London B*, **358**, 1863-1877.

Hawes, C., Haughton, A.J., Osborne, J.L., Roy, D.B., Clark, S.J., Perry, J.N., Rothery, P., Bohan, D.A., Brooks, D.R., Champion, G.T., Dewar, A.M., Heard, M.S., Woiwod, I.P., Daniels, R.E., Young, M.W., Parish, A.M., Scott, R.J., Firbank, L.G. & Squire, G.R. (2003). Responses of plants and invertebrate trophic groups to contrasting herbicide regimes in the Farm Scale Evaluations of genetically modified herbicide-tolerant crops. *Philosophical Transactions of the Royal Society London B*, **358**, 1899-1913.

Haysom, K.A., McCracken, D.I., Foster, G.N. & Sotherton, N.W. (2004). Developing grassland conservation headlands: response of carabid assemblage to different cutting regimes in a silage field edge. *Agriculture, Ecosystems and Environment*, **102**, 263-277.

Heard, M.S., Hawes, C., Champion, G.T., Clark, S.J., Firbank, L.G., Haughton, A.J., Parish, A.M., Perry, J.N., Rothery, P., Roy, D.B., Scott, R.J., Skellern, M.P., Squire, G.R. & Hill, M.O. (2003a). Weeds in fields with contrasting conventional and genetically modified herbicide-tolerant crops. II. Effects on individual species. *Philosophical Transactions of the Royal Society London B*, **358**, 1833-1846.

Heard, M.S., Hawes, C., Champion, G.T., Clark, S.J., Firbank, L.G., Haughton, A.J., Parish, A.M., Perry, J.N., Rothery, P., Scott, R.J., Skellern, M.P., Squire, G.R. & Hill, M.O. (2003b). Weeds in fields with contrasting conventional and genetically modified herbicide-tolerant crops. I. Effects on abundance and diversity. *Philosophical Transactions of the Royal Society London B*, **358**, 1819-1832.

Hepburn, L.V. (2000). Establishment of a register of species-rich hedgerows in Scotland 1998-1999. Unpublished report No. F99AA104. Scottish Natural Heritage, Perth.

Hole, D.G., Whittingham, M.J., Bradbury, R.B., Anderson, G.Q.A., Lee, P.L.M., Wilson, J.D. & Krebs, J.R. (2002). Widespread local house-sparrow extinctions – agricultural intensification is blamed for the plummeting populations of these birds. *Nature*, **418**, 931-932.

Holland, J.M., Cook, S.K., Drysdale, A.D., Hewitt, M.V., Spink, J. & Turley, D.B. (1998). The impact on non-target arthropods of integrated compared to conventional farming: results from the LINK Integrated Farming Systems project. *Proceedings of the 1998 Brighton Conference – Pests and Diseases*. British Crop Protection Council, Farnham. pp. 625-630.

Hutchings, M.R. & Harris, S. (1996). *The Current Status of the Brown Hare* (Lepus europaeus) *in Britain*. Joint Nature Conservation Committee, Peterborough.

Inglis, I.R., Isaacson, A.J., Smith, G.C., Haynes, P.J. & Thearle, R.J.P. (1997). The effect on the woodpigeon (*Columba palumbus*) of the introduction of oilseed rape into Britain. *Agriculture, Ecosystems and Environment*, **61**, 113-121.

Kendall, D.A., Chinn, N.E., Glen, D.M., Wiltshire, C.W., Wintstone, L. & Tidboald, C. (1995). Effects of soil management on cereal pests and their natural enemies. In *Ecology and Integrated Farming Systems*, ed. by D.M. Glen, M.P. Greaves & H.M. Anderson. John Wiley & Sons, Chichester. pp. 83-102.

Kirkham, F.W., Mountford, J.O. & Wilkins, R.J. (1996). The effects of nitrogen, potassium and phosphorus addition on the vegetation of a Somerset peat moor under cutting management. *Journal of Applied Ecology*, **33**, 1013-1029.

Kleijn, D. & Sutherland, W.J. (2003). How effective are European agri-environment schemes in conserving and promoting biodiversity? *Journal of Applied Ecology*, **40**, 947-969.

Leake, A.R. (2002). Biodiversity in different farming systems. *The BCPC Conference – Pests and Diseases 2002*. British Crop Protection Council, Farnham. pp. 949-956.

Marrs, R.H. (1993). Soil fertility and nature conservation in Europe: theoretical considerations and practical management solutions. In *Advances in Ecological Research 24*, ed. by M. Begon & A.H. Fitter. Academic Press, London. pp. 241-300.

Martin, P.A. & Forsyth, D.J. (2003). Occurrence and productivity of songbirds in prairie farmland under conventional versus minimum tillage regimes. *Agriculture, Ecosystems and Environment*, **96**, 107-117.

Moorcroft, D., Bradbury, R.B. & Wilson, J.D. (1997). The diet of nestling linnets *Carduelis cannabina* before and after agricultural intensification. *Brighton Crop Protection Conference – Weeds*. British Crop Protection Council, Farnham. pp. 923-928.

Moreby, S.J., Aebischer, N.J., Southway, S.E. & Sotherton, N.W. (1994). A comparison of the flora and arthropod fauna of organically and conventionally grown winter wheat in southern England. *Annals of Applied Biology*, **125**, 13-27.

Morris, M.G. (2000). The effects of structure and its dynamics on the ecology and conservation of arthropods in British grasslands. *Biological Conservation*, **95**, 129-142.

Morris, A.J., Bradbury, R.B. & Wilson, J.D. (2002). Indirect effects of pesticides on breeding yellowhammers *Emberiza citrinella*. *The BCPC Conference – Pests & Diseases 2002*. British Crop Protection Council, Farnham. pp. 965-970.

Mountford, J.O., Lakhani, K.H. & Kirkham, F.W. (1993). Experimental assessment of the effects of nitrogen addition under hay-cutting and aftermath grazing on the vegetation of meadows on a Somerset peat moor. *Journal of Applied Ecology*, **30**, 321-332.

Newton, I. (1995). The contribution of some recent research on birds to ecological understanding. *Journal of Animal Ecology*, **64**, 675-696.

Pain, D.J., Hill, D. & McCracken, D.I. (1997). Impact of agricultural intensification of pastoral systems on bird distributions in Britain 1970-1990. *Agriculture, Ecosystems and Environment*, **64**, 19-32.

Petit, S., Stuart, R.C., Gillespie, M.K. & Barr, C.J. (2003). Field boundaries in Great Britain: stock and change between 1984, 1990 and 1998. *Journal of Environmental Management*, **67**, 229-238.

Potts, G.R. (1986). *The Partridge: Pesticides, Predation and Conservation*. Collins, London.

Preston, C.D., Pearman, D.A. & Dines, T.D. (2002). *New Atlas of the British and Irish Flora*. Oxford University Press, Oxford.

Raven, M.J., Noble, D.G. & Baillie, S.R. (2003). *The Breeding Bird Survey 2002*. BTO Research Report No. 334. British Trust for Ornithology, Thetford.

Robinson, R.A. & Sutherland, W.J. (2002). Changes in arable farming and biodiversity in Great Britain. *Journal of Applied Ecology*, **39**, 157-176.

Robinson, R.A., Wilson, J.D. & Crick, H.Q.P. (2001). The importance of arable habitat for farmland birds in grassland landscapes. *Journal of Applied Ecology*, **38**, 1059-1069.

Scottish Executive (2001). *A Forward Strategy for Scottish Agriculture.* Scottish Executive, Edinburgh.

Scottish Executive (2002a). *Meeting the Needs ... Priorities, Actions and Targets for Sustainable Development in Scotland.* Scottish Executive Environment Group, Edinburgh.

Scottish Executive (2002b). *Custodians of Change.* Report of the Agriculture and Environment Working Group, Scottish Executive, Edinburgh.

Sharrock, J.T.R. (1976). *The Atlas of Breeding Birds in Britain and Ireland.* Poyser, Berkhamsted.

Shrubb, M.J. (1990). Effects of agricultural change on nesting lapwings *Vanellus vanellus* in England and Wales. *Bird Study*, **37**, 115-127.

Shrubb, M.J. (2003). *Birds, Scythes and Combines.* Cambridge University Press, Cambridge.

Smart, S.M., Clarke, R.T., van de Poll, H.M., Robertson, E.J., Shield, E.R., Bunce, R.G.H. & Maskell, L.C. (2003). National-scale vegetation change across Britain; an analysis of sample-based surveillance data from the Countryside Surveys of 1990 and 1998. *Journal of Environmental Management*, **67**, 239-254.

Soil Association (2000). *The Biodiversity Benefits of Organic Farming.* Soil Association, Bristol.

Sotherton, N.W. & Self, M.J. (2000). Changes in plant and arthropod biodiversity on lowland farmland: an overview. In *Ecology and Conservation of Lowland Farmland Birds*, ed. by N.J. Aebischer, A.D. Evans, P.V. Grice & J.A. Vickery. British Ornithologists' Union, Tring. pp. 26-35.

Stoate, C. (1995). The changing face of lowland farming and wildlife part 1 1845-1945. *British Wildlife*, **6**, 341-350.

Stoate, C. (1996). The changing face of lowland farming and wildlife part 2 1945-1995. *British Wildlife*, **7**, 162-172.

Tapper, S.C. & Barnes, R.F.W. (1986). Influence of farming practice on the ecology of the brown hare (*Lepus europaeus*). *Journal of Applied Ecology*, **23**, 39-52.

Taylor, K. (2000). Machair – a land with a flower-sweet taste. *British Wildlife*, **11**, 414-422.

Tilzey, M. (1998). Sustainable development and agriculture. *English Nature Research Report No. 278.* English Nature, Peterborough.

Tucker, G (2004). The burning of uplands and its effect on wildlife. *British Wildlife*, **15**, 251-257.

Vaughan, N., Lucas, E.-A., Harris, S. & White, P.C.L. (2003). Habitat associations of European hares *Lepus europaeus* in England and Wales: implications for farmland management. *Journal of Applied Ecology*, **40**, 163-175.

Vickery, J.A., Tallowin, J.R., Feber, R.E., Asteraki, E.J., Atkinson, P.J., Fuller, R.J. & Brown, V.K. (2001). The management of lowland neutral grasslands in Britain: effects of agricultural practices on birds and their food resources. *Journal of Applied Ecology*, **38**, 647-664.

Watson, A. & Rae, R. (1998). Use by birds of rape fields in east Scotland. *British Birds*, **91**, 144-145.

Wickramasinghe, L.P., Harris, S., Jones, G. & Vaughan, N. (2003). Bat activity and species richness on organic and conventional farms: impact of agricultural intensification. *Journal of Applied Ecology*, **40**, 984-1007.

Wilson, P.J. (1992). Britain's arable weeds. *British Wildlife*, **3**, 149-161.

Wilson, P.J. (1999). The effect of nitrogen on populations of rare arable weeds in Britain. In *Field Margins and Buffer Zones: Ecology, Management and Policy. Aspects of Applied Biology*, **54**. Association of Applied Biologists, Wellesbourne. pp. 93-100.

Wilson, P.J. (2000). Management for the conservation of arable plant communities. In *Fields of Vision – a Future for Britain's Arable Plants*, ed. by P. Wilson & M. King. Royal Society for the Protection of Birds, Sandy and English Nature, Peterborough. pp. 38-47.

4 Farming, Forestry and the Natural Heritage: Towards a More Integrated Future

5 The influence of forest and woodland management on biodiversity in Scotland: recent findings and future prospects

Jonathan Humphrey, Chris Quine & Kevin Watts

Summary

1. Over the past two decades there have been considerable changes in UK forestry, driven by the desire to adopt sustainable forest management practices. The maintenance and enhancement of biodiversity remain high on the public agenda, while the need to deliver improved social and community benefits from forests has increased in importance.

2. Forest and woodland management practices have evolved to embrace wider objectives, and to deal with the increasing complexity inherent in managing habitats and species at a variety of spatial scales. This chapter reviews recent developments in woodland management, focusing on the likely outcomes for biodiversity and exploring ways of integrating forest management with other land-use activities at the landscape scale.

3. The role of research in informing practice is also reviewed. Examples of recent research relating to the improvement of habitat conditions in both semi-natural and planted woodland are highlighted, together with a consideration of current methods for achieving habitat restoration and expansion.

4. Approaches to assessing the effects of landscape change on biodiversity are discussed with special reference to the role of focal (or key) species modelling. Future strategic priorities for forest biodiversity research are likely to include: obtaining more information on the ecology and genetics of key species; improving knowledge of the process of habitat and biodiversity development; developing better methods for modelling; and monitoring (through indicators) the effects of climate and landscape changes on biodiversity.

5. To provide effective input into land management decision-making, research and advice needs to be packaged in a user-friendly way, such as through computer-based decision support tools which integrate large amounts of complex information and provide accessible options for managers.

5.1 Introduction

Past land management has had a profound effect on the biodiversity of Scottish woodlands (Smout, 1997). Extensive areas of native woodland were cleared or fragmented for agricultural purposes, resulting in substantial loss, and contraction in the ranges, of habitats and species. Many notable members of the woodland fauna such as wolf (*Canis lupus*), bear (*Ursus arctos*) and beaver (*Castor fiber*) became extinct in the period 900-1800 A.D. (Corbet

& Yalden, 2001). However, management of the woods themselves had both positive and negative impacts on biodiversity. For example, the vibrant coppice industry of the 18th and 19th centuries ensured the survival of extensive areas of Atlantic oakwoods, but depleted their biodiversity by simplifying woodland structure and composition (Lindsay, 1975). In contrast, overgrazing by domestic stock, deer (*Cervus elaphus*) and the removal and non-replacement of timber had serious effects on habitat quality, in particular the loss of veteran trees, and removal of deadwood (Humphrey *et al.*, 2002c).

At the start of the 20th century, Scotland's forest cover was at an all-time low and the government of the day initiated an ambitious programme of afforestation to expand the strategic timber reserve. As a result, there were dramatic changes in forest area and composition, with cover increasing from 3.5% of total land area at the start of the 20th century to over 17% by 2000 (Forestry Commission, 2002a). Many of the new forests were planted using exotic species of North West American origin. For example, in the most recent National Forest Inventory for Scotland (Forestry Commission, 2002b), approximately 69% of the forest area was classed as coniferous, and of this 59% was of Sitka spruce (*Picea sitchensis*). About two-thirds of new planting took place on open land (Kirby, 1993), the remainder within existing semi-natural woodland (Roberts *et al.*, 1992).

In the latter part of the 20th century, considerable attention was given to the deleterious effects of coniferous afforestation on the flora and fauna of open habitats (Ratcliffe & Thompson, 1989). Most concerns related to loss and fragmentation of habitat for birds of open country and wetland such as merlin (*Falco columbarius*), greenshank (*Tringa nebularia*) and golden plover (*Pluvialis apricaria*) (Avery, 1989), although there was also pressure to restore plantations on ancient woodland sites back to semi-natural woodland (Mitchell & Kirby, 1989). Increasing public pressure and a suite of national and international government commitments to biodiversity conservation, such as the UK Biodiversity Action Plan (Anon., 1995), led to some marked changes in forestry policy and practice. A number of measures were put in place, including:

- a reduction in the rate of upland afforestation with non-native conifers (due also to changes in tax regime), directing further planting onto land of low conservation value, and avoiding habitats of existing high conservation value such as heath and mire (Anon., 1995);
- expansion, restoration, and improvement in the condition of priority native woodland habitats and conservation of associated species (Humphrey & Nixon, 1999; Thompson *et al.*, 2003);
- conservation of genetic resources (Herbert *et al.*, 1999);
- restructuring of forests at the landscape scale in line with design guidelines (Forestry Commission, 1994) to reduce visual impact of clear-felling and provide structural diversity (McIntosh, 1995);
- increasing open space and broadleaved planting both within new woodland and during restocking after felling (Patterson, 1993; Ferris & Carter, 2000); and
- removal of dense conifer stands along streams to reduce shading (Forestry Commission, 2000) and restoration of priority habitats such as peatland and mire (Patterson & Anderson, 2000).

In the last few years there have been further changes in forestry policy with important implications for the way that forests and woodlands will be managed for biodiversity in the future. Firstly, there is a much greater focus on management at the landscape scale, with increasing recognition that there should be better integration of forestry and agriculture to help rationalize public expenditure, and to produce diverse landscapes of greater value to wildlife (Scottish Executive, 2000). Secondly, there is raised awareness of the value of plantations of introduced conifer species as habitats for native flora and fauna including some endangered species (Humphrey *et al.*, 2003), and the need for improved integration with native woodland (Mason *et al.*, 1999) and other habitats at the landscape scale, such as through Forest Habitat Networks (Peterken, 2003). Lastly, there is a sense in which biodiversity conservation has matured, become less dogmatic as a concept and is now routinely included in management planning. With the principles now widely understood and embedded in government policy (Anon., 2003), there is much more emphasis on the 'how' rather than 'why'.

Consequently, the focus is increasingly on forest managers to deliver agreed biodiversity targets. This requires management at a range of spatial scales, incorporating the needs of key species and habitats at the stand scale (Hansson, 2001) within wider objectives for the design and management of forest landscapes (Lindenmayer & Franklin, 2002).

In this chapter we review the likely outcomes for biodiversity of current and future forest management, highlight the role of research in informing the development of practice and set out future research priorities.

5.2 Implementing the Biodiversity Action Plan – priority woodland habitats

5.2.1 Progress to date

Following on from the United Nations Conference on Environment and Development (UNCED) in Rio de Janeiro in 1992, the UK Government produced a Biodiversity Action Plan setting out plans and targets for the conservation of priority habitats (Habitat Action Plans – HAPs) and species (Species Action Plans – SAPs) (Anon., 1995, 1998). Although it also has a role in non-woodland HAPs, the Forestry Commission accepted UK responsibility for the woodland priority habitats. Those relevant to Scotland are: native pinewoods; upland oakwoods; upland ashwoods; wet woodlands; upland birchwoods; and lowland mixed broadleaved woodlands (Jones *et al.*, 2002). Targets and milestones for maintaining area, maintaining and improving condition (habitat quality), and expansion and restoration (removal of non-native trees on ancient woodland sites and re-establishment of a native woodland ecosystem) have been published for most of the HAP types (Table 5.1). Progress towards achieving these targets was evaluated recently through a sample survey (MacKenzie & Worrell, 2003). It was concluded that, although the improvement, expansion and restoration of native woodlands in Britain was progressing well, when compared to the relatively small amount of activity prior to 1992, many of the challenging targets set in the native woodland HAPs may not be achieved by the target dates, unless there is a significant increase in activity.

MacKenzie & Worrell (2003) found that successful progress is being made towards maintaining and improving the condition of upland ashwoods, native pinewoods (Figure 5.1) and upland oakwoods; progress towards achieving expansion targets for the latter two types is also on track. However, very little progress has been made towards the targets for expansion and restoration of upland ashwoods, with only slightly better progress with the restoration of upland oakwoods (Table 5.1).

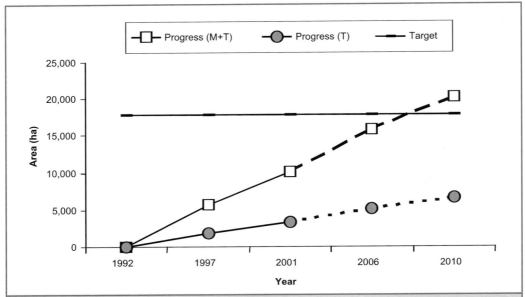

Figure 5.1 Native Pinewoods 'Maintain & Improve Condition' (targets only 'T'; targets/milestones' 'M+T'). The HAP target is 17,882 ha to be achieved by 2010. Adapted with permission from Figure 3.16 in MacKenzie & Worrell (2003). This projection relies on milestones achieving target status. This may not occur in all cases, or may take a longer period of time.

Table 5.1 The area of native woodland recorded as 'Restore' and 'Expand' targets and milestones as a proportion of the total Habitat Action Plan (HAP) area targets for Scotland. Adapted with permission from MacKenzie & Worrell (2003). Targets for Upland Birchwoods and Lowland Mixed Deciduous woodland were published (Forestry Commission Scotland, 2004) but were not included in this study.

HAP type	Restoration %		Expansion %	
	Target (ha)	Target/Milestone Achieved	Target (ha)	Target/Milestone Achieved
Native Pinewoods	11,200	Achieved ?	25,000	89
Upland Oakwoods	3,000	11	3,000	51
Upland Mixed Ashwoods	800	0	2,000	0
Wet Woods	1,000	1	2,200	119-187
Upland Birchwoods	1,800	No data	4,500	No data
Lowland Mixed Deciduous woodland	1,120	No data	2,000	No data

There is a variety of possible reasons for the disparity in progress towards HAP targets between the different woodland types. Arguably, the key reason is that targets were developed from the 'top down' with little account taken of the feasibility of achieving these on the ground. Recent research (Thompson *et al.*, 2003) has shown that some woodland types are simply easier to restore than others. An essential element in the restoration process is the establishment of native tree regeneration after or during removal of the non-native tree canopy, usually conifers such as Sitka spruce or Douglas fir (*Pseudotsuga menziesii*). Regeneration is most successful on sites which have well-drained mineral soils where the ground flora is only weakly competitive, such as native pinewoods. Sites which have fertile moisture-retaining clays and loams, such as lowland mixed broadleaved woodland, support vigorous weed growth which restricts seedling development.

5.2.2 Research and advice in support of restoration objectives

Current best practice for restoration has recently been brought together in a guidance booklet (Thompson *et al.*, 2003) which includes advice on prioritizing sites, evaluating constraints and opportunities for restoration, and identifying appropriate methods. There is increasing interest in developing a more gradual approach to restoration, avoiding the use of clear felling and reducing the environmental stress on remnant native veteran trees, ground flora and epiphytic communities. Maintaining a tree canopy over the site also reduces weed vigour. Methods which are being tested in upland oak and ashwoods include thinning and creating small gaps in non-native conifers and 'haloing' around ancient trees to gradually open them up to light (Read, 2000).

5.2.3 Research and advice in support of improving woodland condition

Kirby *et al.* (2002) have produced guidance on methods for setting objectives for woodland condition within Sites of Special Scientific Interest (SSSIs), but much of this is applicable to native and semi-natural woodland in general. Criteria relating to the 'features of interest', or attributes upon which the original SSSI designation was founded, are proposed to help with setting objectives and targets. One of the key features of interest is woodland quality. An aspect of quality which is pertinent to most woodland types is the survival of veteran/ancient trees and the occurrence of dead and decaying wood habitats. Current knowledge relating to the creation and maintenance of deadwood habitats in woodland has been brought together in the form of a guide (Humphrey *et al.*, 2002c). The guide takes a strategic approach, stratifying woodlands into different types in terms of the potential importance of deadwood habitats within each type. Benchmark quantities/volumes of deadwood are proposed, together with illustrations of key features (Figure 5.2) specific to different types. Guidance on management methods and monitoring is also included.

Research is also continuing on predicting the impacts of deer on woodland quality and biodiversity (Gill, 2000; Fuller & Gill, 2001) through modelling of population dynamics and distribution. There is also increasing interest in using controlled cattle grazing in woodlands to promote biodiversity and tree regeneration (Mayle, 1999) and there is evidence on the ground of beneficial effects (Armstrong *et al.*, 2003).

Figure 5.2 Stylised view of a forest landscape showing distribution of different deadwood habitats (numbered 1-20). 1) Retention within developing stand. 2) Snags provide raptor perches on clear-fells. 3) Retention on felling coupe. 4) Retained cut snags. 5) Partially submerged deadwood in ponds and streams provides habitat for bryophytes and cranefly larvae. 6) Stumps provide important niches for lichens. 7) Retained stand encourages structural diversity and deadwood accumulation. 8) Windthrow is an important source of deadwood. 9) Surroundings of veteran native trees opened up within plantations on ancient woodland sites. 10) Large logs retained to benefit bryophytes. 11) Snags retained as habitat for hole-nesting birds. 12) Brash piles provide temporary shelter for invertebrates. 13) Natural reserves of old conifers located near to semi-natural woodland. 14) Young dense conifers cleared from riverbank. 15) Permanent shelterbelts provide continuity of deadwood. 16) De-stumping should be avoided as it removes below ground deadwood important for invertebrates. 17) Small groups of veteran larch provide nesting sites for goshawks. 18) Snags retained as they mature. 19) Stump piles provide shelter for adders and perches for woodlark. 20) Various species of clear-wing moths use stumps for egg-laying. Reproduced with permission from Humphrey *et al.* (2002c).

5.2.4 Research and advice in support of woodland expansion

There has been a considerable increase in the area of native pine and broadleaved woodland in Scotland over the last two decades, particularly in the Highlands (Rollinson, 2003). Much of this has been achieved through planting, although natural colonisation (regeneration) is the preferred option from a nature conservation perspective (Peterken, 1993). Research in northern Scotland (Thompson, 2004) has shown that natural colonisation of birch (*Betula pendula*) and pine (*Pinus sylvestris*) can be very successful, particularly on dry, relatively infertile sites in the north and east where there is a good seed supply. Colonisation was found to be inhibited by rank competitive vegetation, a combination of wet and very nutrient poor soils, and by excessive browsing (Thompson, 2004). Where planting is the only option for woodland establishment, guidance on the selection of trees and shrubs appropriate to different site types is available through the Ecological Site Classification Decision Support System (ESC-DSS) (Ray, 2001). The species information within ESC-DSS is revised continually as and when new research findings become available. For example, information on establishment techniques for juniper (*Juniperus communis*) was published in 2003 (Broome, 2003).

5.3 Implementing the Biodiversity Action Plan – priority species

There are 391 species which have published action plans within the UK BAP, 62 of which have research actions for the Forestry Commission (FC) to consider with other partners. A review was undertaken (Broome, 1999) to identify which of these species were priorities for FC research activity. The criteria for prioritising were:

1) degree of dependence on woodland habitats;
2) potential severity of impact from forestry operation;
3) public perception of importance;
4) formal role for the FC either as lead partner or contact point; and
5) tractability of research problem, e.g. where answers to questions could be obtained in a reasonable time-scale.

Thirteen priority species were identified and research work plans drawn-up for each. Those relevant to Scotland are listed in Table 5.2, together with the main management issues and research priorities for each species.

Table 5.2 UK Biodiversity Action Plan (BAP) species identified as priorities for research in Scotland.

Species	Management issues	Research priorities
Capercaillie (*Tetrao urogallus*)	Fence strikes; predation; climate change; lack of pinewood habitat	Habitat usage in managed forests; silviculture; bird-friendly fences
Red squirrel (*Sciurus vulgaris*)	Spread of grey squirrels; protection of core areas; identifying appropriate management of conifer stands	Modelling habitat requirements; identifying core areas; thinning trials
Scottish crossbill (*Loxia scotica*)	Stand management to promote food supply	Establishing links between bird abundance and cone supplies; taxonomic identification
Twinflower (*Linnaea borealis*)	Genetic impoverishment; appropriate stand conditions	Population genetics; testing effects of different canopy densities in mature Scots pine stands
Juniper (*Juniperus communis*)	Fragmentation and loss of habitat; methods of regeneration	Status and condition surveys carried out; experimentation on methods for propagation and establishment
Butterflies pearl bordered fritillary (*Boloria euphrosyne*) and chequered skipper (*Carterocephalus palaemon*)	Maintaining quality and provision of open ground and scrub habitats	Habitat usage in spruce forests; effects of creating new open ground

Of the activities listed in Table 5.2, research on capercaillie (*Tetrao urogallus*) continues to be a high priority given the rapid decline in numbers (Kortland, 2003). Considerable resources are being invested in the testing of different approaches to thinning in pole stage Scots pine plantations in Strathspey and Easter Ross. The objectives here are to create light conditions suitable for the growth of blaeberry (*Vaccinium myrtillus*) (one of the main food plants for the capercaillie), and to speed up the process of stand development to the old-growth stage (*sensu* Franklin *et al.*, 1986).

5.4 Managing plantation forestry to provide biodiversity benefits

Increasingly, planted forests are seen as having potential value as a habitat for native flora and fauna. Biodiversity assessments have been carried out in a range of different planted forests types in Britain (Humphrey *et al.*, 2003); 28 stands of different age in Sitka spruce, Scots pine and semi-natural oak-birch woodland were surveyed in Scotland. Assessments were made of vascular plants, bryophytes, lichens, fungi, songbirds, invertebrates and deadwood, and relationships established between the diversity of these groups and habitat, climate, locational and historical factors.

The broad conclusions were that for the majority of species groups there were few differences in diversity between planted and semi-natural woodland, although considerable differences in species groups and guilds were recorded (Quine *et al.*, 2003). Figure 5.3 shows that the total count of invertebrate species was similar in pine and spruce stands, but lower in oak (although there were fewer plots in this woodland type).

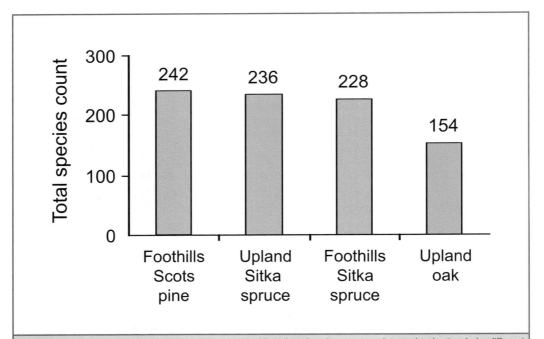

Figure 5.3 Count of invertebrate species recorded in 1 ha plots in spruce, pine and oak stands in different climate zones (Uplands: precipitation >1,500 mm annum^{-1}; Foothills: 800-1,500 mm annum^{-1}). There were eight foothills and upland spruce plots respectively; eight pine plots and four oak plots. Reproduced with permission from Humphrey & Quine (2001).

Species-richness and diversity is highest in the early successional stages of the forest cycle (for 10-20 years after planting) or in stands retained beyond economic maturity (over 60 years in the case of Sitka and Norway spruce (*Picea abies*)). Early-successional forest stands provide niches for open ground and scrub species of particular conservation importance in upland Britain (Humphrey *et al.*, 2002a; Fuller & Browne, 2003). For example, hoverflies (Syrphidae) were found to be most abundant in young open stands with well developed ground vegetation structure (Figure 5.4).

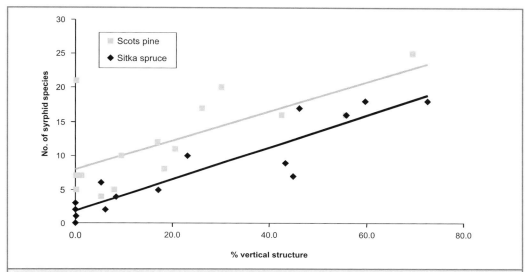

Figure 5.4 The relationship between field layer (0.2-1.0 m) vertical structure and the numbers of syrphid species sampled by Malaise traps in Sitka spruce and Scots pine biodiversity assessment plots. Reproduced with permission from Jukes *et al.* (2003).

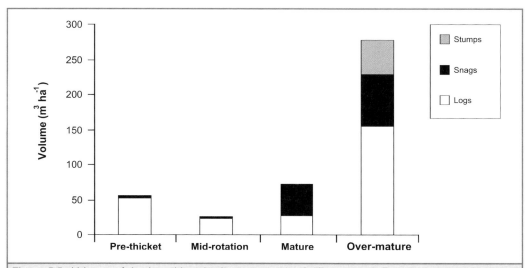

Figure 5.5 Volumes of deadwood in upland spruce stands of different ages. Pre-thicket stands were 8-10 years in age; mid-rotation 20-30 years, mature 40-50 years, and over-mature 60-80 years. Reproduced with permission from Humphrey & Peace (2003).

Late successional stands in plantations provide niches for woodland specialist species such as herbs, bryophytes, and species associated with deadwood (Humphrey et al., 2002). Old conifer stands can accumulate large volumes of deadwood (Figure 5.5) of considerable importance to lower plants, and hole-nesting birds.

Current research is focused on characterizing the structure of late-successional stands; establishing their value for biodiversity, identifying minimum patch sizes, and evaluating the importance of connectivity and distribution in the landscape. Emerging findings suggest that, based on an assessment of wind-risk, over 50-70% of the landscape in Scotland could harbour old-growth stands with large trees (Humphrey, 2005). To maximize their value for biodiversity, these stands should be in the order of 50-100 ha in size (to accommodate species with large minimum patch size requirements such as capercaillie) and no more than 1-2 km from other mature and semi-natural woodland to ensure adequate dispersal between patches.

Within Forestry Commission forests in Scotland, England and Wales there are plans to create large (200+ ha) areas of continuous cover in low wind risk zones. These areas provide scope for developing mosaics of non-intervention old-growth, and stands managed by continuous cover; the latter stands acting as 'buffer' for the old growth stands in terms of maintaining a more stable microenvironment (Humphrey, 2005).

5.5 Putting the pieces together: management at the landscape scale

5.5.1 Development of Forest Habitat Networks (FHNs)

With the publication of the Scottish Forestry Strategy (Scottish Executive, 2000), more emphasis has been placed on the role of woodland in diversifying and improving the quality of the wider landscape, with Forest Habitat Networks (FHNs) identified as the key strategic tool for guiding woodland restoration at the landscape scale for biodiversity enhancement. The main ingredients of FHNs are the development of large core forest areas, and corridors and links between woodlands (Peterken, 2003). The new Scottish Forestry Grant Scheme (SFGS) supports the development of FHNs by encouraging the expansion and connection of existing woodlands in preference to creation of new, isolated woodlands (Forestry Commission, 2003). Both semi-natural and planted woodland can form parts of a network.

Humphrey (2003) stresses the importance of considering within individual FHNs the relative balance between the restoration of planted native woodland, the expansion of existing native woods, the modification of plantations and the establishment of linkages and corridors between woodlands. Current thinking suggests that in order to maximize woodland biodiversity, the restoration and buffering of existing woodland ought to take precedence over connecting isolated woodlands. However, in many landscapes (particularly in the Highlands) there is also a need to consider the appropriate or 'reasonable' balance between open and wooded ground (Scottish Natural Heritage, 2000). This includes consideration of both the absolute amounts of wooded and non-wooded habitat patches as well as their spatial configuration. Habitat Action Plans for woodland and open ground encourage the restoration and expansion of both, and there is a potential land-use conflict where open and woodland habitats compete for the same ground.

There will be different answers for different regions and a range of alternative spatial options which will have varying consequences for wildlife. There is a danger that open ground species will lose out if FHNs are not planned with consideration for both woodland and open ground biodiversity at larger spatial scales. Landscape evaluation is complex, and there is a temptation to simplify the implications of landscape change for species by developing general rules such as 'more woodland is better' and 'isolated woodlands should be connected'. A human view of connectedness is not always the species view. Connecting woodland patches for one species may fragment areas of open ground for another and *vice versa*. By evaluating the effects of landscape change on species with varying habitat requirements and dispersal abilities (focal species), the implications of those changes for biodiversity can be estimated.

5.5.2 Focal species modelling

The focal species approach builds on the concept of umbrella species whose requirements are believed to encapsulate the needs of other species (Lambeck, 1997). Focal species are used to define key spatial and compositional attributes of the landscape. For each relevant landscape parameter, the species with the most demanding requirements for that parameter is used to define its minimum acceptable value. Parameters include: habitat preferences; ability to disperse through different habitat types; and minimum patch size requirements. Forest Research is developing a prototype GIS-based model called BEETLE (Biological & Environmental Evaluation Tools for Landscape Ecology) to undertake focal species modelling. BEETLE is based upon the concept of ecologically scaled landscape indices (Vos *et al.*, 2001) which, when combined with spatial data on the distribution and extent of different habitat types, allows evaluation of the quality of different landscape scenarios for a range of focal species.

The basic outputs from the model are maps of functional networks for individual species. These networks comprise areas of suitable habitat which are functionally linked so that the species in question are able to disperse successfully between the patches. The degree of dispersal is determined by the 'permeability' of the matrix or area between suitable habitats. Landscape quality for each focal species is assessed in terms of the number of networks the landscape supports, the size of those networks, and the size of the habitat patches within each network. Overall biodiversity value is assessed by comparing network quality across the range of focal species. The BEETLE approach is currently being tested in a number of locations. The example in Figure 5.6 from the Scottish Borders, shows the effects of adding additional woodland on the number and size of networks for red squirrel (*Sciurus vulgaris*) – a core woodland species. Further work is planned to assess the effects of this change in woodland on open ground species. The value of the focal species approach is that a range of different landscape scenarios can be tested and evaluated over a short time scale, allowing more effective strategic planning.

5.6 Conclusions: priority areas for future research

The conservation and enhancement of biodiversity is firmly enshrined in government policy in Scotland, and there is now increasing recognition that policy implementation rather than policy development is the key challenge for the future. Currently the most important aspect of implementation is the need to provide integrated solutions at a variety of spatial scales, whilst at the same time involving a much wider range of stakeholders in the decision-making process (Anon., 2003). Getting the ecology 'right' is a complex task requiring the

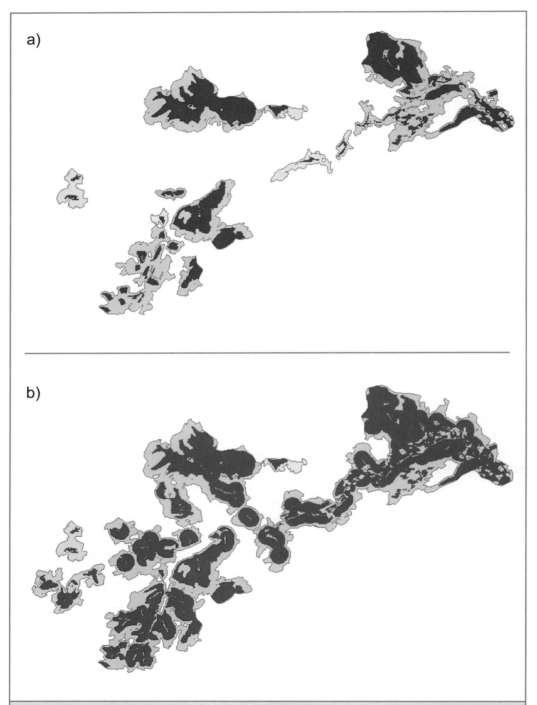

Figure 5.6 Habitat networks for red squirrel in part of the Scottish Borders: a) before woodland expansion – 11 networks; b) after woodland expansion – five networks. Woodland patches are shown in dark green, the lighter coloured areas represent the 'non-habitat' parts of the network; different colours representing different networks.

innovative use of both existing guidance and emerging research findings. Future research needs can be categorized under three broad headings: basic understanding; modelling and monitoring; and decision support.

5.6.1 Basic understanding

A basic understanding of ecological processes and species requirements underpins all modelling, monitoring and decision support tools. With the implementation of the UK Biodiversity Action Plan, there have been increasing efforts to obtain more information on the ecology and genetics of key species. But much more work is needed if we are to develop even an 'entry-level' understanding of how species use and disperse through landscapes. The effects of climate change on species' distributions, and the composition of communities and assemblages, also poses a considerable research challenge (Riley *et al.*, 2003).

The Scottish Forestry Grant Scheme uses a number of rules to help with better targeting of incentives for the creation of new woodlands. Currently, these rules are based on a very imperfect understanding of the process of habitat and biodiversity development. In addition, policy dictates that large areas of state-owned conifer forests should be converted to continuous cover (non clear-fell management systems) with little knowledge about how biodiversity might develop in these novel environments. Opportunities for research will arise with the designation of a number of large continuous cover areas as 'research forests', but resources are needed to carry out biodiversity assessments.

5.6.2 Monitoring and modelling

Developments in forestry and farming practices coupled with climate change will continue to drive changes in the structure and dynamics of wooded landscapes with resulting consequences for biodiversity. The focal species modelling approach has the potential to predict the impact of landscape change on biodiversity, but monitoring is required to validate those predictions. The Countryside Survey 2008, the National Inventory of Woodlands and Trees and the proposed new survey by Forestry Commission Scotland of the native woodland resource will provide information at the broad scale, but there is also a need to develop indicators of biodiversity based on species (e.g. woodland birds), habitat (e.g. deadwood) and landscape (e.g. functional networks) to provide more specific information. Research is needed to help identify these potential indicators and test their robustness in different situations. A preliminary review of potential landscape-scale indicators of biodiversity has been carried out (Quine & Sellars, 2003) as part of the EU Research and Technical Development 5th Framework project VISULANDS – Visualization Tools for Public Participation in the Management of Landscape Change. Such indicators may have an important role in allowing biodiversity values to be assessed in the same decision framework as economic and social values.

5.6.3 Decision support

To provide effective input into land management decision-making, research and advice needs to be packaged in a user-friendly way, such as through computer-based decision support tools which are capable of integrating large amounts of complex information whilst at the same time presenting accessible options for management. The focal species modelling approach offers potential as a decision-support tool for strategic planners, but will always be limited by the amount of species-related information available. Thus there are ongoing initiatives to try and make existing knowledge available to end-users. Work is ongoing on a database system called HARPPS (Habitats and Rare, Priority and Protected Species) linked to the National Biodiversity Network (www.nbn.org.uk) which will provide managers with readily accessible ecological information on species recorded within their forests, and help them to evaluate the effects of current and future management. A prototype of the HARPPS database will be available in late 2004.

Acknowledgements

The authors are grateful to staff in Forest Research who contributed information on their research to this chapter: Alice Broome, Duncan Ray and Richard Thompson. GIS work reported on in this chapter was undertaken by Matthew Griffiths, Louise Sing and Joe Hope. Rick Worrell and Neil MacKenzie gave permission for the reproduction of some of their work. Richard Thompson and Chris Quine made valuable comments on earlier drafts.

References

Anonymous (1995). *Biodiversity: the UK Steering Group Report. Volume 2. Action Plans*. HMSO, London.

Anonymous (1998). *Tranche 2 Action Plans. Volume II – Terrestrial and Freshwater Habitats*. English Nature, Peterborough.

Anonymous (2003). *Towards a Strategy for Scotland's Biodiversity: Biodiversity Matters. Strategy proposals*. Scottish Executive, Edinburgh.

Armstrong, H.M., Poulsom, E.L., Connolly, T. & Peace, A.J. (2003). *A Survey of Cattle-Grazed Woodlands in Britain*. Forest Research, Roslin.

Avery, M.I. (1989). Effects of upland afforestation on some birds of the adjacent moorlands. *Journal of Applied Ecology*, **26**, 957-966.

Broome, A.C. (1999). Review of Species Action Plan research needs for the Forestry Commission. Unpublished report. Forestry Commission. Forest Research, Roslin.

Broome, A.C. (2003). *Growing juniper: propagation and establishment practices. Forestry Commission Information Note No.* 50. Forestry Commission, Edinburgh.

Corbet, G.B. & Yalden, D.W. (2001). Mammals. In *The Changing Wildlife of Great Britain and Ireland*, ed. by D.L. Hawksworth. Taylor and Francis, London and New York. pp. 399-409.

Ferris, R. & Carter, C. (2000). *Managing rides, roadsides and edge habitats in lowland forests. Forestry Commission Bulletin* 123. Forestry Commission, Edinburgh.

Forestry Commission (1994). *Forest Landscape Design Guidelines*. Forestry Commission, Edinburgh.

Forestry Commission (2000). *Forests and Water Guidelines*. Forestry Commission, Edinburgh.

Forestry Commission (2002a). *Forestry Statistics 2002*. Economics and Statistics Unit, Forestry Commission. Edinburgh.

Forestry Commission (2002b). *National Inventory of Woodland and Trees: Scotland*. Forestry Commission, Edinburgh.

Forestry Commission (2003). *Scottish Forestry Grants Scheme*. Scottish Executive, Forestry Commission, Edinburgh.

Forestry Commission Scotland (2004). Native woodland habitat action plans in Scotland. Draft Guidance Note. Forestry Commission Scotland, Edinburgh.

Franklin, J.E., Hall, F., Laudenslayer, W., Maser, C., Nunan, J., Poppino, J., Ralph, C.J. & Spies, T. (1986). Interim definitions for old-growth Douglas fir and mixed conifer forests in the Pacific Northwest and California. *Pacific Northwest Research Station Research Note* PNW-447. USDA Forest Service, Portland, Oregon.

Fuller, R.J. & Browne, S.J. (2003). Effects of plantation structure and management on birds. In *Biodiversity in Britain's Forests: Results from the Forestry Commission's Biodiversity Assessment Project*, ed. by J.W. Humphrey, R. Ferris & C.P. Quine. Forestry Commission, Edinburgh. pp. 93-102.

Fuller, R.J. & Gill, R.M.A. (2001). Ecological impacts of increasing numbers of deer in British woodland. *Forestry*, **74**, 193-199.

Gill, R.M.A. (2000). The impact of deer on woodland biodiversity. *Forestry Commission Information Note* 36. Forestry Commission. Edinburgh.

Hansson, L. (2001). Key habitats in Swedish managed forests. *Scandinavian Journal of Forest Research*, **3** (Supplement), 52-61.

Herbert, R., Samuel, S. & Patterson, G. (1999). Using local stock for planting native trees and shrubs. *Forestry Commission Practice Note No. 8*. Forestry Commission. Edinburgh.

Humphrey, J.W. (2003). The restoration of wooded landscapes: future priorities. In *The Restoration of Wooded Landscapes*, ed. by J.W. Humphrey, A.C. Newton, J. Latham, H. Gray, K.J. Kirby, C.P. Quine & E. Poulsom. Forestry Commission, Edinburgh. pp. 153-157.

Humphrey, J.W. (2005). Benefits to biodiversity from developing old-growth conditions in British upland conifer plantations: a review. *Forestry*, **78**, 33-53.

Humphrey, J.W. & Nixon, C.J. (1999). The restoration of upland oakwoods following the removal of conifers: general principles. *Scottish Forestry*, **53**, 68-76.

Humphrey, J.W. & Quine, C.P. (2001). Sitka spruce plantations in Scotland: friend or foe to biodiversity? *Glasgow Naturalist*, **23** (Supplement), 66-76.

Humphrey, J.W. & Peace, A.J. (2003). Deadwood. In *Biodiversity in Britain's Forests: Results from the Forestry Commission's Biodiversity Assessment Project*, ed. by J.W. Humphrey, R. Ferris & C.P. Quine. Forestry Commission, Edinburgh. pp. 41-50.

Humphrey, J.W., Davey, S., Peace, A.J., Ferris, R. & Harding, K. (2002a). Lichen and bryophyte communities of planted and semi-natural forests in Britain: the influence of site type, stand structure and deadwood. *Biological Conservation*, **107**, 165-180.

Humphrey, J.W., Ferris, R., Jukes, M.R. & Peace, A.J. (2002b). The potential contribution of conifer plantations to the UK Biodiversity Action Plan. *Botanical Journal of Scotland*, **54**, 49-62.

Humphrey, J.W., Stevenson, A. & Swailes, J. (2002c). *Life in the Deadwood: a Guide to the Management of Deadwood in Forestry Commission Forests*. Forest Enterprise Living Forests Series. Forest Enterprise, Edinburgh.

Humphrey, J.W., Ferris, R. & Quine, C.P. (Eds) (2003). *Biodiversity in Britain's Planted Forests: Results from the Forestry Commission's Biodiversity Assessment Project*. Forestry Commission, Edinburgh.

Jones, A.T., Gray, H. & Ray, D. (2002). Strategic application of modelling forest potential: calculating local targets for native woodland habitat action plans in Scotland. *Scottish Forestry*, **56**, 81-89.

Jukes, M.R., Peace, A.J. & Ferris, R. (2003). The invertebrate community of plantation forests. In *Biodiversity in Britain's Planted Forests: Results from the Forestry Commission's Biodiversity Assessment Project*, ed. by J.W. Humphrey, R. Ferris & C.P. Quine. Forestry Commission, Edinburgh. pp 75-92.

Kirby, K.J. (1993). The effects of plantation management on wildlife in Great Britain: lessons from ancient woodland for the development of afforestation sites. In *Ecological Effects of Afforestation*, ed. By C. Watkins. CAB International, Wallingford. pp. 15-30.

Kirby, K.J., Latham, J., Holl, K., Bryce, J., Corbett, P. & Watson, R. (2002). *Objective setting and condition monitoring within woodland Sites of Special Scientific Interest*. English Nature Research Report No. 472. English Nature, Peterborough.

Kortland, K. (2003). Multi-scale forest habitat management for Capercaillie. *Scottish Forestry*, **57**, 91-95.

Lambeck, R.J. (1997). Focal species: a multi-species umbrella for nature conservation. *Conservation Biology*, **11**, 849-856.

Lindenmayer, D.B. & Franklin, J.F. (2002). *Conserving Forest Biodiversity: a Comprehensive Multi-Scaled Approach.* Island Press, Washington DC.

Lindsay, J.M. (1975). The history of oak coppice in Scotland. *Scottish Forestry*, **29**, 87-95.

MacKenzie, N. & Worrell, R. (2003). Contributions to native woodland habitat action plan targets in private and Forestry Commission woodlands. Unpublished report. Forestry Commission. Edinburgh.

Mason, W.L., Hardie, D., Quelch, P., Ratcliffe, P.R., Ross, I., Stevenson, A.W. & Soutar, R. (1999). "Beyond the two solitudes": the use of native species in plantation forests. *Scottish Forestry*, **53**, 135-144.

Mayle, B. (1999). Domestic stock grazing to enhance woodland biodiversity. *Forestry Commission Information Note* 28. Forestry Commission, Edinburgh.

McIntosh, R. (1995). The history and multi-purpose management of Kielder Forest. *Forest Ecology and Management*, **79**, 1-11.

Mitchell, P.L. & Kirby, K.J. (1989). *Ecological Effects of Forestry Practices in Long-Established Woodland and Their Implications for Nature Conservation.* Oxford Forestry Institute, University of Oxford, Oxford.

Patterson, G.S. (1993). *The value of birch in upland forests for wildlife conservation. Forestry Commission Bulletin* 109. Forestry Commission, Edinburgh.

Patterson, G.S. & Anderson, A.R. (2000). *Forests and peatland habitats. Forestry Commission Guideline Note No.* 1. Forestry Commission. Edinburgh.

Peterken, G.F. (1993). *Woodland Conservation and Management.* Chapman & Hall, London.

Peterken, G.F. (2003). Developing forest habitat networks in Scotland. In *The Restoration of Wooded Landscapes*, ed. by J.W. Humphrey, A.C. Newton, J. Latham, H. Gray, K.J. Kirby, C.P. Quine & E. Poulsom. Forestry Commission, Edinburgh. pp. 85-92.

Quine, C.P. & Sellars, H. (2003). *A Review of Non-Visual Indicators of Landscape Change.* Forest Research. Roslin.

Quine, C.P., Humphrey, J.W. & Ferris. R. (2003). The future management of plantation forests for biodiversity. In *Biodiversity in Britain's Planted Forests: Results from the Forestry Commission's Biodiversity Assessment Project*, ed. by J.W. Humphrey, R. Ferris & C.P. Quine. Forestry Commission, Edinburgh. pp. 103-114.

Ratcliffe, D.A. & Thompson, D.B.A. (1989). The British uplands: their ecological character and international significance. In *Ecological Change in the Uplands*, ed. by M.B. Usher & D.B.A. Thompson. Blackwell, Oxford. pp. 9-36.

Ray, D. (2001). *Ecological Site Classification Decision Support System (ESC-DSS).* Forestry Commission, Edinburgh.

Read, H.J. (2000). *Veteran Trees: a Guide to Good Management.* English Nature, Peterborough.

Riley, J., Kirby, J., Linsley, M. & Gardiner, G. (2003). Review of UK and Scottish surveillance and monitoring schemes for the detection of climate-induced changes in biodiversity. Unpublished report. Flexible Fund Project JEY-001-01. DEFRA, London and Scottish Executive, Edinburgh.

Roberts, A.J., Russell, C., Walker, G.J. & Kirby, K.J. (1992). Regional variation in the origin, extent and composition of Scottish woodland. *Botanical Journal of Scotland*, **46**, 167-189.

Rollinson, T. (2003). The UK policy context. In *The Restoration of Wooded Landscapes*, ed by J.W. Humphrey, A.C. Newton, J. Latham, H. Gray, K.J. Kirby, C.P. Quine & E. Poulsom. Forestry Commission, Edinburgh. pp. 3-6.

Scottish Executive (2000). *Forests for Scotland: the Scottish Forestry Strategy.* Scottish Executive, Edinburgh

Scottish Natural Heritage (2000). *Scottish Natural Heritage's position on the reasonable balance of land uses.* Scottish Natural Heritage, Perth.

Smout, T.C. (1997). *Scottish Woodland History.* Scottish Cultural Press, Edinburgh.

Thompson, R.N. (2004). *Predicting site suitability for natural colonisation: upland birchwoods and native pinewoods in northern Scotland. Forestry Commission Information Note* 54. Forestry Commission. Edinburgh.

Thompson, R.N., Humphrey, J.W., Harmer, R. & Ferris, R. (2003). *Restoration of native woodland on ancient woodland sites. Forestry Commission Practice Guide.* Forestry Commission. Edinburgh.

Vos, C.C., Verboom, J., Opdam, P.F.M. & Ter Braak, C.J.F. (2001). Toward ecologically scaled landscape indices. *The American Naturalist*, **183**, 24-41.

5 Farming, Forestry and the Natural Heritage: Towards a More Integrated Future

6 The Scottish Biodiversity Strategy: developing the Rural Land Use Implementation Plan

John Henderson

Summary

1. This chapter describes how the Biodiversity Strategy for Scotland has been drafted and how the Rural Land Use Working Group of the Scottish Biodiversity Forum has contributed towards the development of an implementation plan for that Strategy.
2. The process used to develop the draft Strategy and the implementation plan has involved, and continues to involve, a wide range of people, all of them stakeholders in the delivery of biodiversity, landscape and other environmental objectives. The chapter describes how key stakeholders and others have been encouraged to take ownership of the Strategy.
3. Using the development of the Rural Land Use Implementation Plan as an example, the chapter demonstrates how the different sectoral interests have been able to identify and agree both sector-specific and common issues, and to take an integrated approach to the planning and delivery of benefits to the rural environment and thus also to both rural and urban communities.

6.1 Introduction

The Scottish Biodiversity Strategy provides a major opportunity to think strategically and holistically about the interactions between biodiversity, land use and people. This chapter describes how the Strategy has been drafted and how work on producing implementation plans is being taken forward. The chapter also describes how interested organisations and individuals have been involved in the process.

6.2 Developing the draft Strategy for Scotland's Biodiversity

The starting point for producing a biodiversity strategy for Scotland was the need to fulfil the United Kingdom Government's obligation under the Convention on Biological Diversity. A biodiversity action plan for the UK was published in 1994 (Anon., 1994) and work has been proceeding through a Scottish Biodiversity Forum (SBF) to develop a detailed biodiversity strategy for Scotland.

At the first meeting of the Scottish Biodiversity Forum in February 2002, the Deputy Environment Minister, Allan Wilson, asked the Forum to develop a draft Strategy for Scotland's Biodiversity by February 2003. The development of the draft Strategy was overseen by a steering group, representing a wide range of organisations, and was chaired by the Scottish Executive. Figure 6.1 illustrates the information-gathering and consultation process undertaken between the first and second meetings of the Forum.

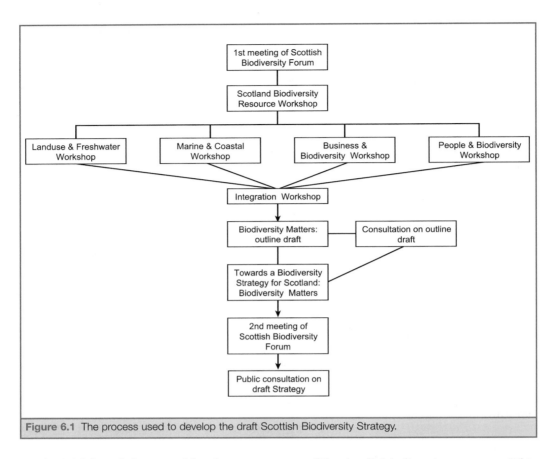

Figure 6.1 The process used to develop the draft Scottish Biodiversity Strategy.

An initial workshop considered an assessment of Scotland's biodiversity resource. This was followed by four sectoral workshops, at which stakeholders established the necessary elements for each sector in the Strategy. The workshops also helped to highlight key actions needed in the Implementation Plan for the Strategy, and identified priority issues in the Strategy. An integration workshop was run in October 2002 to draw these various issues and actions together. The resulting outline draft was distributed to workshop participants, and to other individuals and organisations for comment.

At the second meeting of the Scottish Biodiversity Forum, in February 2003, the Deputy Environment Minister, Allan Wilson, was presented with the draft Strategy (Scottish Executive, 2003). At the meeting, the Deputy Minister launched a public consultation on the document, which closed on 6 June 2003. With the draft Biodiversity Strategy in place, three working groups (Urban, chaired by Steve Sankey; Marine, chaired by Michael Scott; and Rural Land Use, chaired by John Henderson) were set up to contribute towards the development of an Implementation Plan.

A total of 108 consultation responses was received on the draft Strategy from a wide range of agencies, non-governmental organisations and other bodies (Hocknell et al., 2003). Responses to the consultation on the draft Strategy were given full consideration by the three working groups, and a number of amendments to the text of the draft strategy documents made.

The draft Strategy had two key aims. First, to halt the loss of Scotland's biodiversity, and to continue to reverse previous losses by targeted action for species and habitats. Second, to raise awareness of the many benefits of biodiversity, by significantly increasing the number, and range, of people contributing to its conservation and enhancement.

The draft Strategy committed the Scottish Biodiversity Forum to developing a three-year implementation plan for the Strategy. This implementation plan will be a working document complementing the Strategy and will outline existing and new actions that will satisfy the three key principles upon which action for biodiversity in Scotland should be founded. These principles were:

- providing a framework for action;
- promoting the conservation, enhancement and sustainable use of Scotland's biodiversity by placing people at the heart of the strategy; and
- gathering, developing and applying the best available knowledge to assist people in understanding, caring for, enjoying and making wise use of Scotland's biodiversity.

6.3 The Rural Land Use Working Group

What has since become the Rural Land Use Working Group of the Scottish Biodiversity Forum was known as the Agriculture Working Group when it was established in June 1998. It included representatives from a wide range of organisations and its remit included action to raise awareness of biodiversity within the farming and landowning communities, and to encourage the use of 'biodiversity-friendly' practices. The Working Group met several times a year, usually in association with a farm visit, and tasked smaller sub-groups with taking forward certain specific projects.

Outputs from the Working Group included leaflets targeting different types of farms and crofts, the promotion of key biodiversity messages on calendars, information notes for agricultural machinery contractors, and biodiversity awareness seminars. The Group also initiated research into genetic diversity in traditional breeds of livestock and plant varieties, and indirectly contributed to a number of initiatives that will deliver benefits to biodiversity.

Following the release of the draft Strategy in February 2003 (Scottish Executive, 2003), the Agriculture Working Group widened its remit when it took on the development of a detailed three-year Scottish Biodiversity Strategy Implementation Plan for the rural land use sector. The Group's membership was expanded to include people with an expert knowledge of rural environments, such as freshwater and mountains, and of specialist areas, such as access and planning. In recognition of its wider remit, the title of the Working Group was changed to the Rural Land Use Working Group (RLUWG).

6.4 Developing the Implementation Plan

As with the development of the draft Strategy, the active participation of key stakeholders in developing the implementation plan has been essential. The RLUWG was asked to prepare an implementation plan covering rural land use aspects of the Scottish Biodiversity Strategy, which was to be submitted to the Strategy Steering Group early in 2004. At its first meeting, RLUWG members were allocated to one of five sectoral sub-groups: agriculture, uplands and mountains, forestry, freshwater, and sustainable development. Each sub-group was encouraged to co-opt additional expertise as necessary. To ensure consistency of approach, the Group agreed to work to a standard Implementation Plan template (see Table 6.1).

Each sub-group agreed the main issues for their sector, identifying what actions, if any, were already being carried out to address each issue. Where 'gaps' were found in the 'Current Contributing Actions' list, new actions with clear targets were proposed. Sub-groups were tasked with establishing a lead organisation for each new action, identifying stakeholders with an interest in outcomes, identifying milestones towards each target and 'success' indicators, and estimating the resource implications. A number of gaps still require to be filled, in particular identifying organisations prepared to take the lead in delivering the new actions. In many situations the resource implications are not yet fully appreciated. A summary of the entry for 'assisting local delivery of biodiversity' is given in Table 6.2.

The RLUWG organised a workshop in October 2003. The objective was to provide the key rural stakeholders with an opportunity to examine what the Working Group had produced, and to provide feedback, suggestions and contributions which would allow the Group to refine and finalise its implementation plan. This objective was largely achieved: further priority targets and actions to be tackled in the next three years were identified and there was agreement on what targets and actions were non sector-specific or cross-cutting.

The Group agreed which targets and actions were specific to each of the five rural land use sectors, and which targets and actions cut across sectoral boundaries. The cross-cutting, or non sector-specific, targets and actions would later be grouped under a number of strategic cross-cutting or common issues in the implementation plan. These common issues would provide opportunities for further collaboration and integration and included:

- research, monitoring and evaluation;
- landscape-scale planning and management;
- assistance of local delivery of biodiversity;
- the agri-environment programme;
- awareness, understanding and participation;
- advice, best practice and capacity building;
- enhancing the implementation of Habitat and Species Action Plans;
- the need to audit biodiversity enhancement and to incorporate this into policies, plans and mechanisms that impact upon or have potential to enhance biodiversity;
- minimise the threat to biodiversity from climate change, pollution and alien (or non-native) species;
- promote sustainable uses which assist, or are compatible with, biodiversity conservation and help meet Scottish Executive policies for sustainable development; and
- develop and implement a programme for the conservation of genetic diversity in Scotland.

Examples of sector-specific targets and actions for Agriculture and Freshwater include the following.

- Agriculture: facilitate the delivery by land managers of biodiversity benefits through current contributing actions, such as advisors and consultants promoting the Rural Stewardship Scheme. Encourage participation in the Scheme, and the Scottish

The Scottish Biodiversity Strategy: developing the Rural Land Use Implementation Plan

Table 6.1 The template used in developing the implementation plans.

Issue	Current Policy	3-Year Target	Current contributing actions	New Action	Lead Stakeholder	Milestones	Timescale	Monitoring Indicators	Resources

Table 6.2 A completed entry for 'assisting local delivery of biodiversity'.

Issue	Current Policy	3-Year Target	Current contributing actions	New Action	Lead	Stakeholder	Milestones	Timescale	Monitoring Indicators	Resources
Assist local delivery of biodiversity	Local and National Government Responsibilities LBAP Network	Effective co-ordination of LBAP actions across local government boundaries	Cairngorms LBAP; Rural Stewardship and Countryside Premium Schemes; Habitat networks; Water Framework catchment management plans; LBAP network, LBAP Partnership meetings and LBAP newsletter	Regional co-ordination with regard to shared habitats/species, especially with regard to management practices e.g. grazing management (deer, sheep) including common grazings	SNH, LAs	LBAP Officers, Local Authorities, SEERAD, Deer Management Groups, Deer Commission for Scotland, Crofters' Commission, Scottish Crofting Foundation, SNH, Local land managers, Scottish Environment Link	By Autumn 2004, agreed scope of local collaborative planning and organisations and people to be involved. By Autumn 2005, develop proposals for implementing local collaborative planning for biodiversity.	January 2007	Questionnaires seeking feedback from LBAP Officers on degree of co-ordination	Finding resources to extend LBAP Officers' contracts

Executive to produce a CD-ROM training package for use by Area Agricultural staff and others involved in the Rural Stewardship Scheme procedures. New actions, such as piloting collaborative applications to the Scheme and developing the training package further to cover, for example, collaborative applications, were also identified.
- Freshwater: ensure that freshwater biodiversity issues are fully integrated into catchment planning and management, through current contributing actions such as setting up River Basin and Sub-Basin Management Plans, and through new actions such as ensuring that all Local Biodiversity Action Plan freshwater and wetland targets are fully integrated into River Basin and Sub-basin Management Plans by 2007.

6.5 What are the next steps?

Findings from the October 2003 workshop will be considered further and passed to the Scottish Biodiversity Forum's Steering Group for collating with the feedback from similar workshops run by the Urban and Marine Working Groups. The three Working Groups are expected to submit their implementation plans to the Steering Group early in 2004. The final implementation plans for the Strategy should be completed in time to be presented to the Minister for approval and general issue in May 2004.

Crucially, this is an ongoing process. The announcement of the Strategy, together with the three-year implementation plans, will not be the end. These plans will need to be implemented. The wide range of organisations represented in the various Working Groups, together with other stakeholders, will have a responsibility to monitor progress to ensure a successful outcome, and to be involved in reviewing the implementation plans in three years' time.

Editors' note

Since this chapter was written in 2003, there has been significant progress in taking this work forward. The Scottish Biodiversity Strategy (Scottish Executive, 2004a) was published in May 2004, along with an overview of the Implementation Plans 2005-2008 (Scottish Executive, 2004b), a report on developing a range of indicators for monitoring progress with implementing the Strategy (Scottish Executive, 2004c) and a draft research strategy (Scottish Executive, 2004d). The implementation plans, proposed indicators and draft research strategy were put out to consultation in May 2004. Further information can be found on the following website: www.scotland.gov.uk/biodiversity.

References

Anonymous (1994). *Biodiversity: the UK Biodiversity Action Plan.* HMSO, London.

Hocknell, S., Lenthall, C. & Quigley, C. (2003). *Towards a Strategy for Scotland's Biodiversity: Biodiversity Matters! Analysis of Consultation Responses.* Scottish Biodiversity Forum and Royal Society for the Protection of Birds, Edinburgh.

Scottish Executive (2003). *Towards a Strategy for Scotland's Biodiversity: Biodiversity Matters!* Scottish Biodiversity Forum, Scottish Executive, Edinburgh.

Scottish Executive (2004a). *Scotland's Biodiversity: It's In Your Hands. A Strategy for the Conservation and Enhancement of Biodiversity in Scotland.* Scottish Biodiversity Forum, Scottish Executive, Edinburgh.

Scottish Executive (2004b). *Scotland's Biodiversity: It's In Your Hands. A Strategy for the Conservation and*

Enhancement of Biodiversity in Scotland. An Overview of the Implementation Plans 2005-2008. Scottish Biodiversity Forum, Scottish Executive, Edinburgh.

Scottish Executive (2004c). *Scotland's Biodiversity: It's In Your Hands. A Strategy for the Conservation and Enhancement of Biodiversity in Scotland. Developing an Indicator Set.* Scottish Biodiversity Forum, Scottish Executive, Edinburgh.

Scottish Executive (2004d). *Scotland's Biodiversity: It's In Your Hands. A Strategy for the Conservation and Enhancement of Biodiversity in Scotland. Draft Research Strategy.* Scottish Biodiversity Forum, Scottish Executive, Edinburgh.

6 Farming, Forestry and the Natural Heritage: Towards a More Integrated Future

7 Advisory and planning tools to inform natural heritage management

Daniel Gotts

Summary

1. The range of sources of knowledge which inform natural heritage management are outlined. These include research, demonstration sites, and practical experience. The need to gather knowledge from experienced land managers is emphasised.
2. The importance of the motivation of land managers is highlighted, as is the value of land managers understanding the objectives and outcomes of what they are being encouraged to do. This understanding is a key aspect of effective advisory work. A process of awareness – interest – desire – action will help create a demand for advice.
3. The main types of advisory tools are discussed, as well as how these tools might be developed. It is valuable to engage as much as possible with practical land managers and make advice and information accessible and relevant.
4. Key aspects for the development of advice are highlighted: adding to the knowledge base; the need to balance priorities; the importance of practical demonstrations; and the central role of people in planning land management.

7.1 Introduction

Over the past two decades there has been a significant shift in land use policy to encourage land managers – both farmers and foresters – to manage their holdings for the benefit of the natural heritage, as well as for the production of agricultural and timber outputs. This has partly been driven by a change in emphasis in European policy, especially on agriculture. This has been implemented in the UK mainly through programmes such as agri-environment schemes and incentives for the management and expansion of native woodlands, as well as support for the provision of better public access.

The evolution in policy emphasis has led to a change in what farmers and foresters are expected to manage for on their land. In order to meet this expectation, land managers need to understand what they are being asked to deliver, and to know how to deliver the desired outcomes. If these policy aims are to be delivered, there needs to be a commitment to providing land managers with this knowledge.

This chapter gives an overview of how knowledge about protecting and managing the natural heritage is generated, gathered and distributed. It considers the importance of the motivation behind what land managers do. It then looks at the main ways by which this knowledge is transferred to land managers. Finally, it comments on key areas for further work and issues to be addressed.

In this chapter, the term 'natural heritage' is used to encompass biodiversity, access and landscape interests. 'Natural heritage objectives' refer to what is to be achieved through changes in land management, and 'outcomes' are the measurable results of changes in land management.

7.2 Sources of knowledge

A wide variety of sources of knowledge is used to inform the management of the natural heritage. This knowledge is gathered, and then disseminated, through a range of different media, both formal and informal.

7.2.1 Ecological research

Fundamental research on the ecological needs of species and habitats has helped in the development of prescriptions in incentive schemes, and of management advice to farmers and foresters. Examples here include work on the corncrake (*Crex crex*) (Green & Williams, 1994), the development of beetlebanks (Boatman, 1994), creation of species–rich grassland (Scottish Agricultural College, 2003), and capercaillie (*Tetrao urogallus*) (Summers & Dugan, 2001; Proctor & Summers, 2002). However, there is a need for more information on many topics, such as the management of grazing in woodlands, combining management for different interests such as access and nesting birds, managing for conservation interests in new land use systems that emerge as a result of the reform of the Common Agricultural Policy (CAP), and how intensive grassland systems can provide a habitat for wildlife.

7.2.2 Nature reserves and demonstration sites

Nature reserves and demonstration sites are often used as testing areas for refining research findings as well as for developing management approaches that can then be applied more widely (Gibbons, 2002). This allows trialling and refining techniques to iron out practical problems – much as experienced users test beta-versions of new computer software. For example, Scottish Natural Heritage's (SNH) work at Cairnsmore of Fleet National Nature Reserve to trial the use of heavy-duty flail mowers for swiping heather (*Calluna vulgaris*) as an alternative to burning, and the Royal Society for the Protection of Birds' (RSPB) management of grazing and woodland regeneration at Abernethy (Taylor, 1995; Royal Society for the Protection of Birds, 2003).

7.2.3 The acid test of the real world

As management techniques for access, conservation and the landscape are more widely adopted outside research sites and reserves, the practical issues faced, and solutions found, can be used to help tailor management prescriptions and advice to a wider range of situations. Farmers, foresters and other land managers, for whom managing the natural heritage for conservation purposes is not their main activity, will be able to give a user's view on whether or not something is effective and practical to implement. This may be a better 'beta test' than simply trying something out on a reserve. For example, the development of both the 4 Point Plan (Scottish Agricultural College, 2002; Audsley, this volume) and SNHs practical advice on managing public access have involved drawing on the experience of practical land managers.

7.2.4 Ideas from the ground

Growers have always been developing improved ways of managing their crops and livestock; farmers, crofters and foresters who are involved in managing the natural heritage are also doing this. Where land managers have good links to networks, such as research and advisory organisations, and to non-governmental organisations (NGOs), this information gets incorporated into knowledge and advice about managing the natural heritage. For example, there are many farmers and foresters who have developed effective ways to manage public access in combination with their land management operations. Similarly, a farmer-client of the Game Conservancy Trust (GCT) worked out that conservation headlands are most effective for game and other birds if the nitrogen applied is reduced compared with the rest of the crop. This can be done relatively easily when using liquid fertiliser (E. Baxter, pers. comm.).

7.2.5 Completing the feedback loop

Native woodland schemes, agri-environment schemes and management agreements have been in operation for a number of years (Scottish Natural Heritage, 2001; Forestry Commission Scotland, 2003; Scottish Executive, 2003b), and many land managers throughout Scotland are now being paid to manage land through these incentives. Many practical lessons will have been learned about the effective delivery of management objectives. It is important that these people are involved, and for their experience and knowledge to be gathered and fed back to refine management objectives and prescriptions, and incorporated into future advice. The use of stakeholder fora for schemes goes some way towards this, where a range of participants and other interested parties are involved in developing schemes and then agreeing their refinement. For example, this was proposed for goose management schemes by the National Goose Forum (Scottish Executive, 2000b) and subsequently implemented for goose schemes in operation from 2000 (Scottish Executive, 2005). In addition, the Scottish Executive Environment and Rural Affairs Department (SEERAD) and SNH are working with the Scottish Agricultural College (SAC) and the Farming and Wildlife Advisory Group (FWAG) and others to pilot a way of harnessing practical experience to improve agri-environment advice and prescriptions (Scottish Agricultural College, 2004).

7.3 Motivating people to act

In order to get the best management of the natural heritage, people should be motivated to do this work and to deliver the desired outcomes. Whilst providing knowledge from a range of sources will help encourage positive management of the countryside, for management to be most effective practitioners should understand the overall objectives and the desired outcomes.

Clearly stated objectives allow land mangers to adapt their management if necessary. For example, in a late spring birds may nest later. If so, livestock may need to be kept out of a key area for nesting birds for an extra two or three weeks. When a woodland is not regenerating as much as expected, then scarification or short-term grazing may be needed to provide germination sites. Farmers and foresters are aware of what they want their crops, livestock or forests to be like at a certain point and they manage for this accordingly; this should be the approach for all those involved in managing the countryside for conservation as well.

7.3.1 Creating a demand for knowledge

There needs to be a demand for knowledge and understanding which ultimately leads to positive action. This can be achieved by working through steps – formally or informally – in a process of awareness – interest – desire – action (J. Roberts, pers. comm.):

- raising the **awareness** of what land managers have or could have on their holdings;
- generating an **interest** in what is there or a holding's potential;
- creating a **desire** to do something about it; and
- ultimately, some **action** will be taken on the ground.

The appropriate advisory tools should be used to work through this process to add value at each stage by building on what those involved already know and understand. There may be lessons that can be learned from experience or techniques elsewhere, such as approaches in product marketing, where a process is followed from consumer involvement in design through product awareness to eventual purchasing action and use.

7.3.2 Understanding existing motivation

If land managers are to be encouraged to care for the natural heritage, then those involved in developing schemes and planning advice need to be aware of land managers' existing motivations or drivers. These might encourage them to act or react in a particular way, or provide a base interest which can be built on and encouraged. Key drivers might include:

- the need to make a profit to live on and fund future investment;
- a desire to live in an attractive setting;
- the desire for privacy;
- wanting to maintain the productivity and capital value of the land;
- a wish to provide habitat to support shooting;
- the need to comply with legislative requirements;
- a strategic business aim to produce a quality or niche product, whether from a farm or a forest; and
- an interest in managing for wildlife and the landscape.

While many of these drivers will clearly encourage people to look after the countryside, some of these might be seen as working against an interest in protecting and managing the natural heritage – whether it be for wildlife, the landscape or public access. These have implications for the design and development of advisory programmes.

Advisory programmes will be most effective where they are in 'strategic alignment' with broader land use policy. If both the policy messages and on-the-ground advice and information are much the same, then land managers will be much more likely to act on the advice provided. However, if, as in the past, the policy message has been to maintain or increase production, then it is very difficult for land managers to take account of advice on protecting and managing the natural heritage unless this makes some business sense for them (I. Kenny, pers. comm.). Ideally, the reform of the CAP and the move to the Single Farm Payment (Council of the European Union, 2003) will remove much, but probably not all, of this tension.

In addition, it is important that those planning advisory programmes recognise that the target audience is not uniform, as not everyone is starting from the same level of knowledge and interest. Some will have little knowledge of what wildlife they have on their land or how to manage for it, while others will have been working on access and conservation for many years and will be looking for specialist advice on fine-tuning what they are currently doing.

The most effective advisory programmes will be developed and put in place if these different motivations and different levels of knowledge are taken into account from the start.

7.4 Practical advisory tools

There is a range of tools, both tried and tested, that can be used to share knowledge on land management for the natural heritage. It is important to recognise that different tools are appropriate in different situations and at different stages in the awareness-to-action process. In order to plan its support for future advisory work, the Department for Environment, Food and Rural Affairs commissioned a review of the ways in which land managers obtain advice and information (Dampney *et al.*, 2001). This review found that farmers still value face-to-face advice as one of the most effective ways of understanding new ideas. In addition, farmers put great weight on their own experience and often value this above other sources of information.

7.4.1 Printed and other distributed advisory material

Printed advisory material ranges from comprehensive advisory manuals to simple one-page leaflets, making the lessons from research reports and demonstration areas accessible to both advisers and land managers. Computer-based and interactive tools are being used in many areas of advisory work; these have their place. However, not everyone has access to a computer with a CD-ROM drive or to a fast internet connection. In order to be accessible to all, material must still be available in paper form in some way, even if it is just printed from the internet by an adviser. It must be promoted and distributed through the right channels in order to be effective.

It is essential to think carefully about who is the target audience. With the prevalence of contractors in both farming and forestry, many practical operations are no longer carried out directly by the owner. Persuading a farmer or forest manager to plan to look after the natural heritage is one thing. However, it is very important to help ensure that farm and estate staff and agricultural and forestry contractors are aware of what needs to be done on the ground to meet scheme requirements or to avoid causing pollution. For example, the Scottish Biodiversity Forum's Rural Land Use Working Group has produced laminated 'cab cards', specifically aimed at machinery ring members and contractors, highlighting environmental issues to consider when working on clients' farms.

SNH's Targeted Inputs for a Better Rural Environment (TIBRE) initiative aims to promote the use of technology in intensive agriculture to reduce impacts on the environment, and the arable handbook (Scottish Natural Heritage, 1997) – developed through discussions with farmers and launched in 1997 – has been updated (Scottish Natural Heritage, 2004). The techniques and technologies highlighted by TIBRE include computer-based decision support tools, precision farming equipment, and diagnostic tests to help determine the need to apply nutrients or pesticides. Although a web-based version will be developed, the handbook will still be available in printed form for ease of reference by all those using it. In addition, SNH is surveying farmers to investigate how they hear about new developments, so that the promotion and distribution of TIBRE material can be through the most effective channels.

In 2003, the Forestry Commission and SNH published *Habitat Networks for Wildlife and People* (Fowler & Stiven, 2003) and this sets out to illustrate broadly how the concept of forest habitat networks can be implemented on the ground. The challenge now is to translate this into action on the ground through more detailed local guidance and practical demonstrations.

7.4.2 Advice on the ground

Face-to-face advice is crucial in encouraging people to read advisory material and to adopt new practices. On-the-ground advice can help by showing land managers both what is of interest on their land and how they can protect and enhance it. Advice can also help by offering practical solutions, whether it is on where best to plant a new woodland or how best to manage public access. Advice does not come cheap, but if government, agencies and others expect land managers to adopt new techniques to best effect, then resources must be invested in this.

Forestry Commission Scotland, SNH and SEERAD have been supporting organisations such as FWAG, SAC, Scottish Native Woods and Reforesting Scotland to raise awareness of environmental issues among land managers, as well as to help them implement schemes and agreements. Recently, initiatives such as Greenspace for Communities and Paths For All (Scottish Natural Heritage, 2003a) have been supported to provide on-the-ground advice, especially in support of new legislation and policy initiatives.

7.4.3 Training and awareness events

Training and awareness courses and group-based exercises can be very useful in helping the adoption of new ways of doing things and the development of new skills. However, simply going on a training course will not deliver change overnight. Subsequent exposure to the issue or practicing a new skill is required if the new knowledge is to be retained, developed and eventually become part of a person's routine way of doing things.

Planning to deliver the right level of course for the audience is important. Training on the ecology of individual species is of use to conservation advisers and agency staff. However, land managers are unlikely to have time, or want, to attend many courses on managing for specific habitats or species, unless they have a particular interest, such as shooting.

Many farmers and woodland managers are now coming into incentive schemes. However, it is not clear how many are aware of what is expected on the ground in terms of the objectives towards which they should be managing, and the outcomes which will indicate that the objectives have been achieved. Awareness days on agri-environment schemes for farmers joining these were piloted a few years ago to provide an overview of what is expected. Consideration should be given to introducing these more widely to give farmers and crofters a better understanding of what the schemes are expected to deliver on the ground, so that they have 'ownership' of the outputs. This is especially important at a time when many more land managers will be expected to be aware of environmental conditions following the introduction of the Single Farm Payment (Council of the European Union, 2003).

SNHs Sharing Good Practice programme (Scottish Natural Heritage, 2003b) has been running since 1999. In recent years this has included events on both farming and forestry. SNH could work with others to plan to include more practical events and involve farmers and woodland managers more in the delivery, for example on scheme objectives and cross-compliance conditions.

7.4.4 Showing it working on the ground

Many new techniques are developed and tried out initially at research centres, and this can be a good way of demonstrating their initial application. Countryside management approaches can be promoted in this way with nature reserves being some of the places where these are tried and refined.

SACs Hill Sheep and Native Woodland project at Kirkton, near Crianlarich, has been established to look at a new approach to managing sheep and woodland on the same area of ground. This work aims to test whether combining sheep husbandry with woodlands can increase the sustainability of upland farming systems, and to test systems of hill sheep production with off-wintering. Given the nature and scale of the research – effectively looking at a complete system change – this investigation can only be undertaken at a 'farm scale'. This provides the opportunity to tackle a wide range of issues that arise in the implementation of the idea.

For many years, SNH has been managing Creag Meagaidh National Nature Reserve to reduce red deer (*Cervus elaphus*) numbers and to encourage native woodland regeneration. In this situation the priority is to promote a more diverse landscape which includes habitat for deer and other wildlife, rather than focusing on producing wool or lamb, or maintaining a deer stalking enterprise.

There is still some scepticism about whether the Kirkton and Creag Meagaidh approaches would work in less 'supported' situations. However, trying this management in these situations is likely to plant a germ of an idea in people's minds which comes to fruition much later on.

Demonstrating a new land management technique or system at a research site might encourage some people to try it out. However, many will need to be convinced that it can work – practically or economically, or both – in more general situations. Also, as policy develops so that management of the natural heritage is set to become a more central aspect of land management, it is important that the relevant systems and techniques can be seen in practice in more commercial farms and forests.

A good example of this is the work being done by some commercial farms, with the support of LEAF (Linking Environment And Farming) and FWAG to show how environmental considerations can be built into farm management. SNH is also supporting Scottish Native Woods to establish demonstrations of co-ordinated riparian management in six catchments in the Highlands, and the aim is to get land managers to work together to manage riparian woodlands. The European Union-financed LIFE projects, such as the work on Atlantic oakwoods in Sunart, have shown how co-ordinated work can deal with an issue on a large scale (Caledonian Partnership, 2003).

7.4.5 Getting land managers involved

Discussion groups have long been used in both farming and forestry. These provide a good opportunity to put forward new ideas and have them discussed constructively, with an opportunity for many of the potential pitfalls to be examined. Given the evolution of land use policy, it is important for SNH, the Scottish Environment Protection Agency and conservation NGOs to engage with these groups to help farmers and foresters better understand the practical implications of policy changes.

FWAG, for example, regularly organises farm walks to look at examples of how conservation management is done on farms, and generate discussion about this. However, this might be seen as 'preaching to the converted', so there is a need to make more of a push to contact those who would not come on a FWAG farm walk. The introduction of the environmental conditions attached to the Single Farm Payment (Council of the European Union, 2003) will encourage more farmers and crofters to find out how to put things into practice.

7.5 Conclusion – key areas for future work

A number of key areas need to be addressed and tackling these will help meet the challenges ahead to achieve more effective integration of the range of land management objectives.

7.5.1 Adding to the knowledge base

There are some areas where more data or information are needed to inform advice and planning. There is a need to identify gaps and help prioritise future research. However, expanding the knowledge base includes gathering practical knowledge and experience, as well as improving the ways in which information is communicated among land managers, researchers and advisers.

7.5.2 Better co-ordination on the ground

Integration of interests is necessary to identify and to plan how biodiversity, access and landscape objectives can be combined. However, in many areas of the country both land and funding resources will be limited. Management activity needs to be targeted and local management priorities agreed.

Both the *Scottish Forestry Strategy* (Scottish Executive, 2000a) and *Custodians of Change* (Scottish Executive, 2002) emphasised the importance of habitat networks. Forestry Commission Scotland and SNH are taking forward work to look at how to balance requirements for woodland and open ground habitats, especially in the uplands. In the lowlands, where land is more productive, the challenge is more about how best to use the resources available to tackle habitat fragmentation and degradation. There is a need to identify, agree and put in place management that takes into account the range of natural heritage interests.

SNH and SEERAD are currently funding initial work to develop ways of putting this into practice (Humphrey *et al.*, 2005). This will provide guidance and planning tools to help with the implementation of lowland habitat networks. By using a more planned approach, people with a range of interests can be brought together to look at the wider scale and how management for nature conservation might also contribute to enhancing the landscape and providing access, as well as to wider rural development. These agreed objectives should then be used to help prioritise the use of funding.

7.5.3 Showing things *in situ*

Research work is only part of the picture. The ideas generated can appear abstract to many land managers. Seeing things happening on the ground in a situation similar to their own – both economic and physical – is most likely to influence them to take action. SEERAD has proposed 'monitor' farms and benchmarking as a way of helping farmers improve business performance (Scottish Executive, 2003a). In parallel, a network of locally relevant environmental 'monitor' farms and environmental benchmarking is needed, showing what is good practice to emulate.

The Scottish Biodiversity Strategy draft implementation plans include a number of key actions on the development of demonstration sites (Scottish Executive, 2004). These need to be taken forward to show how land management can provide for conservation interests. The development of demonstrations of long-term management changes can take a number of years before they come to fruition and resources need to be committed to this. However, demonstrations of smaller-scale changes, new approaches and novel techniques need to remain flexible to adapt to developing knowledge, needs and markets.

7.5.4 The importance of people

People are key to helping deliver many natural heritage objectives, as the draft Scottish Biodiversity Strategy emphasises. If land managers are expected to play their part in this, they must know what to do and be motivated to act.

In planning advisory programmes, it is essential to take account of the audience for advice and accept that the audience is varied – from those who have been doing conservation management for many years, to the new entrants to land management. Many land managers have been receiving advice and guidance for a number of years, and applying this and learning from the experience. It is essential that the value of their experience is recognised and that their knowledge is captured.

People need to be involved in planning what is delivered and how it is delivered at a local level. Objectives for incentive schemes and other initiatives need to reflect both national and local priorities. This is not always an easy balance to achieve. However, the development of planning and advisory tools can help with this by involving people on the ground in the process as early as is practical. Any plans for land use in an area – whether for nature conservation, access or the landscape – are going to be more acceptable and easier to implement if those who it affects directly have been involved in their development.

7.5.5 Commitment to advice

Finally, it is important to remember that when things change and when government, agencies and others expect people to work in new ways, there is a need to provide good, consistent advice and support to enable this. The nature of the advice and the ways in which this advice is provided to farmers, foresters and other land managers also needs to adapt in order to remain effective. That is an important challenge to us all.

References

Boatman, N.D. (Ed.) (1994). *Field margins – integrating agriculture and conservation.* BCPC Monograph No. 58. British Crop Protection Council, Farnham.

Caledonian Partnership (2003). *Restoring Natura Forest Habitats.* Highland Birchwoods, Munlochy.

Council of the European Union (2003). *CAP Reform – Presidency Compromise (in agreement with the Commission).* Council of the European Union, Brussels.

Dampney, P., Winter, M. & Jones, D. (2001). Communication methods to persuade agricultural land managers to adopt practices that will benefit environmental protection and conservation management (AgriComms). Unpublished report. Department of Environment, Food and Rural Affairs, London.

Forestry Commission Scotland (2003). *Scottish Forestry Grants Scheme.* Forestry Commission, Edinburgh.

Fowler, J. & Stiven, R. (2003). *Habitat Networks for Wildlife and People.* Forestry Commission Scotland, Edinburgh and Scottish Natural Heritage, Perth.

Gibbons, D. (2002). The role of science in the RSPB. In *Conservation Science in the RSPB, 2002.* Royal Society for the Protection of Birds, Sandy.

Green, R.E. & Williams, G. (1994). The ecology of the corncrake *Crex crex* and action for its conservation in Britain and Ireland. In *Nature Conservation and Pastoralism in Europe*, ed. by E. Bignal, D.I. McCracken & D.J. Curtis. Joint Nature Conservation Committee, Peterborough. pp. 69-74.

Humphrey, J.W., Watts, K., McCracken, D., Shepherd, N., Sing, L., Poulsom, E.G. & Ray, D. (2005). A review of approches to developing lowland habitat networks in Scotland. Unpublished report. Scottish Natural Heritage contract AB(02AA10212) 040549. Forest Research, Roslin.

Proctor, R. & Summers, R.W. (2002). Nesting habitat, clutch size and nest failure of capercaillie *Tetrao urogallus* in Scotland. *Bird Study*, **49**, 190-192.

Royal Society for the Protection of Birds (2003). *Reserves Annual Report 2001/02.* Royal Society for the Protection of Birds, Sandy.

Scottish Agricultural College (2002). *The 4 Point Plan.* Scottish Agricultural College, Edinburgh.

Scottish Agricultural College (2003). The management of long-term set-aside for nature conservation/the creation of species rich grassland: summary 1993–2001. *Scottish Natural Heritage Commissioned Report F99AA105.*

Scottish Agricultural College (2004). *Agri-environment Case Studies – Improving Advice on Practical Habitat Management.* Scottish Executive, Edinburgh. Available from www.scotland.gov.uk/library5/environment/aecsiaph-00.asp.

Scottish Executive (2000a). *Forests for Scotland: the Scottish Forestry Strategy.* Scottish Executive Environment and Rural Affairs Department, Edinburgh.

Scottish Executive (2000b). *Policy Report and Recommendations of the National Goose Forum.* Scottish Executive, Edinburgh.

Scottish Executive (2002). *Custodians of Change.* Report of the Agriculture and Environment Working Group, Scottish Executive, Edinburgh.

Scottish Executive (2003a). *National Strategy for Farm Business Advice and Skills: Whole Farm Review Pilot Scheme Explanatory Booklet.* Scottish Executive, Edinburgh.

Scottish Executive (2003b). *Rural Stewardship Scheme.* Scottish Executive, Edinburgh.

Scottish Executive (2004). *Scotland's Biodiversity: It's In Your Hands. A Stratagy for the Conservation and Enhancement of Biodiversity in Scotland. An Overview of the Implementation Plans 2005-2008.* Scottish Biodiversity Forum, Scottish Executive, Edinburgh.

Scottish Executive (2005). *Report of the National Goose Management Review Group: Review of the National Policy Framework for Goose Management in Scotland. Response by the Scottish Executive.* Scottish Executive, Edinburgh.

Scottish Natural Heritage (1997). *TIBRE – New Options for Arable Farming.* Scottish Natural Heritage, Perth.

Scottish Natural Heritage (2001). *Natural Care Strategy.* Scottish Natural Heritage, Perth.

Scottish Natural Heritage (2003a). *Facts & Figures 2002/03*. Scottish Natural Heritage, Perth.

Scottish Natural Heritage (2003b). *Sharing Good Practice – Programme of Events 2003-2004*. Scottish Natural Heritage, Perth.

Scottish Natural Heritage (2004). *TIBRE Arable Handbook – 2nd edition*. Scottish Natural Heritage, Perth.

Summers, R.W. & Dugan, D (2001). An assessment of methods used to mark fences to reduce bird collisions in pinewoods. *Scottish Forestry*, **55**, 23-29.

Taylor, S. (1995). Pinewood restoration at the RSPB's Abernethy Forest reserve. In *Our Pinewood Heritage*, ed. by J.R. Aldhous. Forestry Commission, Edinburgh, The Royal Society for the Protection of Birds, Edinburgh and Scottish Natural Heritage, Perth.

7 Farming, Forestry and the Natural Heritage: Towards a More Integrated Future

PART 3:
Farming, Forestry and Access

View north over Loch Lomond from Duncryne near Gartocharn © Lorne Gill, Scottish Natural Heritage

Farming, Forestry and the Natural Heritage: Towards a More Integrated Future

PART 3:
Farming, Forestry and Access

The last few years have seen significant changes to the legal arrangements for access to Scotland's countryside. Wide-ranging statutory access rights have been introduced through the Land Reform (Scotland) Act 2003. Explaining how these access rights are to be exercised and managed responsibly is the job of a new Scottish Outdoor Access Code. At the time of the conference, in late 2003, there was still considerable uncertainty about what the new statutory access rights would mean for farmers and other land managers.

In Chapter 8, Silvester describes this uncertainty from the perspective of the landowner. Although there are undoubtedly some uncertainties, Silvester highlights the need to make the new legislation work on the ground. She argues that a common sense approach to managing and facilitating access is the way forward. This approach must include core path networks, education, effective access management and adequate resources.

Hunt (Chapter 9) picks up the thread of resources, looking at the ways in which existing agri-environment and forestry schemes might help or hinder the successful integration of access and land management. Echoing some of the points made by Gotts (Chapter 7), Hunt emphasises the importance of understanding the motivations of farmers and other land managers if funding schemes are to work well. He argues the case for Local Farming Plans, developed at an appropriate landscape scale, that set out how farming, biodiversity, landscape, access, culture and flood defence can be integrated at the local level.

In Chapter 10, Taylor *et al.* report on research undertaken for Scottish Natural Heritage. This research looked at the concerns of land managers and identified the solutions that are already available for addressing these. The chapter provides several practical examples and concludes that although there are a wide range of solutions available, adequate funding needs to be provided to make these become fully effective.

Farming, Forestry and the Natural Heritage: Towards a More Integrated Future

8 The Land Reform (Scotland) Act 2003 and the draft Scottish Outdoor Access Code: what does it mean for land managers?

Marian Silvester

Summary

1. The Land Reform (Scotland) Act 2003 will introduce statutory access rights at a time of unprecedented change in the land management sector. The final content of the legislation has left landowners and managers with several concerns and these are outlined in the chapter.
2. The impact of the new access arrangements will vary across Scotland, with it likely to be most significant in areas close to towns and cities.
3. Landowners and managers want to make the new access arrangements work on the ground. This can be helped by them being involved in the planning of core path networks and in the management of access. An effective education programme and adequate funding are also important.

8.1 Introduction

This chapter reviews the statutory access rights created by the Land Reform (Scotland) Act 2003, and the draft Scottish Outdoor Access Code (Scottish Executive, 2001), and what this means for land managers. Land managers are varied and the impact of the legislation and the Code will be different for different people.

The land manager who has an open access policy will feel different about the legislation than someone whose land has 'Keep Out' and 'No Entry' signs at all possible access points. The land manager who lives in remote Galloway will be less affected, and possibly less concerned, than someone who farms on the edge of Livingston. The land manager with 4 ha may find it more difficult to integrate access with their current activities than the land manager with 4,000 ha. Current and past experience of access will also influence how individual landowners and managers react to statutory access rights and the draft Code.

This chapter takes stock of the new access arrangements and the concerns of land managers. The chapter then looks forward at how the new arrangements might be implemented on the ground.

8.2 Background
8.2.1 Access rights in a wider context

It is important to look at access rights in the wider context in order to gain a better understanding of how land managers may view it. The last few years have seen a wide range of legislation and related issues that impinge on landowners and managers, including:

- the Land Reform (Scotland) Act 2003 – this covers community and crofting rights to buy as well as access rights;
- the foot and mouth crisis – from which the south of Scotland is still recovering;
- the reform of the Common Agriculture Policy (CAP) – which brings considerable uncertainty; and
- the Agricultural Holdings (Scotland) Act 2003.

These are just some of the headline issues that landowners and managers currently have to deal with. Any one of these on their own would be demanding and difficult, but taken together they create an environment in which land managers can feel overwhelmed, even threatened. This is the context in which statutory access rights will be rolled out, along with a code of responsible behaviour, bringing new obligations and ways of working to land managers. It is little wonder then that this is a stressful time for landowners and managers, with access rights just one more issue to come to grips with in an environment that is ever-changing, with new rules and regulations to learn, understand and implement.

8.2.2 Feedback from the members of the Scottish Rural Property & Business Association

Over the last few years, Scottish Rural Property & Business Association (SRPBA) (formerly known as the Scottish Landowners' Federation) has spoken to hundreds of its members at access roadshows, local meetings and during one-to-one site visits. As a result of this, the SRPBA has received considerable feedback from members about their feelings towards the access legislation and their expectations of its impact on land management. This feedback forms the basis for the opinions expressed in this chapter.

8.2.3 Scottish Rural Property & Business Association and access

The SRPBA has an impressive track record of working in partnership to improve and facilitate access dating back over many years. Examples of this include the Hillphones scheme, *Heading for the Scottish Hills* (Scottish Mountaineering Club, 1996) and the Concordat on Access to Scotland's Hills & Mountains (Scottish Natural Heritage, 1996). The Hillphones scheme provides daily information on deer stalking operations to enable hillwalkers and climbers to find out where stag stalking is taking place and to plan routes avoiding these operations. More information about Hillphones can be found at www.snh.org.uk/hillphones/index.asp. At a local level, many SRPBA members are involved with developing path networks and participating on local access forums. Through its work on the National Access Forum, the SRPBA has supported a right of access, exercised responsibly, for informal recreation and passage.

In preparing for the access legislation, the SRPBA wanted to see the key obstacles to better public access resolved. The SRPBA saw these obstacles as:

- a lack of duties on local authorities;
- the issue of liability; and
- the need for funding.

Most of all, the SRPBA wanted the legislation to provide clarity and balance.

The Land Reform (Scotland) Act 2003, however, extends the right of access beyond recreation to include commercial activity, and fails to give local authorities clear duties to create, and importantly, to maintain path networks, or to address the issue of occupier's liability. It is unlikely, therefore, to lead to the significant increases in public funding which are essential to making the legislation work well.

So, there is a general feeling of uncertainty for land managers at the moment, and probably for others such as local authority staff and access officers. The legislation has been passed but it will not come into force until a new Code of responsible behaviour has been agreed and published. During this interregnum whilst we wait for the Code to be published, there is a lack of certainty about how various elements of the Act will be interpreted. Furthermore, important questions about the role of local access forums, dispute resolution and core path planning have not been answered because guidance has not yet been provided to local authorities.

Unfortunately, therefore, we are in a state of limbo, and many people, recreational users and land managers alike, are either unaware of the legislation or unsure of what it will mean in practice. This extended period of uncertainty is already causing problems and leading to unnecessary conflict.

Feedback from landowners suggests that some access-takers are behaving as though they now have a statutory right of access, even though without the Code they cannot fully understand what this means and what responsibilities the Code will place upon them. Attitudes among some people are already changing, and generally this is felt by land managers to have been for the worse. For example, on one estate where previously university groups would contact the landowner before a visit, this is now not happening. Common courtesy, it would seem, may be the first casualty of the new legislation.

8.3 Current concerns

The legislation has now been passed by the Scottish Parliament but there remain some important concerns for landowners and managers. These are outlined below.

8.3.1 Public liability

A right of access is likely to result in more people going to places they have not visited before. The legislation broadly means that people can be on most land or water, at any time of day or night. Consequently, there may be an increase in accidents involving the public whilst taking access. As society becomes more litigious, there are very real fears that this will lead to an increase in claims against the occupier of the land, and this, in turn, may lead to an increase in insurance premiums and other costs. A key need, therefore, is for those exercising access rights to be aware of the risks and to take responsibility for their actions.

8.3.2 Risk assessment

There will be an increased burden of risk assessment on land managers. Land managers are no strangers to the issue of duty of care, but there is a difference between the general risk assessment that has applied so far, and the increased burden of risk assessment that applies when people have a statutory right to be on most land and inland water at any time. Furthermore, the right of access also applies to children and this means that land managers have to consider whether children, accompanied or otherwise, might be on land or water. As many farms are worked by just one or two people, the practical implications of this increased burden of risk assessment and management are serious and, potentially, unmanageable.

Concerns over liability, which seem to be unresolved by the legislation, mean that some land managers fear that having sign-posted paths on their land will increase their duty of care to those taking access. This could lead to some land managers being resistant to having new sign-posted routes on their land, unless public liability can be assumed by another party. This possible effect of the uncertainty around liability runs counter to the intention of increasing responsible access to the countryside.

8.3.3 Biosecurity

The guidance provided to land managers on biosecurity measures by the Scottish Executive (Scottish Executive, 2003) does not fit well with the new right of access and the guidance in the draft Access Code. For example, land managers are encouraged to reduce the number of business visitors to their farms to reduce the potential for importing infection, and this does not sit comfortably with facilitating public access. Many land managers are understandably concerned about maintaining biosecurity on their farms when the new access rights come into effect.

8.3.4 Farmyards

One difficult issue is that of access to farmyards. The legislation is clear: farmyards are excluded from the right of access and the SRPBA welcomes this. Yet recreational users argue that farm tracks provide access routes to the land beyond and that many of these tracks pass through farmyards. On a practical level the SRPBA has suggested that, as a priority, local authorities should work with land managers to fund new routes around farmyards where there is a demand for access, and where access through the farmyard is not appropriate.

8.3.5 Commercial access

The SRPBA supported the right of access for informal recreation and passage, but only as part of a balanced package. The extension of access rights to include activities run by commercial providers has left many land managers angry and bewildered. They question how such commercial access can be considered informal recreation and passage.

It is not easy to estimate how many new businesses will take advantage of this right to carry out commercial recreation activities: only time will tell. However, the SRPBA, believes it is completely unreasonable that a pony-trekking business or a bike-hire company, for example, could lead tours over private land by exercising access rights, without consulting with the land manager. There is also a concern about the level of liability arising from such access. The SRPBA has suggested that commercial operators should have some form of agreement with the land manager which could clarify issues such as liability, maintenance and reparation for any damage.

8.3.6 Privacy and night-time access

Many land managers, and other rural residents, are confused by the lack of clarity in determining what is reasonable in the protection of people's privacy. Residents need guidance on what they can do to clarify the extent of their 'sufficient adjacent land', as the Land Reform (Scotland) Act 2003 puts it, to ensure that people enjoying access rights do so in the most responsible way and that people's privacy is protected.

Connected to the issue of privacy is concern over night-time access. Seen against the backdrop of an increase in rural crime rates, it is understandable that rural residents are worried about safety and security. Night-time access can also cause unnecessary disturbance to wildlife, disturbance to livestock and provide additional problems for land managers when carrying out land management activities at night.

8.3.7 Conservation

There is considerable concern about increased access to field margins, especially where these areas have been managed for the conservation of wildlife. While some interests might argue that there is little evidence to suggest that public access along field margins results in the loss of wildlife, land managers believe that their own experience contradicts this and that regular disturbance along a wildlife habitat, such as a field margin, will have an impact on the amount and range of wildlife that is supported there. The SRPBA believes there needs to be recognition that where field margins are managed for wildlife there should be an opportunity to protect these areas from disturbance. In practice, this would mean being able to sign-post alternative routes to avoid disturbance to sensitive areas.

8.4 Looking forward

This section seeks to suggest where, and in what ways, the impact of the Act will be most felt. While this is inevitably a crystal-ball gazing exercise, SRPBA experience would suggest that the following areas are likely to be where the greatest impacts will be felt.

8.4.1 Areas close to settlements

Areas close to where people live are likely to be most affected by the right of access. Farms close to settlements tend to be smaller landholdings than those in more rural settings and, because of this, land managers have less land to work with, while generally having higher demands for access, both in quantity and variety. In a remote upland area, the main demand will be for walking. On the edge of a village or town, people will want to walk, cycle, horse-ride, picnic, walk their dogs, fly kites, kick a ball around and so on. Managing access pressure along with economic land management activities on smaller parcels of land can be very difficult.

The SRPBA recognises that farms around settlements are also more prone to anti-social problems such as vandalism, litter and drugs. While these are not intrinsically access problems, they add to the difficulties experienced by land managers around settlements. The SRPBA believes that funding for path creation and maintenance, and for employing staff to help manage access, should be targeted at areas around settlements.

8.4.2 Range of activities

As the legislation gives new rights for access to inland water, and for activities such as horse-riding and cycling, there are likely to be changes in the patterns of recreational use. For example, there is likely to be increased levels of off-road cycling, an activity which is not well provided for at present.

8.4.3 Possible conflicts

Undoubtedly there will be a few high-profile 'events' staged to draw attention to the new right of access, and these will probably focus on those areas where there have been perceived, or actual, problems with access.

Anecdotal evidence suggests that conflict between different recreational users is already a bigger problem in some cases than conflict between land managers and recreational users. The right of access applies to a whole range of activities. Walkers, horse-riders and cyclists could all expect to use the same path and without there being a tradition of shared-use, courtesy and respect this can give rise to a number of problems. Unfortunately, the land manager may well be drawn into these conflicts or be expected to prevent or resolve problems.

8.5 Working with land managers

8.5.1 Making the legislation work on the ground

An attitude of consideration that respects the concerns of landowners and managers is important. Many concerns relating to the management of land use activities alongside a right of access are genuine and are not merely trumped-up to discourage access. Worries about the management of livestock, timber extraction, and concern for public safety are all real. Many of the so called 'obstructions' mentioned in the legislation, such as barbed wire, animals at large and electric fences, are simply elements of the Scottish countryside; they are not there for the purpose of preventing access, they are facets of land management. It is important, therefore, that those involved with access to the outdoors understand how the legislation will affect landowners and managers.

A co-operative approach which includes consultation with landowners and managers and joint projects which find shared solutions to problems can be very effective, as has been demonstrated by the Paths for All Partnership. Working together is probably the only way to make the right of access work on the ground.

Finally, a common sense approach to managing and facilitating access opportunities and solving problems, rather than one dominated by politics and points of principle, is much more likely to be successful. After all, it may not make sense for horse-riders and walkers to be on the same path in some circumstances, nor for a core path to go through a field regularly used for suckler cows. If we take a common sense and practical approach, access will work much better for everyone.

8.5.2 Integrating access rights with land management

In order to help integrate access rights with land management, there are a number of key areas where the SRPBA would like to see real effort by all those involved in access to work in a practical, realistic and co-operative way.

Core Path Networks

The SRPBA believes that the planning of these networks should be inclusive, with local authorities working with landowners and managers from the start of the process. Consideration needs to be given to the various land use activities undertaken and common sense applied to identifying core paths. This should be carried out with the local access forum, which should have a balanced representation of local land managers and recreational users.

Education

A targeted, effective and well-resourced education programme is essential to increase awareness and understanding of both the right of access and what is expected in terms of responsible behaviour. SNH has begun the preparatory work for this process, and the SRPBA looks forward to playing its part. As part of its commitment to this, the SRPBA has strengthened its access team with the appointment of an Access Officer: this post is part-funded by SNH.

Access management
The mechanisms for the practical management of access, through reasonable temporary closures or diversions, should be simple and flexible. The SRPBA believes that if a common sense approach is taken, then day-to-day land management activities can continue and that access can be integrated with them.

Benefits to land managers
It is easy to see the benefits of the access legislation to the wider public which includes the creation of better opportunities for people to enjoy the outdoors. There are also potential benefits to land managers: positive access management can help to integrate access with land management and can prevent or resolve problems. In some situations there will be opportunities for diversification, green tourism initiatives and other business links to access to the countryside. There could also be opportunities for land managers to be contracted to carry out path creation and maintenance, thus bringing additional income to some land managers. Making access relevant to land managers in this way benefits everyone.

Funding
All of this will require the investment of money and other resources to make it happen.

8.6 Conclusions
For land managers, the implementation of the legislation will herald changes in the day-to-day management of land to accommodate access in new and, hopefully, practical ways. It will cost land managers time, effort, and energy to plan and implement access opportunities and to assess and manage risks.

The legislation and the final Code, taken together, must provide a balanced framework for the right of access whilst ensuring that land managers can continue to manage the land. The SRPBA argues that without balance, landowners and managers will be poorly served, and the public, while having a 'right' of access, will not enjoy the 'welcome' in the countryside that they desire.

It is therefore vital that land managers themselves are welcomed as part of the solution and not seen to be part of the problem. If people feel ignored, or side-lined, or that their views are not given due consideration, they tend to become unco-operative. This is as true of land managers as it is of everyone else.

However, if there is a genuine desire to use access rights to provide better opportunities for the public to enjoy the outdoors, rather than to settle old scores, then everyone involved with access should take personal responsibility for working in a way that respects the rights of others, listens to their views, and chooses co-operation over conflict.

These are exciting times for some; interesting times for others. The challenge is to have the courage to make access rights work for everyone.

Editors' note

Since this paper was written, the new Scottish Outdoor Access Code (Scottish Natural Heritage, 2005) was approved by the Scottish Parliament on 1 July 2004 and came into effect, along with the statutory access rights, in February 2005. For more information, go to www.outdooraccess-scotland.com.

References

Agricultural Holdings (Scotland) Act 2003. www.opsi.gov.uk/legislation/scotland/acts2003/20030011.htm
Land Reform (Scotland) Act 2003. www.opsi.gov.uk/legislation/scotland/acts2003/20030002.htm.
Scottish Executive (2001). *The Draft Scottish Outdoor Access Code.* Scottish Executive, Edinburgh.
Scottish Executive (2003). *Biosecurity.* www.scotland.gov.uk/Topics/Agriculture/animal-welfare/15721/2959.
Scottish Mountaineering Club (1996). *Heading for the Scottish Hills.* Scottish Mountaineering Trust, Glasgow.
Scottish Natural Heritage (1996). *Concordat on Access to Scotland's Hills & Mountains.* Scottish Natural Heritage, Perth.
Scottish Natural Heritage (2005). *Scottish Outdoor Access Code.* Scottish Natural Heritage, Perth.

9 Agri-environment schemes, forestry schemes and access – are these policies working together?

Steve Hunt

Summary

1. The existing agri-environment, forestry and access policies are not well integrated.
2. Funding schemes only work and deliver policy goals if the people on the ground implement them properly, and in sufficient numbers.
3. Farmers and crofters are essential to the delivery of these schemes, but to ensure their active involvement a number of changes need to take place. These include: greater recognition of farmers' and crofters' roles in providing environmental benefits by government, local communities and the general public; a clearer overall vision for a sustainable Scotland; more accredited advice; the use of approaches such as local farming plans; and better funding to underpin such work.

9.1 Introduction

Scotland has been farmed for over 6,000 years. Farming has helped define the environment, managed wildlife and habitats, and shaped the landscape (Smout, 2000). Over 70% of Scotland is farmed, so agriculture will continue to exert a huge influence on the natural heritage in the future. Farmers and crofters manage the very fabric of the countryside. In turn, this affects the well-being of the people and wildlife that live there or pass through. Farming today has the opportunity to provide a greater understanding of the importance of the farmed environment and holds the key to access – both physical and emotional – to much of the countryside.

The success of agri-environment, forestry and access policies and their implementation on the ground will have a major influence on whether the natural heritage is truly established as a priority on the agricultural, economic and rural agendas in Scotland, or is marginalized in favour of other demands. Such schemes and policies need to be an integral part of farm businesses.

9.2 Policy

UK and European agricultural policies of the last 50 years have had a profound influence on the Scottish landscape and its wildlife (Smout, 2000). These policies reached a point where they became unsustainable in agricultural and environmental terms. Fortunately, the last decade has seen a swing towards more sustainable policies which are beginning to address wider environmental and social responsibilities.

However, as highlighted by Scottish Natural Heritage (SNH) in *Natural Heritage Futures: Farmland* (Scottish Natural Heritage, 2002), there is a trend towards polarisation where there are relatively few, highly productive farms capable of surviving without subsidies, and a much larger number of smaller farms unable to make a decent income from food production alone.

The Scottish Executive's vision for the environment and rural development (Scottish Executive, 2003) included objectives for delivering the *Forward Strategy for Scottish Agriculture* (Scottish Executive, 2001b) and *The Scottish Forestry Strategy* (Scottish Executive, 2000). Access threads its way through these at certain key points. Delivery on the ground is through the Rural Stewardship Scheme (RSS) (Scottish Executive, 2001c), Scottish Forestry Grants Scheme (SFGS) (Scottish Executive & Forestry Commission Scotland, 2003) and the draft Scottish Outdoor Access Code (Scottish Executive, 2001a).

Delivery of these strategies is acknowledged to be a massive undertaking, but there is now a clear commitment to, and understanding of, agricultural, forestry, biodiversity and sustainability needs. However, the success of these policies, individually and collectively, is highly dependent on the goodwill, abilities and imagination of the farmers, crofters and other land managers on the ground, and on the quality of the advice and information that is available to them should they need it.

9.3 What happens on the ground?

The *Forward Strategy for Scottish Agriculture* (Scottish Executive, 2001b) stated that rural Scotland is not a single entity and that the type of farming varies from place to place. The role of agriculture, and the balance of economic, social and environmental benefits expected from it, will also differ from one part of Scotland to another. Schemes such as the RSS and SFGS must be attractive and flexible enough to account for the myriad environmental conditions, landscapes, local biodiversity and types of farming. They need to enable farmers and crofters to both protect the rare and exceptional, the jewels in the crown, and to enhance local habitats in general. The provision of paths and other recreation facilities also needs to reflect the variations across Scotland.

The number of successful RSS projects increased significantly in 2002 and interest in the scheme has continued to grow. More money will be needed to meet this increased level of interest. The SFGS provides land management alternatives for farmers and a sustainable extra income. Provided that new planting does not compromise other habitats, this can be an excellent environmental and economic option for the farmer or crofter and can enhance biodiversity.

There are examples where these schemes have successfully gone hand in hand and instances where this has not worked well. Above all, the success of individual schemes and the extent to which they can be integrated are extremely dependent on the attitude and abilities of the individual farmer or crofter.

Another issue is how these schemes can be integrated with other needs, such as good access to land. The potential for conflict is often seen as greatest around access. There has been a mixed response to the proposed Scottish Outdoor Access Code, with some farmers worried by issues such as the possible trampling of crops and field margins, and possible disturbance to game birds, livestock and wildlife by dogs. However, some observers doubt that it will result in significantly higher numbers of people using the countryside, or in a radical change in behaviour by those who already use it. Others see it as an opportunity to resolve access problems or to support on-farm enterprises such bed and breakfasts or farm shops. The new access arrangements are now inevitable and it is perhaps best to see these as an opportunity to increase and foster an understanding of the countryside.

Looking at these schemes and the wider changes affecting the countryside, it is unclear what land managers are being asked to do. Even if individual scheme designs are clear in their intended outcomes, they need to sit within an overall vision that the farmer can sign up to. It is a feature of human nature that we all differ and see things in different ways. Provided the overriding policies are right, the success of any scheme or code depends on the people who have to carry it out and the nature of the climate they are operating in. Have they got the right tools? Is there adequate and accessible advice available if needed? Is society at large supporting them in what they do? And is there an overall vision for what we want?

For funding schemes to be successful, they should be an integral part of a farmer's business and an integral part of what is needed for the local area, and should fit clearly within a national picture. This chapter proposes that this is not happening currently and a hypothetical example will help to illustrate this point.

Imagine a farmer whose family has worked hard over many generations to create a large and balanced mixed farm. The farm had a lot of wildlife and produced food of high quality, and the farmer had ensured a long tradition of access to his land for others to enjoy. On his retirement the farmer sold the farm and it was divided into five new holdings. The farm could have remained sustainable and the retiring farmer hoped that the tenets of *Custodians of Change* (Scottish Executive, 2002) – that farming and the environment are inextricably linked and that farming has the potential, the responsibility and the opportunity to bring increased environmental and agricultural benefits to the countryside – would continue to be upheld. However, the five farmers who bought the farm had farmed in the area themselves for many years and watched the success of the retiring farmer from afar, but each was the product of their own individual upbringing, and each brought a different approach to, and connection with, the land.

- Farmer 1 was a passionate believer in the intrinsic value of nature and farmed with wildlife and the environment in mind, irrespective of any schemes on offer. He took full responsibility for the stewardship of the land.
- Farmer 2 was a staunch defender of modern agriculture, holding that farming had created the countryside as a by-product of the need to produce food, and so he had created big fields and was not interested in planting hedges or new woods that did not immediately support his crop productivity.
- Farmer 3 was business-orientated and recognised the financial benefits of agri-environment and forestry schemes and the positive impact they could have on his business, but his environmental stewardship did not extend any further than was necessary to be eligible for grants.
- Farmer 4 liked to be independent and was uncomfortable with the idea of rural stewardship regardless of economic considerations – that was not his responsibility.
- Farmer 5 enjoyed farming as a way of life, but was much less business-orientated.

All five farmers might work hard and be relatively successful in their own right without the overall operation being sustainable. Making a good – and sustainable – living from agriculture in current conditions is difficult and the policy framework keeps changing. Given these five farmers, it is likely that the character of the original farm would begin to change. Perhaps some of the flowers and birds would become less common, or there might be flooding downstream from where one of the farmers had straightened a stretch of burn and drained a patch of wetland. With access no longer being particularly encouraged, the community could begin to feel distanced, and perhaps farm stalls might begin attracting fewer customers. Over time, it is likely that the whole character of the landscape would begin to alter.

In this example, the environment is not a priority for Farmers 2, 4 or 5. Farmer 1 gives the environment a high priority but cannot make much money out of it. Farmer 3 could make money but has a limited notion of stewardship, so the biodiversity of his farm outwith the scheme areas began to deteriorate. Overall, no-one is effectively integrating business and environmental objectives. As a result, all are farming less sustainably in environmental and economic terms than they could. They are not taking the opportunities available to them.

9.4 Securing greater adoption of schemes by farmers

A key step to making progress is to encourage more farmers to adopt a particular scheme. Edwards-Jones (2004) explored the minimum factors that would need to be in place to make farmers do this. Farmers would need:

- to be aware of what is on offer;
- to be open to persuasion;
- to want to adopt schemes and sustainable management techniques; and
- to be able to explore and question the options available and then act on them.

The first need is for appropriate information to be sent to farmers to make them aware of the various opportunities. This information has to be presented attractively and accessibly. Time is precious to farmers and some scheme documents can be bulky and take a long time to read. Turning back to the five farmers, Farmer 1 might not read it if he felt he was already doing the right thing whilst the independent Farmer 4 would probably throw it away as he is not seeking any help. Farmer 5 might ignore everything he was not compelled to read and Farmer 2 may remain focused on increasing agricultural outputs. Information needs, therefore, to be interesting and relevant, clear, understandable, concise and to the point, and visually striking enough to make busy farmers read it.

Both scheme administrators and farmers and crofters need to be open-minded and persistent. Those interested in wildlife or in improving their bottom line are relatively easy to engage in the process. But what about others? An opportunity for one person is interference to another. Some may be interested but have been unsuccessful in applying for grant aid in the past and been put off as a result. At present, Farmer 1 would secure no financial gain as there is no money to help those farmers who are already doing good work in support of the environment.

9.5 The added value of farming

Why is it so important to engage the farming and crofting communities in managing their land more sustainably? Furthermore, as part of this process, why should agri-environment and forestry schemes, along with access and any other scheme for that matter, all go hand in hand?

Firstly, as *Custodians of Change* (Scottish Executive, 2002) highlighted, agriculture and the environment are inextricably linked, along with our well-being and that of wildlife. Furthermore, since many farming systems are neither economically nor environmentally sustainable, business and environmental goals should be integrated as a priority.

Secondly, farming can deliver all sorts of targets set out by policy makers. The *Forward Strategy for Scottish Agriculture* (Scottish Executive, 2001b) sets out a range of issues where farmers can make a significant positive contribution, such as reducing water and air pollution and climate change. Moreover, farmers can help deliver the Scottish Executive's objectives and targets for the environment and rural development – such as flood defence, water quality and the development of woodland products. The Scottish Executive is beginning to make some of these links, but needs to find ways to enable farmers to deliver on a much larger scale.

Farming has been underpinned by subsidies for so long that some businesses are untenable without them. If farming is to become a prosperous, market-oriented industry and a responsible custodian of the environment, as is the aim in the *Forward Strategy*, and deliver agri-environment and forestry schemes, along with all of these extra policy goals, it is going to need effective support.

Clearly, this is not a simple task and will take time and resources to implement. Farmers and crofters manage the very fabric of the countryside upon which all these benefits depend. We need to see farmers and crofters, and they need to see themselves, not only managing the land in the conventional sense, for conventional products, but in a much broader way where food is not seen as the only valid or, in some cases, the main product of farming. If we look at the added value of what farmers and crofters do, it is clear that it is also valid to farm for flood protection, for plentiful supplies of clean water, for sustainable forest products, for access, for biodegradable fuel, to protect our cultural heritage, to make nicer places to live and visit, and so on. If farmers and crofters are supported in this way, we are not only ensuring a decent living for farmers, a boost for rural communities, and places for wildlife to thrive, we are enabling the farming community to help us to help everyone.

9.6 Looking to the future

If the goal is to achieve agricultural sustainability, the connection between farming and a broad range of environmental products must become much more firmly established in the minds of the government, the farming community and the general public. Most farmers see the need for change, but if there is to be a decisive move towards producing a broad range of environment products, and if farmers are to be encouraged to do this, then more cash needs to be injected into supporting these environmental products through the Common Agricultural Policy and UK funding.

Public awareness and understanding is also vital. Government needs to be more proactive in promoting what is already being achieved by farmers and in showing the public why their money is being spent this way. Policy must allow the implementation of these interlinked schemes and codes to be significantly increased at the local level, not just through disjointed strands, but as a coherent whole and part of an overall vision. Apart from explaining schemes more clearly, there are only two main ways of securing increased adoption of schemes, codes and all that goes with them. One is enforcement, by making the rules more rigid and strengthening penalties for not adhering to the rules. The difficulty with this approach is that it is likely to generate bureaucracy, resentment and resistance. The other approach is to set out the context very carefully, make it relevant, set up more effective ways of engaging practitioners, give people on the ground the right support in terms of cash, available advice and public understanding, and let farmers and crofters get on with making it work on their land.

9.7 Making policy work on the ground

For policies to work and targets to be delivered, there needs to be a looser and more flexible approach that engages as many people as possible in actively doing something, whether they are a farmer or crofter or part of the local community. Ways need to be identified to make schemes essential to participate in in terms of cash, pride, support and stewardship. Agri-environment and forestry schemes and access are all part of the bigger picture and a better delivery mechanism, based on an overall vision, is needed. Such an approach would need to:

- be less formulaic and rigid;
- devolve as much as possible to the grassroots in a robust but flexible framework;
- engage everyone and let people on the ground have a say;
- explain how wildlife, landscape and access fit together in very specific areas tailored to the local environmental, economic and social needs, and how a farm or croft fits into this and the benefits it will bring;
- provide accredited advice on how to go about things and how they fit the bigger picture;
- encourage neighbour to work with neighbour, or groups of farmers to work together within river catchments;
- build on some of the projects already underway; and
- trust individuals and communities to take the work forward themselves.

One way of achieving this would be to produce Local Farming Plans (LFPs) (Hunt, 2003). These could act as driving mechanisms for delivering our vision for a sustainable Scotland, and for making these targets and schemes more integrated and workable. The LFPs could set the necessary context for the delivery of agricultural 'products' that are appropriate to an area; such as high quality food, biodiversity, access, landscape, archaeological protection and flood defence. Clearly, the LFPs should relate to Local Authority plans, Local Biodiversity Action Plans, forestry and landscape plans, catchment management plans and other established systems that are appropriate or exist for the area. The farmer or crofter can then see how all these aspects fit together and where he or she fits in. Each farm or croft could then have a plan which integrates the farmer's personal business and environmental goals and encompasses agri-environment and forestry schemes and access, based on the Local Farming Plan.

The LFPs should operate at a landscape scale that is appropriate to the area. One of the defining features of Scotland's landscape is its variability, where a distance of a few miles in any direction, or a small change in altitude, can give rise to substantial changes in soils, climate and aesthetics (Scottish Natural Heritage, 2002). Thus, some LFPs should cover a smaller area than others according to the nature of the locality and issues of concern.

These LFPs could build on current good practice and begin to work on a much larger scale. By doing this, the plans should enable many more, if not most, farmers to take sustainability seriously and feel actively involved.

Turning back to the hypothetical example of the five farmers, they need to work together to restore the balance of the farmland. For example, Farmer 2, having worked together with neighbours and discussed issues with outside advisors, might realise that a focus on large fields and increased production had reduced the number of birds, hares and other wildlife, and that this is a loss to him and the local community. He may then go on to do something about it. Each farmer needs the opportunity to contribute their ideas and develop a plan that satisfies all of their needs. It would make even more sense if their neighbouring farmers up and down the strath joined the Plan.

A group such as the five farmers and their neighbours could be proud to be actively getting on with the business of serious farming in its broader sense, and relish the responsibility of helping, for instance, to reduce the flooding which used to inundate the small town downstream of their land. The well-informed townspeople, aware of their Local Farming Plan, might be pleased to support them, especially if they have access to the wetlands and woodlands on the local farms that absorb the rainfall that prevent flooding of their homes. The local community and visitors alike can develop a strong sense of place and enjoy benefits such as locally produced food bought locally.

The moral of this story is that no one thing is right, one size does not fit all. If an environment can be created where farmers believe there is a future for them and their children, and there is a much wider commitment across the board to an overall vision of sustainable farming, then each individual farmer is likely to be able to see his or her part and take it seriously. This is much easier said than done, but it should be one of the main challenges of the coming years.

9.8 Conclusions

In summary, farmers and crofters need to be allowed to take the lead within a context that is secure enough for policies and schemes to be delivered. This will require:

1. greater recognition from government that farmers and crofters hold the key to not only providing high quality food, but also a range of environmental products;
2. more cash to underpin such work;
3. commitment to an overall sustainable vision for Scotland;
4. Local Farming Plans that set out to achieve the integration of farming, biodiversity, access, landscape, culture and flood defence locally, whilst forming part of a national picture;
5. greater understanding within the farming community of their role and its importance;
6. greater understanding by the public of the benefits of farming; and
7. an increased pool of accredited advice to help enable farmers and crofters to deliver.

References

Edwards-Jones, G. (2004). Profiling the serial environmentalist. *FWAG Scotland,* Issue 2 Winter 2004, 30-31.
Hunt, I.S. (2003). Local Farming Plans. Unpublished paper. FWAG Scotland Perth.
Scottish Executive (2000). *Forests for Scotland: the Scottish Forestry Strategy*. Scottish Executive, Edinburgh.

Scottish Executive (2001a). *The Draft Scottish Outdoor Access Code.* Scottish Executive, Edinburgh.

Scottish Executive (2001b). *A Forward Strategy for Scottish Agriculture.* Scottish Executive, Edinburgh.

Scottish Executive. (2001c). *The Rural Stewardship Scheme.* Scottish Executive, Edinburgh.

Scottish Executive (2002). *Custodians of Change.* Report of the Agriculture and Environment Working Group, Scottish Executive, Edinburgh.

Scottish Executive (2003). *Draft Budget 2004-05.* Scottish Executive, Edinburgh.

Scottish Executive & Forestry Commission Scotland (2003). *Scottish Forestry Grants Scheme.* Scottish Executive, Edinburgh.

Scottish Natural Heritage (2002). *Natural Heritage Futures: Farmland.* Scottish Natural Heritage, Perth.

Smout, T.C. (2000). *Nature Contested.* Edinburgh University Press, Edinburgh.

10 Helping farmers and foresters to manage public access

Ken Taylor, Peter Scott & Vyv Wood-Gee

Summary

1. This chapter describes the concerns of farmers and key findings arising as a result of the imminent introduction of Part I of the Land Reform (Scotland) Act 2003.
2. There are many concerns but their significance varies from location to location.
3. Good practice has developed to deal with these concerns and, in most cases, adverse impacts can be reduced to acceptable levels and benefits to the land manager secured.
4. The Land Reform (Scotland) Act 2003 should not significantly affect the use of these management measures.
5. A promotional and educational campaign is needed to raise the public's understanding of what is responsible behaviour.

10.1 Introduction

The objective of the chapter is to highlight the issues that concern farmers, foresters, estate managers, gamekeepers and managers of inland waters (collectively referred to as 'land managers') with respect to public access and to describe possible solutions. The chapter draws on research undertaken on behalf of Scottish Natural Heritage (SNH) to explore the issues of concern. It uses real-life examples to identify measures that can be used to eliminate adverse effects or to bring their impacts to acceptable levels. The chapter also demonstrates the way in which SNH has approached the task of researching the challenges faced by land managers and disseminating the information on good practice.

10.2 Background to the research

In January 2003, the Scottish Parliament passed the Land Reform (Scotland) Act 2003. Part I of the Act introduces a right of public access over most land and water throughout Scotland, but excluding land on which crops are growing. This access is contingent upon the access user behaving responsibly; also, it requires the landowner to manage his/her land in a manner that does not unreasonably interfere with the rights of access. It is difficult to be prescriptive about what constitutes responsible and reasonable behaviour, either for access use or land management. In recognition of this difficulty, the Act requires SNH to produce and, subject to Ministerial and Parliamentary approval, publish a Scottish Outdoor Access Code (Scottish Natural Heritage, 2004).

In this context, SNH has recognised the need to make land managers aware of the management issues that might arise as a result of the change in the law. The management measures available range from informal management techniques to the use of legal measures (such as byelaws imposed by public bodies). SNH also recognises the need to prepare and disseminate this information in the form of advice to assist land managers to adopt positive and sensitive responses to issues and opportunities arising from public access and recreation. Consequently, in 2002, SNH appointed Peter Scott Planning Services Ltd., Asken Ltd. and Vyv Wood-Gee to undertake research into management issues and solutions associated with public access (Scott *et al.*, 2003). The advice arising from this work will apply to current access patterns and issues, and to any predicted issues which may flow from the access provisions of the Land Reform (Scotland) Act 2003.

The research focuses on:

- popular/frequently visited countryside;
- informal recreational activities, including walking, cycling, horse riding and non-motorised watersports;
- impacts of access users on the management of land and inland waters, rather than impacts of land and water management on access users; and
- problems rather than benefits, allowing for the identification of specific solutions to the issues arising.

It is important to recognise that although the research focused on 'problems', there are many benefits associated with the new legislation. For access users, these benefits include opportunities for recreation, health and spiritual uplift, and increased awareness of the countryside and how it is managed. The economies of areas visited by access users are expected to benefit from increased spending. Land managers may benefit from the increased clarity that the Act affords with respect to public access, especially the definitions of responsible and reasonable behaviour that will be set out in the Scottish Outdoor Access Code. The publishing of the Code will be supported by a major, long-term promotion and education programme.

10.3 Methodological considerations

The research involved identifying key issues and potential solutions from:

- a review of selected responses to relevant consultation papers;
- debates in the Scottish Parliament;
- internet searches and visits to specific websites;
- a targeted search of leading organisations concerned with public access and other relevant sources;
- feedback from delegates at two workshops and five focus group meetings, and guidance from the Project Steering Group; and
- a trawl of examples of existing good practice in respect of access management, drawn from across Scotland.

The literature reviewed included government research and policy documents, guidance provided by organisations such as SNH, National Trust for Scotland, Forestry Commission Scotland, and representatives of bodies such as the NFU Scotland and the

Scottish Rural Property & Business Association (formerly known as the Scottish Landowners' Federation). Information from across the whole of the UK was used, although priority was given to Scottish sources.

Attendees at the two workshops and the Project Steering Group comprised representatives of a wide range of sectors, including land managers/owners, user groups and public sector organisations. The five focus groups were held around Scotland, with different themes:

- South Aberdeenshire, South Lanarkshire and East Dunbartonshire: owners and/or managers of farms, estates and woodlands, including some access officers, advisers and users' representatives;
- The Trossachs: managers of access to, on and alongside inland waters and some water users' representatives; and
- Skye: crofters and their representative organisations.

It should be noted that:

- the experience quoted is pre-Land Reform (Scotland) Act 2003 and may need to be modified in the light of experience of access following implementation of the Act;
- it has not been possible to investigate each good practice example 'on the ground'; and
- the good practice examples are offered as ideas, rather than recommendations, that may be of interest to land managers and may be adapted to local circumstances.

Illegal activities (such as fly-tipping and off-road driving) are a major concern to land managers, even though they do not arise from *bona fide* recreational access, and so these were included in the research.

10.4 Issues of concern to land managers
10.4.1 Identification of issues

The first part of the study was to identify the key concerns of land managers. The literature review, focus groups and workshops each produced a list of issues of concern to land managers. The different sources yielded very similar lists. For simplicity, these were grouped into twelve main issues. These are listed in Table 10.1 (not in any order of priority). Many of these concerns have been recognised in the Scottish Outdoor Access Code (in draft, at the time of editing this chapter). If the responsible and reasonable behaviour set out in the Code is followed, the scale of difficulties on the ground should be minimised.

Table 10.1 Issues of concern to land managers.

Issue	Descriptions
Occupiers' liability and public health and safety	These are major concerns for most land managers – especially where there are inherent risks. For example, people with dogs in fields of suckler cows and calves, access to deep water, potentially hazardous operations (such as timber harvesting or crop spraying), and public use of structures, such as bridges. The cost of insurance and the effort needed to avoid risks are further concerns.
Litter, fly-tipping and car dumping	The main issues arising include injury to livestock, damage to machinery and the impact on local scenery. The cost of removing and disposing of waste materials is a concern.
Dogs and their control	The effects of inadequately controlled dogs mostly concern livestock farmers. Their concerns are over impacts such as sheep worrying and spread of disease. Arable farmers and horticulturalists are concerned at the risk of quality assurance scheme rules being compromised. Gamekeepers have concerns over the risk of disturbance to game and wildlife by dogs.
Effects on field boundaries	This concern was felt mainly by livestock farmers, who fear that gates may be left open or the integrity of boundaries compromised. If stock escape from fields, a range of adverse effects may arise, such as the burdens of rounding up escaped stock, livestock breeding in the wrong circumstances, or road traffic accidents.
Damage to infrastructure	Concern was expressed about the damage that may be caused to field boundaries, such as walls and fences, by people trying to cross them. Management time, effort and cost need to be expended to repair the damage.
Damage to crops and forestry	Although the Land Reform (Scotland) Act 2003 does not give a right of access to land on which crops are growing (including hay and silage of a late stage of growth) or young trees, land managers fear that many users may not recognise these crops, resulting in crop losses from trampling.
Illegal activities	A wide range of illegal activities give rise to concern; for example, theft or vandalism. This is not normally a direct consequence of *bona fide* recreational access, but land managers were concerned that the new access rights may make it more difficult to monitor and control the activities of potential miscreants.
Obstructions or interferences caused by visitors	The presence of members of the public may interfere with land management activities. Common examples given were cars blocking entry to fields, interference with stock or game movements (often not deliberate), and well-intentioned, but misplaced, rescuing of 'abandoned' lambs.
Loss of privacy and exclusivity	This concern has two facets. Firstly, managers were concerned that their personal privacy, and those of farm or estate workers, may be infringed by the public. Secondly, there were concerns about the implications for private events and commercial activities, such as wedding receptions or holiday lets whose attraction is their privacy and exclusivity. Importantly, managers and local residents are also concerned that access rights may be used as an excuse by criminals to assess the potential, or undertake, thefts from rural dwellings or farmyards.

Table 10.1 Issues of concern to land managers (cont.)

Issue	Descriptions
Interference with commercial land management activities and utility undertakings	Land is used for many commercial purposes other than farming and forestry. Therefore, interference can take many forms, for example, disruption to stalking, shooting, pest control and muirburn, or canoeists disturbing anglers on high value beats. The potential for commercial activity tourism operators, who make no contribution to its management but take groups onto private land, was a specific issue raised.
Environmental impacts	Public access can lead to damage being caused to the natural and cultural heritage and to field margin and other conservation schemes. Land managers noted that vegetation on popular footpaths can be lost, with resultant soil erosion. Also, water may be contaminated by human or dog faeces.
Inter-user conflicts	Although not necessarily of direct concern to land managers, the risk of inter-user conflict was highlighted. Examples included cyclists and horse riders affecting walkers; canoeists affecting anglers; walkers affecting shooting (and *vice versa*).

10.4.2 Quantification

An attempt was made to assess the actual significance of the issues raised, such as the number of incidents of damage or cases of occupiers' liability taken through the courts. The intention was to rank these issues in terms of actual, rather than perceived, significance. In practice, there were insufficient data to achieve this and issues may be of minor significance to most land managers, but of major significance for others. There are other factors that make the quantification of the issues difficult. These are:

- mitigation: erecting signs and clear waymarking can reduce the risk of people being exposed to hazards;
- under-reporting: many incidents (such as minor injuries to livestock or interference with land management activities) go unreported;
- variations in scale: illegal activities may be virtually insufferable at an urban fringe location, but non-existent in a remote valley;
- uncertainties over the future: for example, land managers have rarely had to pay compensation to access users for failure to meet their duty of care as an occupier, but they are concerned about occupiers' liability in the future, bearing in mind the perceived increase in willingness of people to sue, lawyers operating on a 'no win-no fee' basis, and reported increases in costs of insuring against these risks; and
- unbalanced reporting: reports about access and land management tend to focus on problems, even though these may not reflect the majority of people's experience of countryside access, and ignore the potential benefits.

As a result, it was considered counter-productive to try to rank issues in terms of their severity. The preferred approach is that all issues need to be addressed and to seek good practice examples to provide management solutions for each.

10.5 Solutions
A wide range of different solutions was identified. These can be grouped into three different types: strategic access management; generic solutions; and site-specific solutions. These are discussed below.

10.5.1 Strategic access management
The introduction of the new rights of public access may mean that some land managers encounter public access onto their land for the first time. For others, it may provide a prompt to adopt a more pro-active approach to access management. There is much to commend a strategic, planned approach to access management, rather than piecemeal solutions to specific issues as they arise (i.e. a 'fire-fighting' approach). The aim of a strategic and integrated approach is to manage access positively, rather than simply letting it happen. Such an approach has implications across the whole range of issues and it is considered important that a suitable methodology for an Access Management Appraisal (or some similar process) is developed. Such an approach has been partially developed (Taylor *et al.*, 2001) and key elements are reproduced in Figure 10.1 in a provisional format. It has been incorporated into the Countryside Agency's advice on managing access (Countryside Agency, 2004) and is recommended for the development of advice by SNH.

It is expected that the appraisal may be undertaken by the land manager, particularly if they have previous experience of managing publicly accessible land. An alternative may be to seek advice from a specialist advisor or the local authority. One of the key elements to bear in mind is that, as far as possible, the appraisal should be undertaken with the aspirations of access users in mind, as well as the needs of the land manager.

10.5.2 Generic solutions
Standard techniques can be applied in a wide range of situations and in various forms, and are often interdependent. These techniques are described below.

Signage and information
Signs can be a very good way of getting important information and messages across to the public, although managers report that there is always a minority who ignore them. They should be located carefully, so that the message is delivered where it is most likely to be effective – namely at the point where the information has most meaning. For example, information about a diversion may best be placed at the start of the route, such as where people park their cars. Care should be taken to convey the information in a positive way (e.g. use 'Please do …' rather than 'It is prohibited to …', or 'No …'). Government agencies, including SNH, are developing pictograms using the Common Visual Language. These will assist in conveying messages rapidly and in simple format, and to those who may be unable to read text, including visually impaired visitors. Care will be needed to avoid using too many signs and so degrading the quality of the landscape.

Awareness raising
Visitors who have an awareness and understanding about land and water management are likely to have less impact than those who are unaware. A variety of ways can be used to increase awareness, including personal contact, leaflets, guides, and land managers working

Helping farmers and foresters to manage public access

IS YOUR LAND/WATER, OR IS IT LIKELY TO BE, POPULAR AMONGST VISITORS? IS THE LAND/WATER OF YOUR IMMEDIATE NEIGHBOURS POPULAR?	

If Yes to any/all of the above considerations, then

WHAT TYPES OF VISITOR DO YOU CURRENTLY HAVE AND EXPECT TO HAVE IN THE FUTURE?	
VISITORS ARRIVING BY CAR/PUBLIC TRANSPORT?	VISITORS ARRIVING ON FOOT, BIKE OR HORSEBACK?
IF YOU ARE CLOSE TO A TOWN OR VILLAGE, CONSULT REGULARLY WITH THEM AND TRY TO ARRIVE AT A MUTUAL AGREEMENT TO HOW ACCESS ON YOUR LAND IS BEST MANAGED	
USING A MAP OF YOUR LAND OR WATER :	
Stage 1	MARK THE AREAS AND FEATURES THAT MIGHT BE ATTRACTIVE TO VISITORS AND WHICH THEY MAY WISH TO VISIT
Stage 2	MARK AREAS THAT ARE SENSITIVE TO VISITORS OR MIGHT BE IN THE FUTURE
Stage 3	MARK THE MAIN ENTRY AND EXIT POINTS WHICH PEOPLE USE OR ARE LIKELY TO USE
Stage 4	TRY TO LOCATE ANY CRITICAL POINTS WHERE PEOPLE ARE LIKELY TO VENTURE INTO SENSITIVE AREAS
Stage 5	CONSIDER HOW PEOPLE CAN BE ENCOURAGED TO TAKE ROUTES THAT AVOID SENSITIVE AREAS
Stage 6	ARE THERE ANY PARTICULAR PRESSURE POINTS?
Stage 7	LOOK FOR POSSIBLE NEW LINEAR ROUTES ACROSS YOUR LAND AND PLAN HOW THESE MIGHT BE DEVELOPED Not only might this help minimize impacts on land management, but it will also help improve visitors' enjoyment
Stage 8	ARE THERE ANY WAYS OF BENEFITING FROM VISITORS?
Throughout the process	SEEK ADVICE AND SUPPORT From the Local Authority (access officer, local access forum, local ranger service), SNH, land management and recreation organizations (such as the NFU Scotland, Scottish Rural Property & Business Association (formerly known as Scottish Landowners' Federation), Paths for All Partnership, Ramblers' Association, Mountaineering Council of Scotland) and access advisors/consultants. Grants may be available to provide or upgrade access facilities e.g. replacing stiles with self-closing gates to improve access. TALK TO YOUR NEIGHBOURS How you manage access on your land will be affected by how others manage theirs, and *vice versa*.

Figure 10.1 Outline of a provisional access management appraisal methodology.

with local authority rangers and others. Land managers can find out why people visit their land by talking to visitors and this will help them to decide which measures are most likely to work in practice. The Scottish Outdoor Access Code, and SNHs proposals for long-term support to a campaign of promotion and education, will help to raise awareness generally.

Practical measures

Influencing the choices that walkers and others make about which routes they follow to different parts of the land can help to minimise nuisance or disruption to management activities (although managers must not obstruct access). If paths need to be created, these should follow desired lines to minimise the risk of short cuts developing. Waymarks can be used to encourage people to follow preferred routes.

Zoning

This can be used to influence the overall distribution of visitors on land and water, in both time and space. For example, if managers identify a hazardous area (such as old mine shafts), they may decide not to develop car parking close by. Zoning is based on the identification of specific attributes, activities and sensitivities of the land or water and the impacts different types and levels of recreational use or activities may have on these characteristics (e.g. sensitive zone: comprising important conservation features or principal lambing fields, where access needs to be sensitively managed; robust zone: that can accept higher intensities of recreation without adverse environmental impacts or disruption to farm or other activities).

10.5.3 Issue-specific examples

The research identified 63 different good practice examples, drawn from across Scotland, England and Wales. A small selection is provided below, to illustrate the material gathered. Details of all of the good practice examples can be found in the SNH report on the research featured in this paper (Scott *et al.*, 2003).

Example 1: Promoting understanding of farming and land management

Managers of the Pentland Hills Regional Park, in consultation with local land managers, have devised several means of promoting visitors' awareness of land/water management needs and responsible access. They include:

- welcome boards at key Park 'gateways' (e.g. main car parks), including a user-friendly poster explaining the various roles of key managers (e.g. sheep farmers, gamekeepers, archaeologists, water keepers, the Army) and how visitors can avoid disturbing these interests;
- seasonal posters on sensitive management issues, using cartoon illustrations to draw visitors' attention to information on aspects such as tupping, lambing and bird nesting, and on how visitors can avoid disturbing these; and
- a *Pentland Hills Visitor's Guide* (Pentland Hills Regional Park Authority, undated) leaflet which illustrates the Pentlands' path network, recommends walking, cycling and riding routes, presents a diagram of The Pentlands Cycle of Wildlife and

Farming and indicates specific months when users should keep to paths to avoid disturbing pregnant ewes, lambs, nesting birds, grouse shooting and sheep tupping, provides particular advice to dog walkers and mountain bikers, and promotes the use of public transport to get to the Park.

The Pentland Hills Regional Park has a website that contains interesting information and examples of access management (www.edinburgh.gov.uk/phrp).

Example 2: Controlling fire risk in woodlands

Fires pose potential hazards to crops and woods surrounding property and people. Fires can be started accidentally (such as by dropped cigarettes and matches, or by campfires or barbeques getting out of control) or deliberately. In practice, the Woodland Trust believes that most fires are started by arsonists. The Trust's approach at its urban woodland sites is to:

- seek advice from the fire authorities and let them know of plans for controlled burning;
- compile and maintain a fire plan which contains details of phone and fax numbers of the nearest fire station, vehicle access points, water sources, terrain description and hazard locations, site manager's contact details and a summary of the above;
- avoid potential fire hazards developing, for example by removing litter or garden waste;
- clear any combustible materials, such as bracken or brambles, especially if near to vulnerable areas (for example, land adjacent to properties);
- leave dead wood standing, as this is less combustible than when lying on the ground (and is better for nature conservation), or in large blocks that do not easily catch fire; and
- be particularly alert during fire risk periods (in woods, this can be early in the spring).

This example was drawn from the Woodland Trust's Urban Woodland Management Guide 1 *Damage and Misuse* (Woodland Trust, 2002).

Example 3: Management of risks – Pembrokeshire Coast Path

A full risk assessment for the Pembrokeshire Coast Path has been conducted by the Pembrokeshire Coast National Park Authority. Five levels of risk have been identified and management adjusted to reflect the differing levels. The risk assessment has allocated lengths of the path corridor into the relevant risk zones (Pembrokeshire Coast National Park Authority, 2000). Measures are applied as appropriate and include:

- the line of the route being clearly visible on the ground to reduce the risk of people straying into hazardous areas;
- the tops of cliffs being stabilised so that people are less likely to lose their footing in places where this could have serious or even fatal consequences;
- a minimum width of path being made available so that people can pass other users without being exposed to risk of falling off the path;

- path furniture being maintained at a high standard to reduce the risk of injury;
- a natural surface and good drainage being maintained to minimise erosion; and
- vegetation being used in places to form a natural edge to the path, thus keeping people back from dangerous cliff-tops.

A key aspect of the risk assessment is that the degree to which these measures are applied is varied between different levels of risk identified through zoning. The path is monitored regularly, and the risk assessment is formally reviewed every five years.

Example 4: Risks from dogs and cattle, and behaviour on golf courses

The Fife Coast Path is a strategic route set up and managed by Fife Council's Countryside Services. In places, the route goes through private farmland, including enclosed fields, and across or adjacent to golf courses. In the past there have been access issues relating to people with dogs walking through fields with cattle and across golf courses.

In conjunction with some local land managers, the Council has designed and put up signs at a number of key locations to highlight appropriate standards of behaviour expected from people taking access. The ranger service has been responsible for liaising with the land managers and in putting up and maintaining the notices. These notices adopt a humorous approach.

In general, the signs have worked well and are accepted by the majority of users and managers. Some signs have a contact telephone number to enable people to contact the Council for further information. Figure 10.2 provides an example of the type of sign used.

Guidance on keeping cattle in fields with public access is provided by the Health and Safety Executive (Health & Safety Executive, 2002), although some of this advice needs to be reviewed in the light of the requirements of the Land Reform (Scotland) Act 2003.

10.5.4 Comment on the scope of the good practice examples

Most good practice examples identified in the report are derived from larger estates, or those in the public sector, with a bias towards the management of 'honeypot' sites. Also, many of the solutions have cost implications for the land manager (either financial costs or in management time and effort). Nevertheless, the principles behind the examples are relevant to most situations and it is intended that managers of smaller land-holdings can derive useful ideas from the examples and put them into practice, possibly with support from local authorities.

10.6 Conclusions

The research found that land managers have many concerns as a result of public access to the countryside. There is a wide range of solutions, many of which are already in practice in Scotland, but there are cost implications. It is concluded that there is a requirement for:

- funding to meet the costs of managing access and advisory support to land managers;
- training, awareness-raising and for guidance to be published;
- policy to be developed with respect to access management in areas such as, for example, conflict resolution;

- clarification of issues relating to quality assurance schemes and insurance premiums; and
- support for land managers in cases of wrong-doing.

In all of the above, the needs of managers of inland water should not be neglected.

Figure 10.2 Example of signs used on the Fife Coast Path. Source: courtesy of Fife Council.

Editors' Note

Since this paper was written, the new Scottish Outdoor Access Code (Scottish Natural Heritage, 2005) was approved by the Scottish Parliament on 1 July 2004 and came into affect, along with the statutory access rights, in February 2005. For more information go to www.outdooraccess-scotland.com.

References

Countryside Agency (2004). *Land Managers' Guidance Pack: a Guide for People Who Own and Manage Access Land (Ref CAX 150 F)*. Countryside Agency Publications, Wetherby. See website www.openaccess.gov.uk/wps/portal.

Health & Safety Executive (2002). *Keeping cattle in fields with public access. Agriculture Information Sheet No. 17*. HSE Bookshop, Sudbury. See website www.hse.gov.uk/pubns/ais17.pdf.

Land Reform (Scotland) Act 2003. See website www.opsi.gov.uk/legislation/scotland/acts2003/20030002.htm.

Pembrokeshire Coast National Park Authority (2000). *Pembrokeshire Coast Path National Trail Safety Statement & Risk Assessment March 2000*. Pembrokeshire Coast National Park Authority, Pembroke Dock.

Pentland Hills Regional Park Authority (undated). *Pentland Hills Visitor's Guide: the Best Advice and Information for a Great Visit to the Pentland Hills*. Pentland Hills Regional Park Authority, Edinburgh. See website www.edinburgh.gov.uk/phrp/publications/publications.html.

Scott, P., Taylor, K. & Wood-Gee, V. (2003). Developing advice for land managers on positive access management. Unpublished report. Scottish Natural Heritage, Perth.

Scottish Natural Heritage (2004). *A Proposed Scottish Outdoor Access Code: Report on the Outcome of the Consultation on the Draft Scottish Outdoor Access Code*. Scottish Natural Heritage, Perth.

Scottish Natural Heritage (2005). *Scottish Outdoor Access Code:* Scottish Natural Heritage, Perth.

Taylor, K., Renshaw, W., Ball, H., Hammond, G. & Barlow, J. (2001). *Practical ways of managing access to the countryside*. CD/website prepared for the Countryside Agency and English Nature. Countryside Agency, Cheltenham.

Woodland Trust (2002). *Urban Woodland Management Guide I: Damage and Misuse*. The Woodland Trust, Grantham.

PART 4:
Farming, Forestry and Landscape

View from above Loch Gamhna and Loch an Eilean, Cairngorms © Lorne Gill, Scottish Natural Heritage

PART 4:
Farming, Forestry and Landscape

Given the major policy developments relating to biodiversity and access, it is easy to lose sight of the importance of landscape. The five chapters within this part of the book provide a detailed picture of the current state of play on landscape issues, including landscape character assessment, policy and advice.

Swanwick & Martin (Chapter 11) tell us about the landscape character assessment approach, which helps to describe landscape and to explain what makes landscapes different from each other. This common approach has been applied across Scotland and this enables landscape character assessments to be applied consistently from one end of Scotland to the other in managing the natural heritage. The chapter describes how the assessments are now being used in land management plans and in indicative forestry strategies. However, the authors note that the use of landscape character assessment in agri-environment schemes in Scotland has lagged behind that in England & Wales. The authors also look at some of the challenges ahead in making full use of the landscape character assessment approach.

Bennett (Chapter 12) provides a more detailed look at how landscape character assessment has been used in Dumfries and Galloway to underpin planning policies, landscape guidance and the management of National Scenic Areas. Bennett also points out the need to involve local people in this work and for funding schemes to be developed in ways that encourage better landscape management.

In Chapter 13, Macinnes provides a timely reminder that the landscape character assessment approach only looks at the landscapes that we see today. The historic landscape can often be lost as evidence becomes obscured by natural processes and human activity. Macinnes describes a new approach called Historic Land Use Assessment which is enhancing our understanding of the historic dimension of landscape and informing our approaches to its management.

Although techniques for assessing landscape character have existed for some time now, there is little national policy relating to landscape conservation and management. Brooks (Chapter 14) describes a new discussion paper put out to consultation by Scottish Natural Heritage. This discussion paper focussed on Scotland's lowland landscapes and Brooks describes the main forces for landscape change and their broad consequences. A wider debate is called for on how we want to see these lowland landscapes evolve.

In Chapter 15, Farmer looks at future research and advisory needs. Four case studies help to illustrate the ways in which landscape research and advice are being used to underpin landscape management and the integration of landscape with land management more generally. The need for improved communication and guidance is highlighted.

Farming, Forestry and the Natural Heritage: Towards a More Integrated Future

11 The role of Landscape Character Assessment in farming, forestry and the natural heritage

Carys Swanwick & Julie Martin

Summary

1. Landscape Character Assessment is the process of identifying and describing variation in the character of landscape and using this information to assist in guiding and managing change. It seeks to identify and explain the unique combination of elements and features that make landscapes distinctive.
2. The National Landscape Character Assessment programme in Scotland was initiated in 1994 to meet a number of different objectives, including establishing an inventory, raising awareness and assisting with planning and management casework.
3. The programme aimed to classify and describe the character of Scotland's landscapes – what makes one area 'different' or distinct from another – by examining the various component parts of the landscape and the physical, ecological, and cultural processes that interact to affect the landscape.
4. Outputs from the programme include a series of published reports and a GIS and supporting database, together with a national classification of Scotland's landscape based on combination of the individual assessments.
5. Applications of the Landscape Character Assessment programme in landscape conservation and management are less well-developed than applications in planning, perhaps reflecting more limited awareness of the programme outputs among those involved in this area.
6. Landscape Character Assessment has been used successfully in the formulation of land management proposals and in the forestry arena. It has not been widely used in relation to farming and, in particular, has not yet played a significant role in recent agri-environment scheme targeting and design in Scotland.
7. A recent review concluded that the national programme in Scotland is both visionary and innovative and has a number of important strengths. One of the key recommendations for land management is that there should be greater recognition of the potential for landscape character issues to inform Scotland's Rural Stewardship Scheme and related land management initiatives.

11.1 The meaning and importance of landscape

Although the idea of landscape will always be contested there is today a reasonable consensus about its meaning among those involved in landscape planning, design and management. In simple terms it is generally accepted as meaning the relationship between people and place (Swanwick & Land Use Consultants, 2002). More technically it is a product of the interactions between:

- the physical and natural environment (geology, soils, climate, flora and fauna); and
- social and cultural factors (land use, settlement, enclosure, human activities, both now and in the past).

These interactions can, however, really only describe the land around us. It is people's perceptions that turn land into landscape and people who bring their own complex layers of associations, memories, and values to appreciation of landscape. The importance of landscape in modern society has been widely articulated (see, for example, Swanwick & Land Use Consultants, 2002; Scottish Natural Heritage, 2002b) and it is generally accepted that landscape is one of the most treasured aspects of our natural heritage, alongside natural habitats and wildlife.

Against this background, this chapter describes the role of Landscape Character Assessment in dealing with landscape dimensions of the natural heritage, focussing particularly on the role that it plays in land management in general, and in forestry and farming in particular. It draws on research carried out in 2003 for Scottish Natural Heritage (SNH) to review the programme of Landscape Character Assessment in Scotland (Julie Martin Associates & Swanwick, 2003) and its practical applications.

11.2 The nature of Landscape Character Assessment

Landscape Character Assessment is the process of identifying and describing variation in the character of landscape and using this information to assist in guiding and managing change. It seeks to identify and explain the unique combination of elements and features that make landscapes distinctive. The origins and development of the process have been described elsewhere (Swanwick, 2002) and details of the approach and methods can be found in Swanwick & Land Use Consultants (2002). The key feature of this technique is its emphasis on the concept of landscape character. This concept is now widely understood and is commonly defined as

> *"a distinct and recognisable pattern of elements that occur consistently in a particular type of landscape. Particular combinations of geology, landform, soils, vegetation, land use, field patterns and human settlement create character."*

Thus it can be equated with concepts of distinctiveness or sense of place and is concerned with what makes landscapes different from each other rather than necessarily better or worse.

11.2.1 A common approach

Landscape Character Assessment is now being applied in most parts of the UK. Scotland and England are using a very similar approach (though with some variations) and for this reason have produced common guidance on the method (Swanwick & Land Use Consultants, 2002). This sets out the two key stages of Landscape Character Assessment, namely:

- characterisation – identifying areas of distinctive character, classifying and mapping them and describing their character; and
- making judgements – based on the characterisation and tailored to the particular use to which the assessment is to be put.

Within this two stage process there are six separate steps each with a number of elements, and these are summarised in Figure 11.1.

11.2.2 What are the outputs of Landscape Character Assessment?

The outputs of a Landscape Character Assessment invariably involve the definition, mapping and description of either landscape character types or landscape character areas. Landscape Character Types (LCTs) are generic, allow different landscapes to be compared, and have similar characteristics in different areas. Landscape Character Areas (LCAs) on the other hand are unique, are geographically or space-specific, and have individual identity but share generic character with other areas of the same type.

A Landscape Character Assessment will usually consist of a description of each type or area including: a written description of character with illustrations; a list of key features or key characteristics which make the landscape distinctive; an analysis of what is happening to the landscape in terms of a summary of main pressures for change; and guidelines for the future planning and management of changes that currently, or may in future, affect the landscape.

11.3 The Scottish programme of Landscape Character Assessment

The National Landscape Character Assessment programme in Scotland was initiated in 1994 in recognition of the fact that there was no coherent or comprehensive body of knowledge or inventory in existence on the landscapes of Scotland. Its origins and development are described in detail elsewhere (Hughes & Buchan, 1999). The programme aimed to classify and describe the character of Scotland's landscapes – what makes one area 'different' or distinct from another – by examining the various component parts of the landscape and the physical, ecological and cultural processes that interact to affect the landscape. The specific objectives of the programme were to:

- establish an inventory of all the landscapes of Scotland;
- raise awareness of Scotland's landscapes;
- identify the forces for change in Scotland's landscapes;
- provide information to support various kinds of casework, including development control and other proposals for land use change;
- provide information to help SNH, local authorities and other partners to influence development plans and other land use strategies; and
- help inform national policy on issues relating to landscape interests.

According to the SNH staff involved, one further important objective of the programme was to involve SNHs partners from the outset and to encourage them to make maximum use of all products of the programme.

The studies were prepared mainly by landscape consultants, although two were undertaken 'in-house' by SNH staff, sometimes working jointly with local authority

11 Farming, Forestry and the Natural Heritage: Towards a More Integrated Future

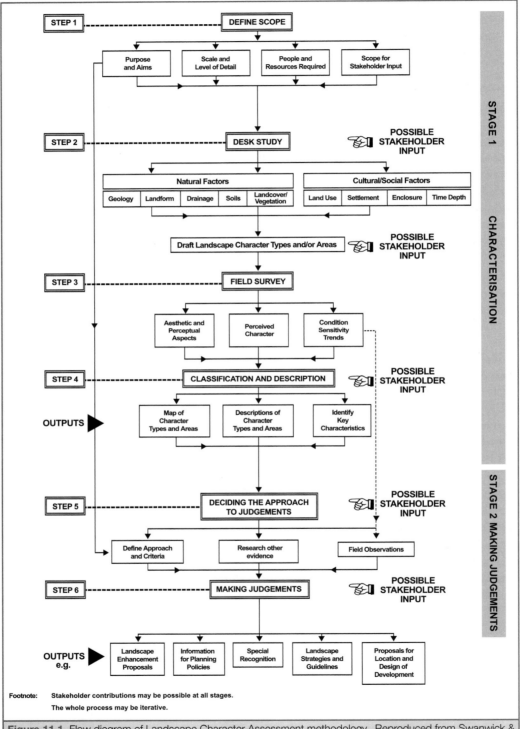

Figure 11.1 Flow diagram of Landscape Character Assessment methodology. Reproduced from Swanwick & Land Use Consultants (2002).

landscape staff. All were undertaken in co-operation with planning authorities and a range of other bodies including Historic Scotland, the Forestry Authority, enterprise companies and local groups, who participated through funding, membership of project steering groups and involvement in consultations.

11.4 Outputs from the programme

The programme resulted in two main types of output. The first output is the series of 29 Landscape Character Assessment reports published between 1994 and 1999. A final report on Loch Lomond and the Trossachs was completed in 2004. Most of the reports cover local authority areas and some covered proposed national park areas, the landscapes around two of Scotland's firths and two specific smaller land areas. Although the briefs were broadly similar, the reports vary widely in content and style. However, there are a number of common components in each, notably:

- a review of the natural and cultural influences that have shaped the landscape;
- a classification and description of landscape character types and/or landscape character areas at a scale appropriate to the study area;
- a review of forces for change affecting the landscape; and
- a set of guidelines designed to assist in future planning, design and management of the landscape.

Although most reports share these elements, the way in which they are structured and presented is individual to each report and the detailed content of the studies varies widely. This variation in the outputs reflects a number of factors. The early studies were viewed as pilot studies for the programme so variety is to be expected. The procurement process meant that each study was carried out by different consultants, each with their own views as to how the work should be carried out. This was inevitable, especially as the method of assessment continued to evolve over the period in which the programme was carried out. In pursuing the full programme, the consultants, SNH project managers and study steering groups were also given considerable latitude in relation to method, content and style. There was no predetermined landscape typology, so the landscape classifications that emerged were specific to the area concerned and not necessarily consistent with other areas. In determining appropriate guidelines the influence of project partners, particularly local authorities, is also evident, with different emphasis in different areas.

At the local level such variation may be a benefit. The reports demonstrate a rich wealth of approaches, all of which may be successful in different ways and so show that variation is helpful in reflecting local requirements and perspectives. However this diversity may become more of a problem when the results of the programme are to be used at the regional and national level.

This is particularly important when considering the second output, namely the Geographic Information System (GIS) and accompanying database which was compiled from the individual assessments. In late 1998, when the national programme of LCA in Scotland was nearing completion, SNH commissioned David Tyldesley & Associates to digitise the mapped landscape units identified during the programme and to develop a national database compiled from the LCAs (David Tyldesley & Associates, 1998a). In summary, 3,967 landscape units (landscape character areas) were plotted, each allocated to one of 366 LCTs. Each unit has an accompanying database entry covering its key characteristics, a brief description of the essence of the character of the landscape type, and a list of pressures for change that are likely to affect the character of the landscape.

In a further, related piece of research (David Tyldesley & Associates, 1998b), David Tyldesley & Associates was commissioned to explore the potential of both the database and the LCAs from which it was derived to develop a national classification of landscape character. To some degree the work was considered to be experimental, examining the database's strengths and weaknesses, but the assumption was that it should be possible to group LCTs into those sharing similar characteristics. It was thought that such groupings might have useful applications in national or regional land use planning and management, and might also help to provide a better general understanding of the landscape resource in Scotland.

Despite the fact that it was never an aim of the LCAs to provide the information for a national classification of landscape character, the consultants considered that the database and the reports (on which they also drew directly) provided the information for a rational and reasonably consistent analysis. In refining the classification and database, the following key steps were taken:

- LCTs, i.e. those areas that are essentially of the same type, were identified and named;
- each of these LCTs was categorised into Highlands and Islands, Uplands, Lowlands or Coastal, depending on its landscape context;
- a decision was taken not to separate out Coastal landscapes, because most also fell within other categories;
- amalgamation then proceeded on the basis of grouping according to similar landform, land cover, land use and other factors to define Level 2 LCTs;
- finally the Level 2 LCTs were further amalgamated on the same basis to define Level 3 LCTs.

Once this work was complete, some further refinement of the GIS and database was undertaken by SNH's Geographic Information Group, in particular to standardise the mapping at the urban edge and the coast, excluding areas for which there was no complete and consistent LCT coverage. The end result of all this work is a classification that defines 275 Level 1 'Same As' LCTs, 121 Level 2 LCTs and 55 Level 3 LCTs within the three broad groupings of Highlands and Islands, Uplands and Lowlands.

The creation of this GIS and database is, in itself, a significant achievement because of the complexities of the task. It does, however, have limitations, not least because the original assessments were not intended to be used in this way and the variations do cause some difficulties at this level. Nevertheless the national classification with its different levels of detail and the accompanying database have proved to be valuable tools.

11.5 Practical applications of Landscape Character Assessment in planning for and managing the natural heritage

In the last 10 to 15 years, Landscape Character Assessment has emerged as a powerful tool to aid the planning, design and management of a wide range of landscapes. It can help to ensure that change and development does not undermine what is characteristic and valued and help to ensure that change to character is informed by a good understanding of its implications. On the other hand, it is not just a tool for conservation and it can be just as useful in helping to inform strategies for the enhancement of existing, or indeed the creation of new, landscapes. Broadly, the main applications of Landscape Character Assessment fall into two categories, namely planning and landscape conservation and management.

- Planning: Landscape Character Assessment may contribute at a variety of levels to the formulation of development plan policies, development control, preparation of development proposals, environmental impact assessment (EIA), development capacity studies and strategies for particular forms of development, and design guidance.
- Landscape conservation and management: Landscape Character Assessment may provide the basis for identification of special landscapes and preparation of landscape management strategies, and may play an important role in a wide range of other initiatives including agri-environment schemes, woodland expansion, strategies to tackle issues of environmental and economic regeneration, and programmes to monitor landscape change (Swanwick & Land Use Consultants, 2002).

The full range of potential applications for Landscape Character Assessment is summarised in Table 11.1. Many applications are relevant from the national or strategic level to the local or project level. Table 11.1 contains general examples of the different applications across these levels in Scotland, and also briefly notes some key examples of completed work where the approach has been used successfully. The potential users of the programme include a very wide range of organisations and individuals working from the national scale down to the scale of an individual landholding.

11.6 Applications of Landscape Character Assessment in landscape conservation and management

Applications of the Landscape Character Assessment programme in landscape conservation and management are to some degree less well-developed than applications in planning, perhaps reflecting more limited awareness of the programme outputs among those involved in landscape conservation and land management. In the context of this chapter the most relevant applications are in the general area of land management and, more specifically, in forestry and farming.

11.6.1 Land management

There is undoubtedly great scope for Landscape Character Assessment to make a significant contribution to the formulation of land management proposals in Scotland. At national level, it has fed into SNH's Natural Heritage Futures programme (Scottish Natural Heritage, 2002a). This programme is a suite of publications prepared by SNH to guide the future management of the natural heritage towards 2025. It considers the natural heritage within six 'settings' across Scotland (coast and seas, farmland, forests and woodlands, fresh waters, hills and moors, and settlements) and also within 21 natural

Table 11.1 The range of potential Landscape Character Assessment applications and examples at the local and strategic levels. Modified from Julie Martin Associates & Swanwick (2003) and based partly on Swanwick & Land Use Consultants (2002)

Potential applications	Local/project level examples	National/strategic level examples
PLANNING		
Development plan policies	• Local plan landscape policies	• Structure plan landscape policies
Development control and consultations	• Planning applications, forestry consultations	• Public inquiries
Preparation of development proposals	• Project development briefs • Project design and EIA	• Development briefs for new settlements
Transport planning and appraisal	• New trunk road routing and design	• Strategic transport appraisal
Development capacity studies	• *Capacity studies for opencast coal, aquaculture, housing, wind farms*[1]	• Policy and locational guidance for wind energy
Urban fringe, townscape and settlement analyses	• Townscape assessments	• Green belt studies
Design guidance	• Housing and other design guides	• Preparation of Planning Advice Notes
LANDSCAPE CONSERVATION AND MANAGEMENT		
Identification of special landscapes	• Boundaries of Areas of Great Landscape Value or (in future) Local Scenic Areas	• Boundaries of National Parks and National Scenic Areas
Landscape management plans and strategies	• *Management plans for estates, country parks*[2]	• Natural Heritage Futures • Management strategies for National Parks and National Scenic Areas
Forestry and agri-environment	• *Forest design guidance*[3]	• Rural Stewardship Scheme design, targeting and evaluation • Indicative forestry strategies
Environmental and economic regeneration	• Central Scotland Forest	
Monitoring landscape change	• Local authority landscape monitoring	• Natural Heritage Trends

EXAMPLES OF COMPLETED PROJECTS

[1] Capacity Studies: David Tyldesley & Associates (1998c); Grant (2002); Land Use Consultants (2002).
[2] Management Plans for Estates: Turnbull Jeffrey Partnership (1996).
[3] Forest Design Guidance: Environmental Resources Management (1998).

heritage areas. For each of the settings and each of the 21 areas, it describes natural heritage character, reviews key influences on that character, and presents a vision, objectives and actions for the future.

These analyses are underpinned by national assessments about specific aspects of the natural heritage, including landscape. The *National Assessment of Scotland's Landscapes* (Scottish Natural Heritage, 2002b) and the 21 individual area prospectuses drew heavily on the outputs of the Landscape Character Assessment programme. In particular, the reports and the forces for change data within the database informed the appraisal of key issues for both the settings and the areas. Key landscape issues were identified for each of the six settings, and for each of the 21 areas a tabular analysis was undertaken of: the diversity, distribution and rarity of the areas' LCTs; the relative importance and value of different aspects of the areas' landscape; and, the pressures affecting the areas' landscape and their degree of significance. These analyses, while very helpful, are complicated by the fact that the boundaries of the LCTs do not always directly relate to the settings or to the Natural Heritage Areas. Unfortunately, there is no 'audit trail' to show where the data for assessment of pressures came from or to allow the analysis to be replicated in future.

At a more detailed level, Landscape Character Assessment is expected to provide a key input to management plans for the Loch Lomond and the Trossachs and the Cairngorms National Parks, and to guide the implementation of management strategies for National Scenic Areas (NSAs). It may also form the basis for local authority landscape management strategies. Such work is still in its relatively early stages, so it is not yet possible to draw any conclusions as to its value or success.

It is already clear, however, that Landscape Character Assessment is proving to be a useful tool at the level of the individual landholding or estate. For example, the LCA for the Mar Lodge Estate within the Cairngorms (Turnbull Jeffrey Partnership, 1996) was jointly sponsored by the landowners (the National Trust for Scotland) and SNH and was specifically designed to inform and shape the contents of the Estate Management Plan. The assessment provided information about different landscape character areas for use by land managers, considered pressures and opportunities for change within the Estate, developed guidelines on how landscape character might be conserved, enhanced or restructured as appropriate, and provided advice on how best to accommodate different forms of change. Guidelines suggested approaches towards native woodland restoration, management of plantations, conservation of wild land quality, treatment of derelict buildings, and moorland and deer management.

11.6.2 Forestry

In forestry, perhaps because of the close involvement of the Forestry Commission in many of the assessments carried out in Scotland, Landscape Character Assessment applications are relatively well developed. *The Scottish Forestry Strategy* (Scottish Executive, 2000) makes particular mention of the role of landscape character in helping to guide decisions about the location and design of new woodland. Major forestry and regeneration initiatives, such as the Central Scotland Forest (Central Scotland Forest Trust, 2004), have made significant use of Landscape Character Assessment to indicate where there is the greatest potential for woodland planting and where expansion is undesirable in terms of landscape character. In addition, at the local authority level, indicative forestry strategies have become increasingly common, for example in Ayrshire where an indicative forestry strategy based on the area's Landscape Character Assessment is being developed with both the industry and the community as partners. In Dumfries and Galloway, *Landscape Design Guidance for Forests and Woodlands in Dumfries and Galloway* has been developed (Environmental Resources Management, 1998), combining landscape character assessment material with Forestry Commission design guidance to provide practical advice on the successful integration within the landscape of new woodlands and forests. Design sheets, intended to help Woodland Grant Scheme applicants develop appropriate planting proposals, have been produced for each of the LCTs where forestry is an issue.

11.6.3 Farming

Many of the Scottish Landscape Character Assessments contain a range of landscape guidelines that relate to farming. For example, for the Pastoral Valleys landscape type in the Dumfries and Galloway assessment the guidelines note that

> *"Dry stone dykes are important characteristic features in establishing well managed pastoral quality ... creating a relationship between human activity and local geology. Restoration and continued maintenance of dykes should be supported."*

In comparison with the widespread adoption of Landscape Character Assessment in relation to forestry, however, there is relatively little evidence of the use of this tool in agri-environment scheme targeting and design in Scotland, except within some of the former Environmentally Sensitive Areas (ESAs) and in the design of crofting demonstration projects in the Highlands and Islands. The information provided on the Rural Stewardship Scheme (Scottish Executive Environment and Rural Affairs Department, 2005), which is now Scotland's main agri-environment scheme (superseding the Environmentally Sensitive Areas Scheme and the Countryside Premium Scheme), makes little or no mention of landscape character issues.

This contrasts with the situation in Wales and in England where Landscape Character Assessment data are now widely used in agri-environment scheme design, targeting, guidance and evaluation. For example, in Wales it has influenced the Tir Gofal Scheme, while in England it has been used in the past in the design and monitoring of the Environmentally Sensitive Areas and Countryside Stewardship Schemes and, most recently, in the development and piloting of the new Entry Level Scheme (ELS), open to all farmers (Department for Environment, Food and Rural Affairs, 2003). In England the guidance provided for farmers in the ELS pilot areas was closely based on the Countryside Character Area descriptions (Countryside Commission & Countryside Agency, 1998-1999) and on county Landscape Character Assessments, and the scheme evaluation will consider whether the measures chosen by farmers offer optimal benefits in terms of landscape character. The Farm Environment Plans (FEPs) which are at the heart of the Higher Level Stewardship element of the new Environmental Stewardship scheme (Department for Environment, Food and Rural Affairs, 2004) require that careful attention be paid to the landscape character of the farm and its surroundings in order to maximise the opportunities for landscape benefit from the scheme. The survey forms for FEPs include information about the key characteristics of the surrounding 'joint character area' (from the national joint character map) to assist in the survey process and the targeting statements are to be based on the management needs of character areas.

The lack of focus on landscape issues within agri-environment initiatives in Scotland has been highlighted in the *Custodians of Change* report (Scottish Executive, 2002). This report notes the importance of agricultural landscapes to quality of life and the tourist economy in Scotland. It points to the recognition given to landscape character issues in national planning policy and guidance and in the forestry sector, and suggests there is a need to develop similar practical advice on agricultural landscape design. The report recommends that the Executive and its agencies should develop policies and guidance for this purpose, and that the Rural Stewardship Scheme should be amended to respond more effectively to environmental (including landscape) concerns.

Both the existing Rural Stewardship Scheme and the proposals for Land Management Contracts, designed to provide economic, environmental and social benefits on whole farms, could undoubtedly be further developed to take account more explicitly of the needs of the landscape. The products from the National Landscape Character Assessment programme would provide an ideal starting point for this.

11.7 Conclusions

A review of the National Programme of Landscape Character Assessment in Scotland (Julie Martin Associates & Swanwick, 2003) which has provided the basis for this chapter, concluded that the programme is both visionary and innovative. Its most significant achievements are summarised below.

- The programme provides a key tool for SNH staff to use in fulfilling SNHs landscape duties and remit. In particular, it has met the agency's urgent need for local level landscape information that can be shared with local authorities and other partners, and has helped to ensure that landscape issues are given due attention in day-to-day casework.
- The LCA programme has achieved formal recognition in policy and advice from central government on natural heritage issues, helping to raise the profile of landscape issues in Scotland.
- It has helped to ensure that the importance and value of *all* landscapes is now given due consideration within the planning system.
- The programme outputs are already being very widely used for a host of different applications.
- The Scottish LCA programme was the first full-coverage, detailed LCA programme to be completed in the UK, and, it is thought, in Europe. Similar but less detailed coverage has now been completed in Northern Ireland, but in England and Wales there is still no full, detailed assessment coverage.
- A special feature of the programme is that by involving all of the local authorities (and other partners) across Scotland in the assessment programme, the programme has fostered a sense of shared ownership and a common understanding of Scotland's landscapes.
- The programme offers an excellent platform from which to implement the European Landscape Convention, providing a detailed landscape characterisation, an analysis

of the forces for change affecting Scotland's landscapes, and information on landscape values and possible landscape objectives.

A special strength is the degree to which the Landscape Character Assessment programme outputs are recognised and used by planners for development planning and development control throughout Scotland. Consultees to the review highlighted its clear, systematic coverage, the fact that it has raised the profile and awareness of landscape issues among other agencies, planners and developers, the legitimacy given to landscape concerns, and the cost-effective use of SNH research funding.

The programme has met many of its objectives, notably in providing an inventory of Scotland's landscapes and information for development control and development planning, and through involvement of SNHs partners. Objectives relating to wider landscape awareness, consistent identification of forces for change, and input to national policy on landscape issues have been less well met. Other weaknesses stem from:

- the variation between the Landscape Character Assessments;
- poor interpretation of historical and ecological characteristics in some Landscape Character Assessments;
- the lack of a national 'top down' landscape perspective;
- the lack of external web access to Landscape Character Assessment outputs;
- limited stakeholder input;
- insufficient separation of the characterisation and judgement stages of the assessment process; and
- a limit to the range of subsequent Landscape Character Assessment applications.

While it has been quite widely used in some aspects of land management, especially with regard to forestry, it has so far been used much less in relation to farming. There is a particular need to see landscape considerations, based on use of Landscape Character Assessment, integrated more fully into the Rural Stewardship Scheme and the emerging proposals for Land Management Contracts.

Some of these weaknesses reflect the fact that good practice in Landscape Character Assessment has moved on since the main suite of assessments making up the programme were prepared. Nevertheless, the review concludes with a number of recommendations. In terms of land management, the key recommendation is that SNH should seek recognition of the potential for landscape character issues to inform Scotland's Rural Stewardship Scheme and related land management initiatives.

References

Central Scotland Forest Trust (2004). *Central Forest Strategy.* Central Scotland Forest Trust, Shotts.

Countryside Commission & Countryside Agency (1998-1999). *Countryside Character.* Eight regional volumes. Countryside Commission and Countryside Agency, Cheltenham.

David Tyldesley & Associates (1998a). Computer Access to Landscape Character Assessment Information. Unpublished report. Scottish Natural Heritage, Perth.

David Tyldesley & Associates (1998b). Analysis of National Landscape Character Types in Scotland. Unpublished report. Scottish Natural Heritage, Perth.

David Tyldesley & Associates (1998c). Clackmannanshire Settlement Landscape Capacity Study. Unpublished report. Clackmannanshire Council, Alloa.

Department for Environment, Food and Rural Affairs (2003). *Entry Level Agri-Environment Scheme Pilot Scheme Guidance Booklet 2003.* Department for Environment, Food and Rural Affairs, London.

Department for Environment, Food and Rural Affairs (2004). *Higher Level Stewardship. 2004 Farm Environment Plan. Application Information Leaflet.* Department for Environment, Food and Rural Affairs, London.

Environmental Resources Management (1998). *Landscape Design Guidance for Forests and Woodlands in Dumfries and Galloway.* Forestry Commission, Dumfries and Galloway Council and Scottish Natural Heritage, Perth.

Grant, A. (2002). Wester Ross Settlement Landscape Capacity Study. Unpublished report. Highland Council, Inverness.

Hughes, R. & Buchan. N. (1999). The Landscape Character Assessment of Scotland. In *Landscape Character – Perspectives on Management and Change*, ed. by M.B. Usher. The Stationery Office, Edinburgh. pp. 1-12.

Julie Martin Associates & Swanwick, C. (2003). Overview of Scotland's National Programme of Landscape Character Assessment. *Scottish Natural Heritage Commissioned Report* F03AA307.

Land Use Consultants (2002). Assessment of the sensitivity of landscapes to windfarm development in Argyll and Bute. Unpublished report. Scottish Natural Heritage, Perth.

Scottish Executive (2000). *Forests for Scotland: The Scottish Forestry Strategy.* Scottish Executive, Edinburgh.

Scottish Executive (2002). *Custodians of Change.* Report of the Agriculture and Environment Working Group, Scottish Executive, Edinburgh.

Scottish Executive Environment and Rural Affairs Department (2005). *The Rural Stewardship Scheme.* Scottish Executive, Edinburgh.

Scottish Natural Heritage (2002a). *Natural Heritage Futures.* Scottish Natural Heritage, Perth.

Scottish Natural Heritage (2002b). *Natural Heritage Zones: A National Assessment of Scotland's Landscape.* Scottish Natural Heritage, Perth.

Scottish Natural Heritage (2003). *Scotland's Future Landscapes? Encouraging a Wider Debate.* Scottish Natural Heritage, Perth.

Swanwick, C. & Land Use Consultants (2002). *Landscape Character Assessment – Guidance for England and Scotland.* Scottish Natural Heritage, Perth and the Countryside Agency, Cheltenham.

Swanwick, C. (2002). *Landscape Character Assessment. Topic Paper 1: Recent Practice and the Evolution of Landscape Character Assessment.* Scottish Natural Heritage, Perth and the Countryside Agency, Cheltenham.

Turnbull Jeffrey Partnership (1996). *Mar Lodge Estate Landscape Assessment. Scottish Natural Heritage Review No. 79.*

11 Farming, Forestry and the Natural Heritage: Towards a More Integrated Future

12 Supporting landscape design through guidance and management strategies

Sue Bennett

Summary

1. The Dumfries and Galloway Landscape Assessment characterises the landscape of Dumfries and Galloway into a series of recurring landscape character types.
2. The assessment informs the statutory land use planning framework for the area, including local landscape designations and policy on wind farms.
3. It has been used to develop partnership management strategies and guidance to influence landscape change, including local Forest and Woodland Landscape Design guidance, Local Forestry Frameworks and National Scenic Area Management Strategies.
4. Such guidance can help to influence landscape change arising through forestry and other land management activities.

12.1 Introduction

The last 10 to 15 years have seen the development of the Landscape Character Assessment approach and the application of this to landscape conservation and planning. This chapter looks at recent work to apply the approach to guidance on forestry through the development of detailed topic-related guidance and broader management strategies.

Dumfries and Galloway Council was the first local authority in Scotland to prepare a regional scale Landscape Character Assessment (LCA) in partnership with Scottish Natural Heritage (SNH). The LCA has since informed a range of local policy and guidance on managing the rural landscape. This chapter introduces the LCA and describes some of the resulting guidance developed to steer landscape changes arising from various forms of land management and development.

12.2 The Dumfries and Galloway Landscape Assessment

Dumfries and Galloway Council first became interested in carrying out a Landscape Character Assessment in the early 1990s, because it was about to prepare a new Structure Plan. In developing its new strategic land use planning framework, the Council wanted to take account of two significant pressures leading to rural landscape change. Forestry already covered over 25% of the region (95% of which was coniferous) and there was continuing interest in forestry expansion. In addition, the area was seeing interest from wind farm developers. In order to guide policy development on these two issues, an understanding of the character of the landscape and its ability to accommodate such land uses was needed.

Meanwhile, SNH was interested in beginning a programme of LCA in Scotland. As a result, a partnership including the Forestry Commission (FC) and the Local Enterprise Company was established to take forward a LCA in Dumfries and Galloway.

Various early 'landscape assessments' had been carried out in the mid to late 1980s. Informed by these, the Countryside Commission had produced initial advice on their approach to landscape assessment in 1987 (Countryside Commission, 1987). In 1991, the Countryside Commission for Scotland published guidance on landscape assessment principles and practice (Countryside Commission for Scotland, 1991). The brief for the Dumfries and Galloway study drew on this early work. Since then, landscape assessment (now termed landscape character assessment) has continued to develop and has expanded to encompass new areas of interest such as historic landscape assessment and landscape capacity. The Countryside Agency and SNH have recently consolidated their advice in new LCA guidance produced in 2002 (Swanwick & Land Use Consultants, 2002).

Land Use Consultants, who had been involved in much of the early work, were appointed to carry out the Dumfries and Galloway Landscape Assessment. The assessment adopts a format that has been broadly followed elsewhere in Scotland (Land Use Consultants, 1998). First, it looks at physical and cultural influences that have shaped the landscape of the region, and key features of the resulting landscape, such as the predominance of improved pasture and rough grazing, the contribution of forests and woodlands, and local variations in building materials that help to create local distinctiveness. It then considers various changes happening in the landscape (including changes in agriculture, forests and woodlands, and various forms of development) and proposes a series of generic guidelines for each agent of change (for example on buildings in the countryside).

The final part of the Council's assessment classifies the landscape of Dumfries and Galloway into four regional character areas, and then into a series of 21 recurring landscape types (see Figure 12.1). Examples include the Southern Uplands and Peninsula landscape types. The landscape character of each landscape type is described and a series of key characteristics is highlighted (see Tables 12.1 and 12.2). Key landscape issues are identified, and an overall strategy is proposed: either to conserve, enhance, or modify the overall landscape character of each landscape type. A series of guidelines is then proposed, with a particular focus on forestry and wind farms. Typical examples of guidance might include supporting the reinstatement of hedgerows and hedgerow trees, supporting appropriate levels of moorland grazing, or identifying the potential for wind farm developments.

The LCA has established a common framework for understanding the landscape, and provides a basis for developing policies and guidance that respect and build on landscape character, as well as a starting point for both preparing and assessing development and land management proposals. It was first published in 1995 by Dumfries and Galloway Council (Land Use Consultants, 1995) and was subsequently incorporated in the Scottish Natural Heritage series of Landscape Assessments (Land Use Consultants, 1998). It is still in regular use. Although the area occupied by some of the forested landscape subtypes has expanded, the descriptions of individual character types have generally stood the test of time. In general the guidance offered remains a valuable starting point for considering landscape issues.

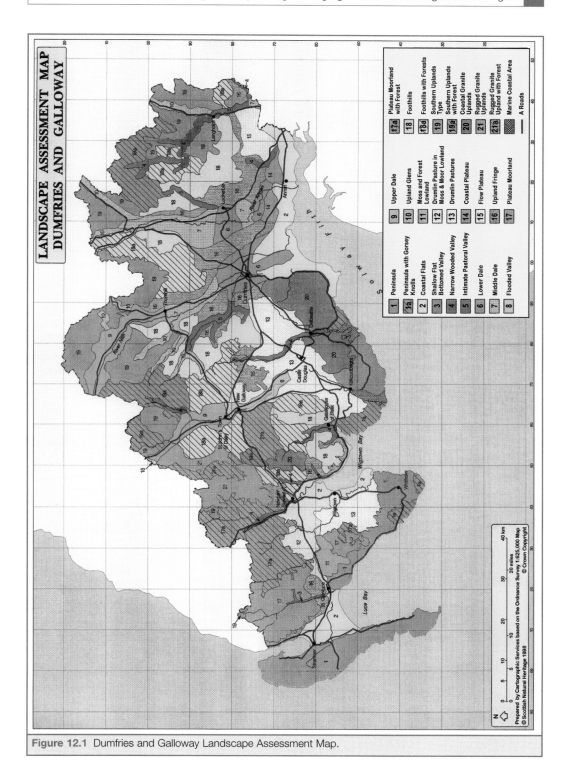

Figure 12.1 Dumfries and Galloway Landscape Assessment Map.

Table 12.1 Southern Uplands Landscape Type. Source: Land Use Consultants (1998).

Key characteristics

- large, smooth dome/conical hills, predominantly grass covered;
- open and exposed character except within incised valleys;
- distinctive dark brown/purple colour of heather on some of the higher areas;
- pockets of woodland in the incised valleys;
- stone dykes occasionally define the lower limits;
- the legacy of lead and other mining activity.

Key landscape issues

- loss or deterioration of heather moorland;
- large-scale forestry expansion;
- demands for wind farms and radio masts.

Table 12.2 Peninsula Landscape Type. Source: Land Use Consultants (1998).

Key characteristics

- medium-scale landscape rising from boggy hollows to rolling pastureland up to Gorsey moorland;
- narrow inter-tidal range with abrupt end to inland land use;
- intimate sheltered bays with stony beaches or occasional narrow strips of exposed flat land, used for transport routes;
- Mull of Galloway distinctive short turf on flat land, no field boundaries, steep cliffs to sea;
- few, but well developed, policy landscapes;
- numerous evenly spaced farmsteads, and few small settlements;
- old forts and castles defend rocky western coast.

Key landscape issues

- farm tipping;
- threat to stone walls as key feature;
- introduction of new woodland planting;
- potential wind farm developments;
- siting and design of tourist facilities.

The assessment was carried out at a scale of 1:50,000, which is particularly relevant in developing strategic policy and guidance. The Council has used the document as a basis for landscape aspects of its statutory planning policy including general landscape policies in the Structure Plan (Dumfries and Galloway Council, 1999a), the review of local landscape designations (Dumfries and Galloway Council, 1999b), and a specific policy framework for wind farms (Dumfries and Galloway Council, 1999c). The LCA has also provided a starting point for a series of partnership guidance documents developed with SNH and the Forestry Commission (Environmental Resource Management, 1998; Dumfries and Galloway Council, Forestry Commission and Scottish Natural Heritage, 2000a,b).

12.3 Statutory planning policy and LCA

The statutory land use planning system is the primary mechanism whereby local authorities influence rural land use and management. The LCA has informed the adopted Dumfries and Galloway Structure Plan (Dumfries and Galloway Council, 1999a) and the finalised Local Plans (Dumfries and Galloway Council, 2002a,b, 2003a,b) in a number of ways that influence the rural landscape. The Structure Plan endorses the overall strategy in the LCA for conserving, enhancing, restoring or modifying the landscape character of different parts of the region, and requires that the LCA is taken into account when assessing development proposals which would have a significant impact on the landscape.

The Structure Plan uses the landscape character assessment to inform a systematic review of local landscape designations (Dumfries and Galloway Council, 1999b). In undertaking the review, a wide range of factors was taken into account, several of which derived directly from the LCA. Certain landscape types are valued more than others for their scenery. Scenic interest also appears to be concentrated where the key characteristics of these valued landscape types are particularly well expressed. Examples include those parts of the Southern Uplands with the most pronounced and sculptural relief, where the characteristic domed shape of the hills is best expressed, and where heather (*Calluna vulgaris*) moorland caps their summits, and, elsewhere, the particularly craggy parts of the Rugged Granite Uplands in Galloway (see Figure 12.1). In addition, the juxtaposition of landscape types such as the Southern Uplands and Upland Glens tends to create dramatic and scenic landscapes. The review resulted in significant changes to the boundaries of the previous Areas of Regional Scenic Significance (ARSS), with some being amalgamated, and the creation of one new designated area.

The Structure Plan policy framework for wind farms (Dumfries and Galloway Council, 1999c) draws on the LCA in several respects. Areas of high wind speed are split into three categories in terms of their suitability for wind farms, according to a range of criteria including landscape. The reviewed local landscape designations are categorised as sensitive to wind farm development. In addition, the advice in the assessment informs the categorisation of individual landscape types according to their potential for wind farm development. The detailed guidance on individual landscape types then forms a starting point for developing specific guidance on the scale and location of wind farms in specific areas.

The statutory planning system has a less direct influence over forestry and, particularly, agriculture. Local Authorities can guide the location and design of forestry at a strategic level through the preparation of Indicative Forestry Strategies (IFS). The strategy for

Dumfries and Galloway was prepared as part of the Structure Plan review (Dumfries and Galloway Council, 1999d), but did not draw directly on the LCA. Additional landscape design guidance was therefore developed subsequently in partnership with relevant agencies (Environmental Resource Management, 1998). This informed consideration of forestry capacity through two Forestry Frameworks (Dumfries and Galloway Council, Forestry Commission and Scottish Natural Heritage, 2000a,b). More recently, Management Plans have been developed for three National Scenic Areas in consultation with land managers and others (Dumfries and Galloway Council, 2003c,d,e). These applications of the LCA approach are described in more detail below.

12.4 Landscape design guidance for forests and woodlands

The first example of the LCA feeding into partnership guidance was a joint project between the Council, SNH and the Forestry Commission to produce local landscape design guidance for forests and woodlands in Dumfries and Galloway (Environmental Resource Management, 1998). The aim was to produce simple, practical, user friendly guidance for forest owners and managers on designing new woodlands, and restructuring existing plantations, to contribute positively to the local landscape character of Dumfries and Galloway. A secondary purpose was to help agencies and community groups to assess proposals.

The Guidance builds directly on that in the LCA and on the Forestry Commission's national Landscape Design Guidelines (Forestry Authority, 1994). A series of individual guidance sheets gives advice for those landscape types where forests and woodlands can contribute positively to the landscape. Each guidance sheet is produced double-sided at A3 size and is designed to be reproducible as an individual, stand-alone document. The first side describes the landscape character and ecology of the particular landscape character type. The second side sets out the forestry opportunities and constraints, and provides design guidance. Both are illustrated with sketches and photographs (see Figure 12.2).

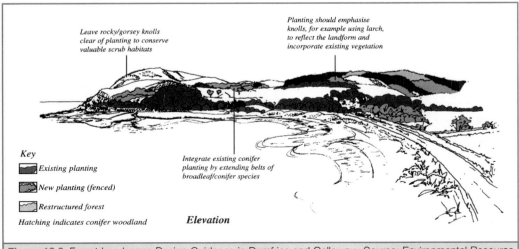

Figure 12.2 Forest Landscape Design Guidance in Dumfries and Galloway. Source: Environmental Resource Management (1998)

The guidance has been helpful in encouraging design proposals that take account of local landscape character. It is available on the Council's website (www.dumgal.gov.uk/dumgal/MiniWeb.aspx?id=91&menuid=1414&openid=1104).

12.5 Local Forestry Frameworks

The above guidance went on to inform two pilot Local Forestry Frameworks (Dumfries and Galloway Council, Forestry Commission and Scottish Natural Heritage, 2000a,b). This initiative developed from a disputed forestry grant scheme application in the Cree Valley in Galloway in 1997. The proposal would have resulted in the loss of remaining open ground within an extensively forested area. It generated strong local opposition and the Council objected to the scheme. The Forestry Commission eventually rejected the grant application on the advice of its Regional Advisory Committee, on the grounds of 'loss of hill farm land and subsequent impact on land use diversity and balance'. This set a national precedent for considering forestry proposals in terms of reasonable balance of land uses.

An unpublished report on *Balancing Forestry with other Land Uses* (ASH Consulting Group, 1996) recommended the use of Local Forestry Frameworks to provide local guidance in sensitive areas that offer limited further scope for forestry. In 1998, the Council, Forestry Commission and SNH agreed to work together to pilot this new approach and to develop more detailed guidance for Langholm/Lockerbie and Galloway (Dumfries and Galloway Council, Forestry Commission and Scottish Natural Heritage, 2000a,b). These were both areas with high concentrations of existing forestry, with continued interest in further forest expansion, and where there were also environmental sensitivities and growing public concern about the extent of forestry. The aim was to provide guidance on future forestry and woodland planting and management, by providing a locally specific interpretation of national and international guidance and commitments, using tools such as the LCA.

A Steering Group advised the project, which also involved extensive community and stakeholder consultation, including eight workshops attended by over 300 people. The study divided the framework areas first into catchments and then into smaller tracts. Each tract was assessed and allocated to one of five broad categories, ranging from 'good potential for forestry' to 'very limited potential'. Detailed guidance was then developed for each individual area, covering the local issues to be addressed (for example archaeology, road extraction) and the appropriate character, distribution and extent of forestry (see example in Box 12.1). The tract boundaries reflect a variety of different factors. Sometimes habitat changes or national nature conservation designations might be the most dominant factor influencing variations in capacity. Elsewhere, changes in landscape character, visual horizons or concentrations of archaeological interest may be more significant.

The Frameworks were developed alongside the Scottish Forestry Strategy (Scottish Executive, 2000) and also informed the current circular on Indicative Forestry Strategies (Scottish Office Development Department, 1999). The documents were formally agreed by the three partner agencies, and are now influencing forestry proposals. Their success has been somewhat difficult to measure because of the recent general slow down in forestry. Since they have been adopted, however, the standard of proposals coming forward has improved and there have been no more contentious forestry grant applications in the region.

> **Box 12.1** Galloway Local Forestry Framework.
>
> **Area 10 Ken – Dalry**
> Moderate Sensitivity
>
> Description
> Water of Ken Valley predominantly moorland area within the Regional Scenic Area, bounded by productive forestry to the north and south and has been identified as important habitat for hen harrier (*Circus cyaneus*) as well as important for tourism. Acid sensitive in western area. Also includes calcium-rich rocks and soils with high botanical interest.
>
> Guidance
> Some new planting may be appropriate particularly where it can provide a more sheltered landscape, provided there is not significant change in the overall balance between pasture and woodland. New planting should seek to protect and enhance the intrinsic landscape character of the area and should have no detrimental effect on the tourism quality of the Ken/Dee valley. Any planting should take the form of irregular interlocking shapes with high species diversity. Views from the Southern Upland Way and A713 should be preserved, and important habitats including geese feeding area in the floodplain should be protected. New planting proposals should provide for access directly onto the A713.

12.6 National Scenic Area Management Strategies

More recently, the LCA has informed National Scenic Area Management Strategies for the three National Scenic Areas (NSAs) in Dumfries and Galloway: the Nith Estuary; Fleet Valley; and East Stewartry Coast (Dumfries and Galloway Council, 2003c,d,e). The current suite of 40 NSAs in Scotland was designated in 1980 for their 'fine scenery', 'unsurpassed attractiveness' and 'natural beauty and amenity'. SNH's advice to Government in 1999 (Scottish Natural Heritage, 1999) recommended that the designation should be made more effective, and a range of measures was proposed, including introducing legislation requiring local authorities to produce and implement Management Strategies. To inform that advice, SNH invited two local authorities to work in partnership with them to pilot the Management Strategy approach. Two contrasting parts of Scotland were selected: Wester Ross NSA in the Highlands; and the three small coastal NSAs in Dumfries and Galloway.

This project provided an opportunity to explore how the designation approach can be complementary to, and informed by, the countryside-wide Landscape Character Assessment approach to managing the landscape. The LCA approach enables us to care for the wider countryside, all of which is important to the people who live and work within it, while the designation approach safeguards particularly valued areas. An analogy can be drawn with the Biodiversity Action Plan approach and wildlife designations. One addresses the wider resource, and the other considers special areas; both contribute to the overall suite of conservation mechanisms.

A project officer was appointed to identify, using the LCA, the special attributes (or 'scenic qualities') of the NSAs and their sensitivities to change, to draw out the forces for change that affect these qualities and to devise a Management Strategy and Action Plan through a participatory process. There is now a fairly standardised approach to Landscape Character Assessment, but at that time there was no standard methodology for selecting special landscapes and for evaluating what makes them special (the SNH/Countryside Agency guidance on Landscape Character Assessment (Swanwick & Land Use Consultants, 2002) includes some initial pointers, and SNH is now developing guidance for local landscape designations). Public preferences play a key role in determining which landscapes may be considered special, so the project officer consulted widely to determine whether people considered the existing designated areas to be special and, if so, why (and also whether the current boundaries were relevant). Through workshops, a common vision was developed for each area and consensus sought on what (if anything) needs special management to deliver that vision.

The resulting strategies (Dumfries and Galloway Council, 2003c,d,e) describe the landscape of each NSA, drawing strongly on the Landscape Character Assessment, and then analyse what is special about the landscape, drawing on both the LCA and the public consultation. Attributes are considered under a range of headings: scale, diversity, harmony, change and movement, light, sensory qualities and landmarks. The strategies then propose a vision for the future, consider issues and opportunities presented by the landscape, and develop Action Plans under a series of themes. Across the three strategies, over 200 actions are identified. The strategies can be viewed on the Council's website (www.dumgal.gov.uk).

The Strategies were adopted as supplementary guidance to the Finalised Local Plans in December 2002 and were launched in May 2003. Advisory Groups have been established to take forward their implementation, supported by the project officer whose post has been extended on a temporary basis. The current priority is to place the project on a more secure footing through a Heritage Lottery Fund bid.

12.7 Conclusions

Management strategies and guidance can provide valuable tools to influence landscape change arising through forestry and other land management activities. To be effective, guidance should be founded on a good understanding of both the processes it is intended to influence and the character of the landscape. LCA offers an appropriate methodology for understanding landscape character. Local communities and stakeholders should be involved in developing any guidance to ensure the outputs reflect their concerns and have wide acceptance and support. This may include people who live and work in the area, those who manage the land, and those who use it for recreation. The advice should be appropriate to the purpose it is intended to serve. Of the two examples of forestry guidance described above, the Forestry Framework approach is most relevant to addressing issues of land use balance in sensitive areas, whereas the Landscape Design Guidance is more useful in influencing detailed design. Mechanisms for implementing the guidance must also be identified, and these can include incorporation in statutory policy and delivery through partnership working or grant schemes. Finally, if the public wants land managers to take special care of landscape character and scenery as well as to produce food and timber, then funding and other mechanisms must be geared to encourage this, such as through appropriately funded agri-environment schemes.

References

ASH Consulting Group (1996). Balancing Forestry with Other Land Uses. Unpublished report. Forestry Commission, Scottish Natural Heritage and the Royal Society for the Protection of Birds, Edinburgh.

Countryside Commission (1987). *Landscape Assessment – a Countryside Commission approach.* CCD18. Countryside Commission, Cheltenham.

Countryside Commission for Scotland (1991). *Landscape Assessment, Principles and Practice.* Countryside Commission for Scotland, Perth.

Dumfries and Galloway Council (1999a). *Dumfries and Galloway Structure Plan.* Dumfries and Galloway Council, Dumfries.

Dumfries and Galloway Council (1999b). *Dumfries and Galloway Structure Plan Technical Paper No. 6, Identification of Regional Scenic Areas.* Dumfries and Galloway Council, Dumfries.

Dumfries and Galloway Council (1999c). *Dumfries and Galloway Structure Plan Technical Paper No. 5, Preparation of Wind Energy Diagram.* Dumfries and Galloway Council, Dumfries.

Dumfries and Galloway Council (1999d). *Dumfries and Galloway Structure Plan Technical Paper No. 4, Preparation of Forestry Strategy Diagram.* Dumfries and Galloway Council, Dumfries.

Dumfries and Galloway Council (2002a). *Finalised Nithsdale Local Plan.* Dumfries and Galloway Council, Dumfries.

Dumfries and Galloway Council (2002b). *Finalised Stewartry Local Plan.* Dumfries and Galloway Council, Dumfries.

Dumfries and Galloway Council (2003a). *Finalised Annandale and Eskdale Local Plan.* Dumfries and Galloway Council, Dumfries.

Dumfries and Galloway Council (2003b). *Finalised Wigtownshire Local Plan.* Dumfries and Galloway Council, Dumfries.

Dumfries and Galloway Council (2003c). *East Stewartry Coast National Scenic Area Management Strategy.* Dumfries and Galloway Council, Dumfries.

Dumfries and Galloway Council (2003d). *Nith Estuary National Scenic Area Management Strategy.* Dumfries and Galloway Council, Dumfries.

Dumfries and Galloway Council (2003e). *Fleet Valley National Scenic Area Management Strategy.* Dumfries and Galloway Council, Dumfries.

Dumfries and Galloway Council, Forestry Commission and Scottish Natural Heritage (2000a). *Langholm/Lockerbie Local Forestry Framework.* Dumfries and Galloway Council, Dumfries.

Dumfries and Galloway Council, Forestry Commission and Scottish Natural Heritage (2000b). *Galloway Local Forestry Framework.* Dumfries and Galloway Council, Dumfries.

Environmental Resource Management (1998). *Landscape Design Guidance for Forests and Woodlands in Dumfries and Galloway.* Forestry Commission, Dumfries and Galloway Council and Scottish Natural Heritage, Dumfries.

Forestry Authority (1994). *Forest Landscape Design Guidelines.* FCGL003. HMSO, London.

Land Use Consultants (1995). *Dumfries and Galloway Landscape Assessment.* Dumfries and Galloway Council, Dumfries.

Land Use Consultants (1998). Dumfries and Galloway Landscape Assessment. *Scottish Natural Heritage Review No. 94.*

Scottish Executive (2000). *Forests for Scotland: The Scottish Forestry Strategy.* Scottish Executive, Edinburgh.

Scottish Natural Heritage (1999). *National Scenic Areas – Scottish Natural Heritage's Advice to Government.* Scottish Natural Heritage, Perth.

Scottish Office Development Department (1999). *Circular 9/1999, Indicative Forestry Strategies.* Scottish Office, Edinburgh.

Swanwick, C. & Land Use Consultants (2002). *Landscape Character Assessment Guidance for England and Scotland.* Countryside Agency, Cheltenham and Scottish Natural Heritage, Perth.

13 From past to present: understanding and managing the historic environment

Lesley Macinnes

Summary

1. The land has long been utilised by people and traces of past use are evident throughout the landscape. These form an important part of its character.
2. Like the natural heritage, the historical dimension of landscape is subject to a range of pressures and needs active management.
3. This chapter describes the new technique of Historic Land Use Assessment and the contribution this is beginning to make to the management of the historic landscape.

13.1 Introduction

Substantial progress has been made in our understanding of present day landscapes through the Landscape Character Assessment (LCA) approach. However, our present day landscapes are a product of complex relationships between people and their environment over a long period of time, even though much evidence is obscured by natural processes and human activity. The Landscape Character Assessment approach does not deal particularly well with the historical aspects of the landscape. This chapter describes a new approach called Historic Land Use Assessment (HLA) which is enhancing our understanding of the historic dimension of landscape and our approaches to its management. As this is a new technique, however, the discussion focuses more on its potential than on its practical application, which is still developing.

13.2 Understanding the historic environment
13.2.1 The historic environment

Our landscape has been utilised by people in one way or another since the end of the last Ice Age. The form of this use has varied, from the fairly light touch of early hunting and gathering communities to the remodelling of the 18th/19th century land improvements and the exploitation of the industrial age. Arguably, the greatest single impact was the introduction of farming in the Neolithic period, as this led to settled communities whose impact on soil and vegetation cover was more focused than before. This broad picture hides a complex relationship between people and their environment, but we can be sure that there has been a great deal of change within patterns of settlement (Glendinning & Mackechnie, 2004), in land use, soils (Davidson & Carter, 1997) and in the nature and extent of tree and other vegetation cover since prehistoric times (Edwards & Whittington, 1997). Some of this has been in response to natural factors, such as climate change, but much has been in response to changing human needs and abilities (Edwards, 1993; Smout, 1993).

Many physical traces of this human impact can be seen in the landscape today. It is perhaps best known through the range of sites and monuments and relict landscapes, which can cover extensive areas on the ground (see, for example, the Royal Commission on the Ancient & Historical Monuments of Scotland (RCAHMS), 1990, 1994). This is particularly true for the post-medieval period, where systematic recording is relatively recent and we are still trying to understand the full pattern of relict land use (RCAHMS & Historic Scotland, 2002a; Lelong, 2003; Dixon, 2003).

But other landscape elements also point to the human past, even though this connection may not always be widely recognised. Designed landscapes were established during the 17th-19th centuries and, in many cases, their natural appearance was the product of deliberate design (Glendinning & Mackechnie, 2004). In many places, particularly upland areas, vegetation cover has been influenced by the grazing of domesticated animals. Semi-natural, or ancient, woodland has usually been managed in the past (Smout, 1993), and that management is often the reason for its survival to the present day. We take field boundaries for granted as part of our modern landscape, but in fact many date back to the 18th century, and in some cases probably even earlier (RCAHMS & Historic Scotland, 2002a). A lot more evidence lies hidden beneath the ground surface, obscured by natural processes and by subsequent human activity (Hanson & Macinnes, 1991; Coles, 2001). In short, traces of human use are evident throughout the landscape and impact on present landscape character. As part of the modern landscape, this evidence of our cultural heritage is subject to the same range of impacts that affects the natural heritage, whether caused by natural processes such as climate change, or by human agency, including development and land management (Macinnes, 1991).

The last few years have seen the development of the LCA approach to landscape conservation and planning (Usher, 1999; Swanwick & Land Use Consultants, 2002). This process, however, did not initially deal particularly well with the historical aspect of the landscape; that is its time-depth. LCAs were generally too limited in scope chronologically, with a focus on the recent past; and, in terms of types of evidence used, with a focus on visually impressive sites and on designed landscapes. Consequently, cultural heritage agencies across the UK developed methodologies to depict the historical character of the landscape in a complementary way, trying to move beyond traditional site-based approaches to a landscape approach to conservation that better reflected historic patterns (Fairclough & Macinnes, 2003; Macinnes, 2004). In Scotland, this is being carried out through the programme of Historic Land Use Assessment (HLA) (Dyson Bruce et al., 1999; Dixon & Hingley, 2002; Stevenson & Dyson Bruce, 2002).

13.2.2 Historic Land Use Assessment

HLA is a partnership project between Historic Scotland and the Royal Commission on the Ancient and Historical Monuments of Scotland (RCAHMS). It maps the historic influences evident in the modern countryside. It is based on the OS 1:25,000 Pathfinder map series and uses topographical and land cover maps and datasets (mainly OS 1st edition and Landline Digital; Macaulay Institute Land Cover maps; Forestry Commission Woodland Data), vertical aerial photographs (1988 Scotland Survey) and archaeological and historical data (mainly National Monuments Record of Scotland, local Sites and Monuments Records, but also the Scottish Burgh Survey, the Inventory of Historic Gardens and Designed Landscapes, Statistical Accounts, and key secondary sources) to identify historic land use patterns and relict landscapes over 1 ha in extent (this size is dictated by the scale of data capture). The data is stored as polygons on GIS, resulting in a map overlay and supporting data-set that can be actively interrogated on screen. The project is essentially desk-based, but there is an element of field-checking.

The basic building blocks of the HLA are historic land use types (see Figure 13.1, which shows the area around Newcastleton, Roxburghshire). These characterise units of land use by their form, function and period of origin: there are 52 historic land use types identified across Scotland to date (a glossary of terms can be accessed through 'Canmore' on the RCAHMS web site – www.rcahms.gov.uk).

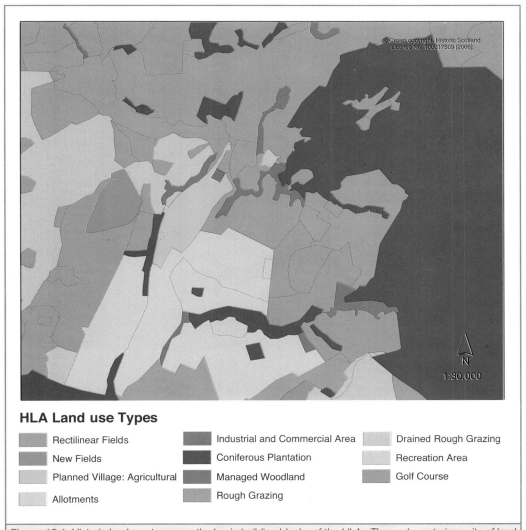

HLA Land use Types

- Rectilinear Fields
- New Fields
- Planned Village: Agricultural
- Allotments
- Industrial and Commercial Area
- Coniferous Plantation
- Managed Woodland
- Rough Grazing
- Drained Rough Grazing
- Recreation Area
- Golf Course

Figure 13.1 Historic land use types are the basic building blocks of the HLA. These characterise units of land use by their form, function and period of origin. © RCAHMS and Historic Scotland. This map is reproduced from Ordnance Survey material with the permission of Ordnance Survey on behalf of the Controller of Her Majesty's Stationery Office © Crown copyright. Unauthorised reproduction infringes Crown copyright and may lead to prosecution or civil proceedings. Historic Scotland Licence No. 100017509 [2005].

Depending on the complexity of the area, the map can be quite detailed, so historic land use types have also been grouped into historic land use categories to provide a simplified – or summary – view of the historic land use pattern which is easier to read and digest (see Figure 13.2). There are currently 14 categories. In effect, the historic land use categories outline the broad framework of historic land use in an area, while the historic land use types break this down into greater detail.

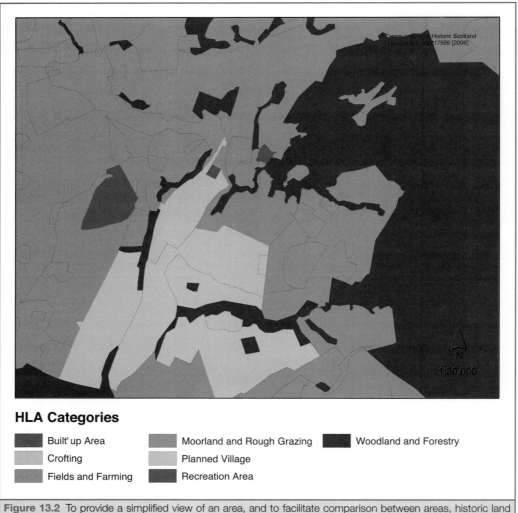

Figure 13.2 To provide a simplified view of an area, and to facilitate comparison between areas, historic land use types have been grouped together into historic land use categories. © RCAHMS and Historic Scotland. This map is reproduced from Ordnance Survey material with the permission of Ordnance Survey on behalf of the Controller of Her Majesty's Stationery Office © Crown copyright. Unauthorised reproduction infringes Crown copyright and may lead to prosecution or civil proceedings. Historic Scotland Licence No. 100017509 [2005].

As the aim is to depict the time-depth in the landscape, each type is also given a period of origin or currency of use, based on the present archaeological and historical understanding of each historic land use type (see Figure 13.3). This provides a chronological framework for the historic land use of an area.

From past to present: understanding and managing the historic environment

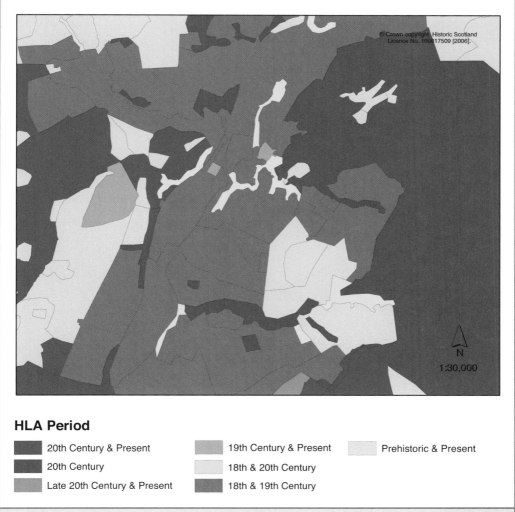

Figure 13.3 Each historic land use type is ascribed a period of origin or currency, showing when the various land uses were established. © RCAHMS and Historic Scotland. This map is reproduced from Ordnance Survey material with the permission of Ordnance Survey on behalf of the Controller of Her Majesty's Stationery Office © Crown copyright. Unauthorised reproduction infringes Crown copyright and may lead to prosecution or civil proceedings. Historic Scotland Licence No. 100017509 [2005].

This period analysis is supported by relict land use types, which depict traces of past land use in the landscape, and archaeological landscapes, where archaeological remains can extend over large areas of land. Like historic land use types, relict types are also assigned periods of use, and can be grouped into categories to provide an overview and aid regional comparison. There can be up to three relict types in any given area, reflecting time-depth within the landscape (see Figure 13.4a-c).

The HLA is a landscape-scale methodology that depicts areas over 1 ha in extent. However, it is complementary to archaeological records and allows smaller areas and individual sites to be brought into the overall picture. Relict types can be overlain on to

13 Farming, Forestry and the Natural Heritage: Towards a More Integrated Future

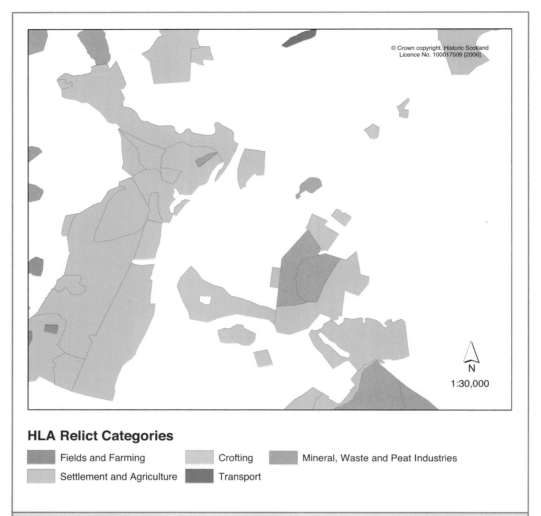

Figure 13.4a Relict land use types highlight further time-depth within the landscape. Relict historic land use types have lost their original use but are still traceable in the landscape; relict archaeological types are concentrations of archaeological features in the landscape (Figure 13.4a). Like historic land use types, relict land use types can also be grouped together by category (Figure 13.4b) or period (Figure 13.4c). © RCAHMS and Historic Scotland. This map is reproduced from Ordnance Survey material with the permission of Ordnance Survey on behalf of the Controller of Her Majesty's Stationery Office © Crown copyright. Unauthorised reproduction infringes Crown copyright and may lead to prosecution or civil proceedings. Historic Scotland Licence No. 100017509 [2005].

historic land use categories, types and periods so that the full time-depth of the landscape can be clearly seen within the modern land use (see Figure 13.5).

Analysis of the HLA data can be carried out by type, category or period for both historic land use and relict land use. This allows a detailed analysis of an area at different levels and shows its distinct characteristics, the origin, form and function of its historic land use pattern, and its development through time. The presentation of data in categories facilitates comparison between areas, enabling regional, and ultimately national, overviews to be achieved. Nevertheless, the full potential of the system is only realised when it is used in its

From past to present: understanding and managing the historic environment | 13

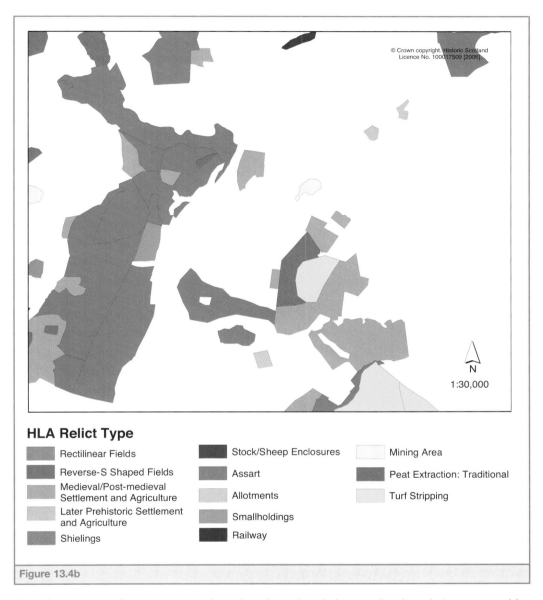

Figure 13.4b

digital interactive form as a research and analytical tool that can be directly interrogated by the user for a particular purpose.

Some 50% of Scotland has now been covered, taking in a reasonable spread of landscape types across the country, and including the National Parks in Scotland and several National Scenic Areas (RCAHMS & Historic Scotland, 2000, 2001, 2002b, 2003). The emerging view is of a dynamic landscape whose main features were established over the 17th-19th centuries, but which has seen considerable change since then, particularly in lowland areas. In upland areas, especially those under rough grazing, there has been less significant change and greater continuity in the nature and intensity of land use, arguably since as far back as the prehistoric period. Regional variations are beginning to emerge which show diversity

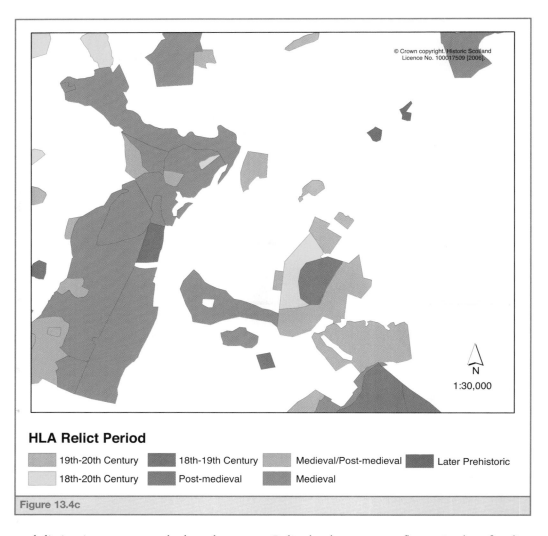

HLA Relict Period

- 19th-20th Century
- 18th-20th Century
- 18th-19th Century
- Post-medieval
- Medieval/Post-medieval
- Medieval
- Later Prehistoric

Figure 13.4c

and distinctiveness across the broad pattern. Relict land use types reflect episodes of earlier use and highlight the time-depth within the landscape. It is clear that very little, if any, of our modern landscape has escaped the impact of human influence.

Together with the detailed recording of relict landscapes through survey, HLA is bringing an additional dimension into landscape conservation. This dimension reflects a major part of the history of the landscape and contributes significantly to our understanding of how the modern landscape has evolved and to our appreciation of landscape character. This knowledge will help us manage the historic landscape more comprehensively. It will also inform discussions about current land management and future land use change by helping us understand, and value, more fully how we got to where we are at present, and how that affects what we might choose to do in future. As with Landscape Character Assessment, the aim of HLA is not to prevent further change, but to understand what the impact of future change will be on the land use pattern inherited from the past – nationally, regionally and locally – and to make more informed decisions as a result.

From past to present: understanding and managing the historic environment

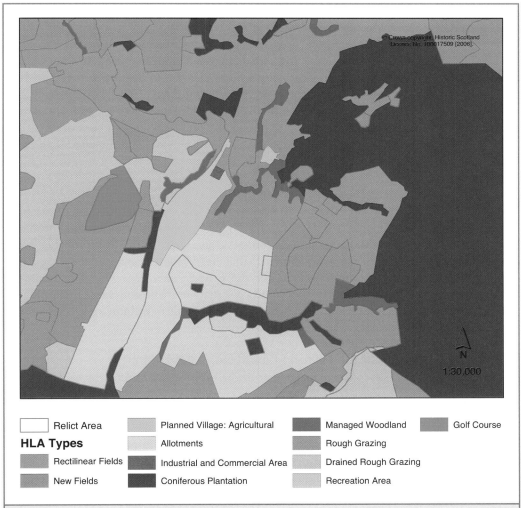

Figure 13.5 Historic land use types and relict land use types can be overlain to give a more complete picture of time-depth in the landscape. Categories and periods can be compared in the same way. © RCAHMS and Historic Scotland. This map is reproduced from Ordnance Survey material with the permission of Ordnance Survey on behalf of the Controller of Her Majesty's Stationery Office © Crown copyright. Unauthorised reproduction infringes Crown copyright and may lead to prosecution or civil proceedings. Historic Scotland Licence No. 100017509 [2005].

13.2.3 Historic Land Use Assessment (HLA) and Landscape Character Assessment (LCA)

Although HLA is a distinct methodology, comparison between HLA and LCA data has shown that they are complementary techniques and that there is often a close correlation between historic land use types and landscape character types, particularly where human influence is most visible (David Tyldesley & Associates, 2001; Fairclough & Macinnes, 2003). HLA also enhances the picture provided by LCA by adding a time-depth that improves our understanding of how changes in the past have affected the modern environment, of the complex interaction between human actions and the natural environment, and of the length of time over which this has occurred. This can help us consider current issues within a longer-term perspective and enable contemporary management decisions to be taken with a fuller understanding of the management practices that have brought us to where we are.

HLA is a stand-alone technique for enhancing our understanding of the historic dimension of landscape, but it can also be integrated into the LCA process (Fairclough & Macinnes, 2003) to give a more comprehensive understanding of how current landscape character developed. However, greater emphasis is placed on completing national coverage of the HLA at present, rather than on integrating the two techniques. Nevertheless, it is feasible for more detailed landscape capacity studies to take an integrated approach, combining the two techniques.

13.3 Managing the historic environment

Traditional approaches to archaeological management are fairly site-focused or, at best, address larger landscape blocks in a discrete way. They generally deal with localised problems relating to agriculture, forestry and natural processes (Macinnes, 1993). It is probably fair to say that the strategic landscape-wide approach that has become the norm for the natural heritage has, until recently, been lacking for the historic environment. The improved understanding of the historic dimension of landscape that we are gaining from archaeological and palaeoenvironmental studies and from the HLA now needs to be built into strategies and policies for the conservation of the rural historic environment more widely.

13.3.1 Key strengths of the Historic Land Use Assessment

The contribution of HLA to the management of the historic environment is likely to lie in three main areas.

The first area is that HLA brings an historical dimension to our knowledge and understanding of current land use, by clarifying the nature and period of land use of a given area, based on historic land use types. We are, consequently, becoming more aware of where current patterns of land use, including field boundaries and vegetation cover, have come from, how these contribute to landscape character and what the impact of losing or changing these will be on the historical depth of the landscape.

The HLA improves our understanding of how past management has helped shape today's natural heritage and land use patterns. Understanding the degree of continuity and change can help assess the impact, and inform the location, of many types of landscape development and of further management and change. It enables us to consider proposals about future change of an area in the context of its historical development, helping decisions to be well-informed about the impact of our actions on the landscape we have inherited from the past.

There is, nevertheless, a need for further development to refine the use of the HLA in practical contexts. Conservation priorities need to be established for the historic land use types, and consideration given to how to address these within land management schemes. We need to clarify the broad management needs of individual historic land use types and their sensitivity to change, addressing issues of limits of acceptable change and identifying priorities for protection and management. Pilot work is already underway to address these issues, in relation to management strategies for National Scenic Areas in Wester Ross (RCAHMS & Historic Scotland, 2003) and the Solway Coast (RCAHMS & Historic

Scotland, 2002b) and the National Park Plan for Loch Lomond and the Trossachs. Further analysis of the HLA data is needed to bring out the diversity of historic land uses across Scotland, to address issues like typicality and rarity on national and regional scales, and to draw out local character and distinctiveness. This will be important for the practical application of the European Landscape Convention (Déjeant-Pons, 2002), assuming that it is ratified by the UK in due course.

The second area where HLA is particularly helpful lies in mapping relict landscapes. It provides a broad overview of the survival of relict landscapes across Scotland, showing major distributions against a land use background which helps us to understand the forces that have impacted on their survival and loss, both in the past and in the present day. The period element of HLA allows us to identify readily those relict areas that are early in date. The survival of landscapes of prehistoric to medieval date is likely to be important simply because they are rare. For later landscapes, though, issues like diversity, distinctiveness and typicality become more important and more detailed work is needed to characterise the remains identified in the HLA. Some of this further work is already underway through initiatives like the First Edition Survey Project (RCAHMS & Historic Scotland, 2002a) and area-based historic landscape projects, such as at Ben Lawers in Tayside (Boyle, 2003). These are beginning to indicate contrasts, for instance between the character of relict post-medieval landscapes in highland and lowland Scotland. Such work will enable us to establish priorities for protection and management within both archaeological and broader land management schemes.

The third role of HLA is to provide a framework for archaeological issues of site survival and the landscape context, or setting, of sites and the impact of development or land use change proposals on these. This will help to establish where the areas of greatest potential survival are likely to be and what kind of sites, or other archaeological evidence, are likely to survive in different land use contexts. It will help us to understand the context of a site in a given area, or regionally, or nationally; how systems of land use have operated in the past; and how sensitive sites and their contexts are to different kinds of change. This will help to prioritise management actions.

13.3.2 Additional benefits

Due to its landscape scale and its land use base, HLA helps to link the historic environment more effectively with other aspects of landscape planning and management, enabling it to be taken into account more easily in integrated approaches (Land Use Consultants, 2003). HLA has been integrated into a landscape capacity study in the World Heritage Site in Orkney (David Tyldesley & Associates, 2001), and used to inform the indicative forestry strategy for Ayrshire. It provides an important baseline for setting and monitoring policies and actions, and for environmental impact assessments. It is being used to help develop management objectives for National Scenic Areas and a landscape strategy for the National Park Plan in Loch Lomond and the Trossachs National Park. It will also be used to help to define the buffer zone of the proposed World Heritage Site of the Antonine Wall. It can, therefore, help to define historic character within designated landscapes from local to international level.

Along with Landscape Character Assessment, HLA enhances our understanding of landscape character and helps to assess the consequences of change, loss and, indeed, enhancement and restoration. There is also scope to explore any correlations between historic land uses and biodiversity. For example, the extent to which semi-natural woodland is linked to past woodland management, or heather moorland to past land uses. This is particularly relevant to the implementation of the Scottish Biodiversity Strategy and associated plans, including Local Biodiversity Action Plans.

HLA will be able to assist in the process of setting national conservation objectives for the sustainable management of the historic environment, in accordance with Historic Scotland's policy, *Passed to the Future* (Historic Scotland, 2002). As it provides an overview of the historic dimension of landscape and landscape change through time, the HLA offers a framework for informing national policy for the historic environment and for assessing the impact of wider land use and land management on it. In giving us a better understanding of regional and local variations, it will provide a broader context for locally based actions. Importantly, HLA also provides a baseline against which change can be monitored and the impact of land use policies on the historic environment assessed.

HLA will also be helpful for drawing in local communities to landscape debates, as it relates to the actions of people on the land: this was found to be case in the consultation process for the pilot management strategies for National Scenic Areas in Wester Ross and the Solway Coast. It will help to enable cultural values of landscape to be included within holistic approaches to the landscape or environment, such as Quality of Life Capital (CAG Consultants & Land Use Consultants, 2001). The Loch Lomond and the Trossachs National Park Authority is beginning to explore this, using LCA and HLA as a base, along with data on the built environment. This is an exciting new development that promises a properly holistic approach to understanding and managing the landscape.

13.4 Conclusion and future directions

Our understanding of the historic environment and its contribution to landscape character is growing. It is beginning to offer an insight into the historic complexities behind the current landscape and to depict regional variation and local character related to this. Like LCA, the focus of HLA is on the wider environment, not on special places; its purpose is to inform and manage the process of landscape change, not to halt it. Although it will initially emphasise the national view, it will show regional variations and can be developed further at a local level. It can also be linked to other site-based data for an even more comprehensive picture. It will help priorities to be set for all policies that impact on the historic environment at a landscape-scale rather than from a site-based focus.

For the future, the immediate aim is to complete national coverage of HLA, to make the data widely available and to support further practical applications. We aim to provide summaries of the main historical patterns and develop guidance on managing change, and ultimately to link HLA and other datasets (archaeological and historical, and other types of landscape data). There is huge potential for forming a more detailed picture of the historic environment and facilitating a more holistic approach to its conservation and management. Along with initiatives on the urban side and more detailed project work, HLA is helping to provide the coherent framework for the long-term sustainable management of the historic environment, in the context of wider land use planning and integrated land management that is envisaged in *Passed to the Future*. This will help to ensure that our future landscape reflects its human influence and its rich time-depth well into the future.

Acknowledgements

I would like to thank several colleagues for their helpful comments on this paper, particularly Dr. D.J. Breeze, Dr. P. Dixon, Dr. N. Fojut, Ms. S. Govan and Mr. R. Turner.

References

Boyle, S. (2003). Ben Lawers: an improvement-period landscape on Lochtayside, Perthshire. In *Medieval or Later Rural Settlement in Scotland: 10 Years On,* ed. by S. Govan. Historic Scotland, Edinburgh. pp. 17-29.

Coles, B. (2001). Britain and Ireland. In *The Heritage Management of Wetlands in Europe,* ed. by B. Coles & A. Olivier. Europae Archaeologiae Consilium Occasional Paper no. 1. Europae Archaeologiae Consilium, Brussels. pp. 23-43.

CAG Consultants & Land Use Consultants (2001). *Quality of Life Capital: Managing Environmental, Social and Economic Benefits.* Countryside Agency, English Heritage, English Nature and Environment Agency, Cheltenham.

David Tyldesley & Associates (2001). Landscape Studies of the Heart of Neolithic Orkney World Heritage Site. Unpublished report. Scottish Natural Heritage, Perth and Historic Scotland, Edinburgh.

Davidson, D.A. & Carter, S.P. (1997). Soils and their evolution. In *Scotland: Environment and Archaeology, 8000BC-AD1000,* ed. by K.J. Edwards & I.B. Ralston. John Wiley, Chichester. pp. 45-62.

Déjeant-Pons, M. (2002). The European Landscape Convention, Florence. In *Europe's Cultural Landscape: Archaeologists and the Management of Change,* ed. by G. Fairclough & S. Rippon. Europae Archaeologiae Consilium Occasional Paper 2. Europae Archaeologiae Consilium, Brussels. pp. 13-24.

Dixon, P. (2003). Champagne Country: a review of medieval settlement in Lowland Scotland. In *Medieval or Later Rural Settlement in Scotland: 10 Years On,* ed. by S. Govan. Historic Scotland, Edinburgh. pp. 53-64.

Dixon, P. & Hingley, R. (2002). Historic Land-use assessment in Scotland. In *Europe's Cultural Landscape: Archaeologists and the Management of Change,* ed. by G. Fairclough & S. Rippon. Europae Archaeologiae Consilium Occasional Paper 2. Europae Archaeologiae Consilium, Brussels. pp. 85-88.

Dyson Bruce, L., Dixon, P., Hingley, R. & Stevenson, J. (1999). *Historic Landuse Assessment (HLA): Development and Potential of a Technique for Assessing Historic Landuse Patterns. Report of the Pilot Project 1996-1998.* Historic Scotland & RCAHMS, Edinburgh.

Edwards, K.J. (1993). Human impact on the prehistoric environment. In *Scotland Since Prehistory*, ed. by T.C. Smout. Scottish Cultural Press, Aberdeen. pp. 17-27.

Edwards, K.J. & Whittington, G. (1997). Vegetation change. In *Scotland: Environment and Archaeology, 8000BC-AD1000,* ed. by K.J. Edwards & I.B. Ralston. John Wiley, Chichester. pp. 63-82.

Fairclough, G. & Macinnes, L. (2003). *Landscape Character Assessment Guidance for England and Scotland. Topic Paper 5: Understanding Historic Landscape Character.* Countryside Agency, English Heritage, Historic Scotland and Scottish Natural Heritage, Cheltenham and Perth. Available from www.ccnetwork.org.uk/ca/LCA_Topic_Paper_5.pdf.

Glendinning, M. & Mackechnie, A. (2004). *Scottish Architecture.* Thames and Hudson, London.

Hanson, W.S. & Macinnes, L. (1991). The archaeology of the Scottish Lowlands: problems and potential. In *Scottish Archaeology: New Perceptions*, ed. by W.S. Hanson & E.A. Slater. Aberdeen University Press, Aberdeen. pp. 153-66.

Historic Scotland (2002). *Passed to the Future: Historic Scotland's Policy for the Sustainable Management of the Historic Environment.* Historic Scotland, Edinburgh.

Land Use Consultants (2003). Application of the Historic Land Use Assessment. Unpublished report. Historic Scotland and RCAHMS, Edinburgh.

Lelong, O. (2003). Finding medieval (or later) settlement in the Highlands and Islands: the case for optimism. In *Medieval or Later Rural Settlement in Scotland: 10 Years On*, ed. by S. Govan. Historic Scotland, Edinburgh. pp. 7-16.

Macinnes, L. (1991). Preserving the past for the future. In *Scottish Archaeology: New Perceptions*, ed. by W.S. Hanson & E.A. Slater. Aberdeen University Press, Aberdeen. pp. 196-217.

Macinnes, L. (1993). Archaeology as land use. In *Archaeological Resource Management in the UK: An Introduction*, ed. by J. Hunter & I. Ralston. Alan Sutton, Stroud. pp. 243-255.

Macinnes. L. (2004). Historic Landscape Characterization. In *Countryside Planning: New Approaches to Management and Conservation*, ed. by K. Bishop & A. Phillips. Earthscan, London. pp.155-169.

RCAHMS (1990). *North-East Perth: an Archaeological Landscape*. HMSO, Edinburgh.

RCAHMS (1994). *South-East Perth: an Archaeological Landscape*. HMSO, Edinburgh.

RCAHMS & Historic Scotland (2000). *The Historic Landscape of Loch Lomond and the Trossachs*. RCAHMS and Historic Scotland, Edinburgh.

RCAHMS & Historic Scotland (2001). *The Historic Landscape of the Cairngorms*. RCAHMS and Historic Scotland, Edinburgh.

RCAHMS & Historic Scotland (2002a). *But the Walls Remained. A Survey of Unroofed Settlement Depicted on the First Edition of the Ordnance Survey 6-Inch Map of Scotland*. RCAHMS and Historic Scotland, Edinburgh.

RCAHMS & Historic Scotland (2002b). *The Historic Land-use Assessment of the Solway Coast National Scenic Areas*. RCAHMS and Historic Scotland, Edinburgh.

RCAHMS & Historic Scotland (2003). *The Historic Land-use Assessment of the Wester Ross National Scenic Area*. RCAHMS and Historic Scotland, Edinburgh.

Smout, T.C. (1993). Woodland history before 1850. In *Scotland Since Prehistory*, ed. by T.C. Smout. Scottish Cultural Press, Aberdeen. pp. 40-49.

Stevenson, J.B. & Dyson Bruce, L. (2002). The Historic Landuse Assessment project and other work. In *Understanding the Historical Landscape in its Environmental Setting*, ed. by T.C. Smout. Scottish Cultural Press, Aberdeen, pp. 51-59.

Swanwick, C. & Land Use Consultants (2002). *Landscape Character Assessment: Guidance for England and Scotland*. Countryside Agency, Cheltenham and Scottish Natural Heritage, Perth.

Usher, M.B. (Ed). (1999) *Landscape Character: Perspectives on Management and Change*. The Stationery Office, Edinburgh.

14 Scotland's future lowland landscapes: encouraging a wider debate

Simon Brooks

Summary

1. SNH issued a discussion paper entitled *Scotland's Future Landscape? Encouraging a Wider Debate* in November 2003. Its aim was to inform and stimulate more discussion about why we need better care for this important national asset and how this might be achieved. This chapter presents an overview of the discussion paper's content, focusing on lowland agricultural and wooded landscapes.
2. The main forces for landscape change are described, and their broad consequences highlighted. Some of the challenges facing our farmed and forested landscapes are reviewed. SNH's broad aim for the care of Scotland's landscape and five factors to be considered in working towards this are set out.

14.1 Introduction

Scotland is renowned for the quality of its distinct and diverse range of landscapes. Many are appreciated and valued as a significant part of the country's natural and cultural heritage, their varied character contributing to a sense of identity, both nationally and regionally (Scottish Office Development Department, 1999). Their contribution to the social and economic well-being of the nation is also increasingly recognised (for example see Scottish Executive, 2004). The landscape provides the setting for where we live, work and choose to recreate, and through this enhances people's quality of lives. The renown of Scotland's landscape is a mainstay of our tourism industry; its image is used to market a range of products, and attractive surroundings can attract investment, resulting in very real economic benefits.

But while Scotland may be famed for the quality of its landscapes, not all are in a condition that we can be proud of. A number have suffered from past exploitation and neglect, some from gradual attrition of valued elements, and others can be at risk if change is not undertaken sensitively. This loss of condition has occurred in spite of a range of mechanisms being put in place to regulate the impact of certain activities on the landscape. There is a history of concern and care for the landscape, led initially by amenity bodies and later in association with statutory agencies such as Scottish Natural Heritage (SNH). If we are to achieve the Scottish Executive's aim "to build environmental capital and pass well-managed, high quality landscapes on to future generations" (Scottish Executive, 2004, para. 94), renewed effort is now required.

That not all landscapes have received appropriate care is of concern to SNH, whose statutory remit for the conservation, enhancement, understanding, enjoyment and sustainable use of the natural heritage includes its 'natural beauty and amenity'. This rather old fashioned term encompasses the landscape's physical fabric and people's appreciation of its beauty and character – what people see, experience and enjoy as they react to their surroundings. This includes all landscapes, regardless of whether they are considered ordinary, degraded or special.

SNHs concern for Scotland's landscape led to a discussion paper being produced which highlighted the challenges facing the landscape (Scottish Natural Heritage, 2003). This chapter outlines why SNH did this, and examines the issues facing our farmed and wooded landscapes that were the focus of the conference. The discussion paper was, however, concerned with the totality of Scotland's landscapes, including the more developed and urban, as well as those degraded by industrial uses.

14.2 Why a discussion paper?

SNH published the discussion paper *Scotland's Future Landscapes? Encouraging a Wider Debate* in November 2003 (Scottish Natural Heritage, 2003), and invited comments by February 2004. The paper set out SNHs thinking on the prospects facing Scotland's landscapes in an effort to stimulate and inform a debate about how we can better care for this important national asset. Through a series of questions, it explored the importance and values that people place on the landscape, discussed the major forces for change over the past 50 years and their consequences, and considered how some of the issues identified might be addressed.

Apart from the designed landscapes consciously laid out to exemplify some aesthetic preference (for example, the setting of a large country house and surrounding estate), the pleasure that many landscapes provide has not come about deliberately. Instead, the appearance of our lowland rural landscapes is largely the incidental consequence of actions undertaken for food and timber production. The decisions of farmers and foresters in their efforts to produce these goods will continue to be the main influence on the character of these landscapes. But the wider public benefit that we have come to expect these landscapes to provide is increasingly seen as relevant to the making of these decisions, illustrated by the ongoing reforms to agricultural and forestry support mechanisms.

Although some may argue that a concern with wider public goods has always been the case, there has at the very least been a lack of clarity about society's intentions for these landscapes. Consequently they have not received the attention and care they deserve, and our surroundings can all too easily be taken for granted. This is no doubt in part a consequence of Scotland retaining many landscapes that are considered of high quality with few impairments, alongside an acceptance that landscapes evolve and need to adapt to changing needs. But the past half century has seen some marked changes, not all of which we might choose to repeat. Some change is sudden, while some is gradual and piecemeal but with cumulative consequences for the appearance of our landscapes.

The discussion paper provided some indicative data on land cover change (Table 14.1) to illustrate some of the ways in which our landscapes are evolving. Four key forces driving this are identified: the transformation of our economy (with the decline of many primary and manufacturing industries); social advances demanding new housing and infrastructure; technological innovation in the land management and development sectors; and substantial public financial support for rural land management. These developments have brought many social and economic benefits to the nation, but we have also seen some detrimental effects to the landscape. These include:

- a loss of landscape diversity and dilution of distinctive character;
- generally poor standards of design;
- the degradation or loss of natural features, a critical component of many landscapes;
- the attrition of remote and wild countryside through the introduction of prominent man-made elements, such as vehicular tracks;
- a reduction in the active management of landscape features such as dykes, hedges and hedgerow trees, and a deterioration in their condition; and
- an erosion of the rural character of lowland Scotland, not least through the spread of noise and light pollution.

Table 14.1 Change in Scotland's landscapes – some illustrative statistics. Sources: data from Mackey *et al.* (2001), except for * Scottish Executive (2000).

Agriculture

- **1940s-1980s** the length of hedgerow is estimated to have decreased by half – a mean annual rate of loss of 560 km.
- **1987-1999** total area of mixed farms (crops and livestock) declined by 23%.

Woodland

- **1924-2000** the area of woodland cover trebled, from 5.6% to 17.2% of Scotland.
- **1924-1980** the area of broadleaved woodland declined by half, while the area of conifer plantation quadrupled.
- **1985-2015** timber production doubled in the past 15 years and is forecast to double again over the next 15 years*.

Natural and semi-natural land cover

- **1940s-1980s** the area of semi-natural habitat decreased by 17% – a mean annual rate of loss of 230 km^2.
- **1940s-1980s** the area of heather moorland decreased by a quarter, and peatland by a tenth.
- **1950-1990** sheep numbers across much of the uplands increased by a third.

Industry and Built Development

- **1940s-1980s** the area of land covered by built development and transport corridors increased by 36% – a mean annual rate of 14 km^2.
- **1993-1999** the area of vacant and derelict land decreased by a fifth.

Are such changes always of concern? Recognising this process of landscape evolution is not to suggest we should seek to fossilise the countryside we see today, as change is intrinsic to our landscape. But the increasing extent and pace of change – which may accelerate further with reform of the Common Agriculture Policy – points to the need for a more pro-active approach to landscape planning and management if we are to retain the character of those landscapes we value, and enhance those less valued. This requires a greater understanding of the values people attach (or might attach in the future) to the landscape, and some consensus on what we are seeking to achieve. The discussion paper aimed to prompt this, by identifying the challenges faced by five broad landscape settings. Those presented by our lowland agricultural and wooded landscapes are discussed here.

14.3 Lowland agricultural landscapes

The diversity of farmed landscapes to be found across Scotland is a reflection of the land's suitability for different agricultural systems and their history of enclosure and improvement. The result is a suite of managed landscapes of distinctive character that impart a sense of regional identity which is valued by society (University of Aberdeen Department of Agriculture and Forestry & Macaulay Land Use Research Institute, 2001). Important contributors to their character are the type and availability of local building stone that is revealed in buildings and dykes and the style of traditional design. In many, the presence of designed landscapes, laid out for the pleasure and satisfaction of past owners, continues to make a significant contribution to the wider scene.

But while the agricultural landscape's broad structure remains intact, there has been a discernable loss of some landscape elements. Intensification has reduced landscape diversity as agricultural systems have been simplified, and field boundaries and hedgerows removed. Economic pressures and mechanisation have reduced the management of woodlands and investment in traditional farm buildings. Decoupling of financial support may result in agricultural restructuring in the longer term, with uncertain results for the landscape (FPD Savills, 2001). Also, although designed landscapes remain clearly discernible, many have long been in decline as the social and economic forces that created and maintained them have changed.

This suggests substantial challenges for the farmed landscape. The Scottish Executive's Agriculture and Environment Working Group recognised the risk of continuing landscape change in its report *Custodians of Change* (Scottish Executive, 2002), and identified it as one of the three priorities to be tackled in the coming decade. The report proposes the preparation of guidance to facilitate local agricultural landscape design. But what should this guidance contain, and how can it be delivered, given the harsh economics of much farming? How do we agree the objectives? While we might expect such guidance to seek to strengthen the distinctiveness of existing agricultural landscapes, are there also opportunities to shape new landscapes where intensification has already removed many of the critical features? The practice of this can learn from similar initiatives, such as ongoing efforts to establish a Central Scotland Forest in the former industrial heartland between Edinburgh and Glasgow (Central Scotland Forest Trust, 2002).

Or, in the case of designed landscapes that are recognised as of particular natural and cultural value, and make such a distinctive contribution to our lowland landscapes, are we looking at reinstating key features? These are landscapes that often remain popular for informal recreation, and hold important habitats. The predecessors to Historic Scotland and SNH jointly commissioned an inventory of designed landscapes in the 1980s, which is currently being extended to capture a more complete picture of the range of designed landscapes. What now seems apparent is that targeted action is necessary if the long-term decline of many of these landscapes is to be halted.

There is also a need to recognise that the achievement of landscape objectives is best tackled at a 'landscape scale'. While farmers will continue to tackle issues on their own land, action is required beyond the level of individual land holdings if significant gains are to be made. Reform of the Rural Stewardship Scheme and its encouragement of collaborative applications by multiple land holdings recognises this (Scottish Executive, 2003). Action at the broader scale also enables the connections between landscape objectives and those for nature conservation or recreation to be better made, such as the creation of core path networks and habitat networks.

14.4 Wooded landscapes

Trees are an important component of many scenes, including our farmed landscapes where hedgerow trees and woodlands can be a defining characteristic. But in the south-west, parts of the Borders, in Argyll and the eastern Highlands, extensive conifer forests have become the defining characteristic. While previous priorities set for forestry did not always give adequate consideration to landscape concerns, we now have new direction provided by *The Scottish Forestry Strategy* (Scottish Executive, 2000) which aims to secure the more sensitive approach developed over the past decade or so. With half of Scotland's forest due to be harvested over the next 25 years, and aspirations to increase woodland cover from 17% to 25% by 2050, now is an important time to consider the opportunities this renewed approach presents.

Although there has already been much effort, we need to recognise that a legacy of past practice remains as a consequence of the length of the forestry cycle. Forest plans and design plans are tackling this, but some inappropriate planting of the past may merit complete removal, particularly where the planting impacts on landscapes recognised as being of particular value. Efforts to create a more diverse forest structure provide opportunities for greater landscape diversity with open ground which can also support biodiversity and recreational objectives.

All of this is recognised by the approach to multi-objective forestry that is now pursued. But if we are to maximise the public benefit that is now sought from forestry, are we clear what and where our priorities for the landscape lie? Should future woodland expansion be more strongly planned and guided? The Scottish Forestry Grants Scheme (Forestry Commission, 2003) has taken a step towards this with the introduction of locational premiums to encourage new planting in five areas, but it remains to be seen whether this needs to be more strongly linked to some spatial planning framework.

14.5 SNHs broad policy approach

This brief discussion of farmed and wooded landscapes highlights the significant challenges facing all of us if we are to achieve better care for Scotland's landscapes. In addressing these challenges, SNHs broad aim will be to safeguard and enhance the distinct identity, the diverse character and special qualities of Scotland's landscapes as a whole. This diversity is illustrated by the results of SNHs national Landscape Character Assessment programme, whose 30 regional studies have identified and described 275 distinct landscape character types for the whole of Scotland, perhaps a third of which can be considered lowland farm and forestry types (Julie Martin Associates & Swanwick, 2004).

But what does SNHs broad aim mean in practice for individual landscapes, and how can we best seek to agree clear landscape objectives at the local level? In the discussion paper, SNH proposes that its approach will be guided by a number of considerations.

First, we need to recognise the reality of change in our landscapes. This is a measure of the dynamic in society and the need to secure social and economic gains. Although we believe most change should seek to fit with and enhance a landscape's existing character, particularly where valued, this is not always appropriate or indeed feasible. New land uses can transform landscapes and as the key characteristics of some are lost new landscapes will emerge. This change requires guiding, through a more proactive approach to landscape planning and management, in order to ensure future landscapes are valued. Local forestry frameworks are already tackling this, and the guidance proposed by *Custodians of Change* could do this, based on a landscape character approach.

Second, we are not just concerned with the special and celebrated landscapes – although these warrant special effort to ensure their qualities are safeguarded. All landscapes deserve care, as the common and familiar are of great value to those who live and work in them. Indeed our lowland and more managed landscapes have perhaps not been given adequate recognition to date by Scotland's national landscape designation – the National Scenic Area (NSA). SNH recognised the emphasis on mountain and wilder landscapes in its advice to Government on NSAs (Scottish Natural Heritage, 1999) and recommended re-examining whether the 40 NSAs adequately cover all of Scotland's best landscapes. Extending the suite to include more managed landscapes merits some attention – perhaps the farmlands of Angus, Buchan and East Lothian might be considered as deserving of this accolade?

Third, while many of Scotland's landscapes are considered to be of high quality with little impairment, SNH believes that a marked improvement in our care for them is required if we are not to lose what we value through continued attrition. Higher standards of design and management are required, along with a more considered approach that takes full and explicit account of the landscape interest.

Fourth, SNH is mindful that Scotland's landscapes are a shared responsibility. Many different stakeholders have an interest in, and influence on, the landscape. Indeed much lies outwith the influence of SNH. If the nation is to effect change, then collective action is required by the many players whose decisions and actions affect this resource. Key to this are the landowners and managers controlling the land, though their decisions are influenced by the priorities and incentives set by Government, and increasingly by world market forces. These are significant forces, but through collective action we believe change can happen. If we are to recognise the public benefits then decisions about our objectives need to be informed by the views of the public – this is not a matter that can be left solely to professionals. An agreed agenda is required and the discussion paper has gone someway to stimulating a debate about what this might be.

Finally, landscape considerations that were the focus of SNHs discussion paper need to be integrated with SNHs other interests, and the objectives of others. Although recreation, biodiversity, earth science and landscape issues are often approached separately by a variety of specialists, we have to consider the strong connections, if not synergy, to be made between these themes. Our surroundings provide the setting if not the main attraction for where we seek recreation, while natural and semi-natural features of conservation interest are an important and sometimes key component of many landscapes. Likewise, the character of many landscapes depends on their continued management. An integrated approach, therefore, is not only required but is inevitable if we are to achieve our objectives.

14.6 Conclusion

If we are to achieve greater care of this important resource, SNH believes that the starting point is a requirement for more debate about how we wish to see these landscapes evolve. The intention was that the discussion paper would stimulate this, but it was only the first step in what needs to be a longer process. SNH will need to continue to provide leadership here, by raising awareness and appreciation of Scotland's landscapes, advising others on their care, promoting good practice, exploring new approaches and encouraging debate. But this needs to be grounded in a clear understanding of what people value about their landscapes, and SNH will continue to listen to the views expressed in the debate about Scotland's future landscapes.

Editor's Note

Following this conference, over 220 responses to SNHs discussion paper were received. Many expressed strong support for the general thrust of the discussion paper, supported the need for more debate about Scotland's landscapes and endorsed efforts to secure greater care. How this could be taken forward was discussed at a seminar in September 2004 (Scottish Natural Heritage, 2004), at which consideration was given to the establishment of a National Landscape Forum. Progress with SNHs landscape work can be followed on the SNH website (www.snh.org.uk).

References

Central Scotland Forest Trust (2002). *Central Scotland Forest Strategy*. Central Scotland Forest Trust, Shotts.

Forestry Commission (2003). *Scottish Forestry Grants Scheme. Applicants Booklet*. Forestry Commission, Edinburgh.

FPD Savills (2001). *Structural change in agriculture and the implications for the countryside*. Paper prepared for the Land Use Policy Group. Executive summary available from www.lupg.org.uk/uploaded_photos/pubs_Structural_Change_Executive_Summary.pdf.

Julie Martin Associates & Swanwick, C. (2004). Overview of Scotland's national programme of landscape character assessment. *Scottish Natural Heritage Commissioned Report No. 029* (ROAME No. F03AA307). Scottish Natural Heritage, Perth.

Mackey, E.C., Shaw, P., Holbrook, J., Shewry, M.C., Saunders, G., Hall, J. & Ellis, N.E. (2001). *Natural Heritage Trends. Scotland 2001*. Scottish Natural Heritage, Perth.

Scottish Executive (2000). *Forests for Scotland: The Scottish Forestry Strategy*. Scottish Executive, Edinburgh.

Scottish Executive (2002). *Custodians of Change*. Report of the Agriculture and Environment Working Group, Scottish Executive, Edinburgh.

Scottish Executive (2003). *Proposals for Changes to Agri-Environment Schemes in Scotland*. Scottish Executive, Edinburgh.

Scottish Executive (2004). *National Planning Framework for Scotland*. Scottish Executive, Edinburgh.Scottish Natural Heritage (1999). *National Scenic Areas. Scottish Natural Heritage's advice to Government*. Scottish Natural Heritage, Perth.

Scottish Natural Heritage (2003). *Scotland's Future Landscapes? Encouraging a Wider Debate*. Scottish Natural Heritage, Perth.

Scottish Natural Heritage (2004). Scotland's Future Landscapes? Moving the Debate Forward. Unpublished report of seminar, Dewars Centre, Perth, 16 September 2004. Scottish Natural Heritage, Perth. Available from www.snh.org.uk/pdfs/strategy/advisingothers/seminarreport.pdf.°

Scottish Office Development Department (1999). *National Planning Policy Guideline NPPG 14. Natural Heritage.* Scottish Office Development Department, Edinburgh.

University of Aberdeen Department of Agriculture and Forestry & Macaulay Land Use Research Institute (2001). *Agriculture's Contribution to Scottish Society, Economy and Environment.* University of Aberdeen, Aberdeen.

15 Towards integrated management: future landscape research and advisory needs

Alison Farmer

Summary

1. Integrated land management, where land is managed for multiple objectives, is an increasing aspiration.
2. Key concerns raised by farmers about current management advice and environmental information include: the poor quality of data available; contradictory advice and priorities from different organisations and advisors; lack of understanding of the economic implications of new initiatives and strategies; and a sense of too much paperwork and not enough individual choice.
3. Advice to farmers and foresters from landscape professionals and policy makers, along with current support systems, needs to be adjusted in order to better support the delivery of an integrated approach.
4. Current landscape projects present some useful ideas on improving advice to the land management community to deliver integrated management and suggest areas for future landscape research.

15.1 Introduction

Without appropriate advice and technical and financial support for farmers, important progress will be compromised and impeded when it comes to the task of delivering various forms of environmental change and improvement, and the realisation of integrated land management. This chapter explores how we can adjust existing advice and available information in order to embrace and deliver this widening concept of land management. It considers the changes which are influencing our approach and attitudes to land management; the concerns raised by farmers about current advice and support; and, through the use of case studies, current work which supports integrated management and informs the direction of future research. This chapter primarily refers to farmers, although the issues addressed relate to others involved in land management, such as foresters and sporting estate owners.

15.2 The context for change

The economic, social and political contexts in which land management has to operate are now undergoing significant changes. For example:

- we live in a society where technological changes are offering, but also demanding, new ways of communicating and problem solving;

- expectations about the 'role' of landscape are no longer seen primarily in terms of production, but increasingly as a resource that should deliver a wider range of benefits or outputs for society as a whole;
- climate change will affect the landscape, but understanding the implications of this at the local level is difficult;
- there are increased expectations for community involvement in decision-making, in order to improve links between land managers and consumers; and
- the mid-term review of the Common Agricultural Policy (CAP) has implications for farm economics and the environment as a result of de-coupling (breaking the link between farm subsidies and production) and cross-compliance (receipt of support payments conditional on maintaining land in good agricultural and environmental condition) (Edwards, 2004).

These changes or trends are influencing our policies, research needs and aspirations. But they also provide a stimulus to review the way in which we offer advice and provide information to farmers, if we are to encourage, support and achieve integrated land management. With so many interrelated demands on the land it is no longer appropriate for advice to come from a number of uncoordinated single-issue sources. The opportunity for conflict and confusion grows with increasing competing demands. The need for disseminating comprehensive, up-to-date and properly weighted advice increases with it.

15.3 What do farmers need?

Custodians of Change (Scottish Executive, 2002) highlighted a number of issues relating to the advice given to land managers in Scotland. It stated that progress in addressing environmental issues was impeded by:

- an advisory system which is not fit-for-purpose;
- poorly integrated and assimilated advice and training; and
- fragmented, and often contradictory, policies and funding schemes from government and its agencies.

These issues are not unique to Scotland. For example, research in the south of England (Dreweatt Neate & Giles Wheeler-Bennett, 2002) examined the needs and attitudes of farmers in the context of local land management plans. This revealed four key areas of frustration for farmers:

- relevant environmental information can be out of date and widely dispersed, instead of being available from just a few sources;
- specialist advice is sometimes offered without proper consideration of other associated topic areas (for example, landscape guidelines may provide advice based primarily on visual character with only minor reference to biodiversity or cultural heritage);
- advice may not be in the land manager's economic interest and so there may be only limited financial incentive; and
- targets and guidelines can be too prescriptive, so that land managers have little room to make their own choices and local decisions.

History tells us that farmers are adept at rising to new challenges and societal needs, such as the provision of more affordable food in the post-war period, and in places farmers already practice multifunctional land management. However, in order to encourage this more comprehensively, there would seem to be two main requirements for the land manager. First, practical advice and information needs to be integrated and readily accessible. Second, the approach must be economically viable. Both of these are essential, and it is the combination of these two requirements that presents the challenge – the need for better, more informed, integrated decision-making that is also economically viable for land managers to implement.

There are initiatives from around the world, such as land management contracts in France and integrated management plans in Western Australia, from which we can learn. But there are also initiatives within the UK, four of which have been selected as case studies below, which illustrate movement towards better integration, and highlight key issues that are directly relevant to the UK.

15.4 Strategic Management Plans

Management plans for protected landscapes such as Areas of Outstanding Natural Beauty (AONBs) or National Scenic Areas (NSAs) offer some interesting lessons (Bennett, this volume). Guidance on the preparation of management plans for AONBs (Countryside Agency, 2001) explains how information about the special qualities, and any key issues, within the area should be gathered and assessed. Management plans should set out a series of aims and objectives, with the intent of delivering an integrated approach to land management. To this end, they bring together all relevant organisations and stakeholders (such as farmers, local businesses, special interest groups and government organisations), collectively set out policies relevant to the particular area, and highlight where policies for different organisations may conflict with the purposes of AONB management.

Protected area management plans, therefore, are of direct relevance to the new integrated land management agenda. They provide farmers with a useful source of baseline data about the local environment, and about integrated and prioritised policy objectives, and involve an element of public consultation. Recently prepared management plans in Scotland (Bennett, this volume) and elsewhere in the UK, such as the following case study, have demonstrated that the process helps to explore the public's expectations about local land management issues and helps to reconcile conflicting aspirations through improved knowledge and transparent prioritisation of action and resources.

15.4.1 Case Study One: Causeway Coast AONB Management Plan, Northern Ireland

The Causeway Coast AONB management plan (Environment & Heritage Service, 2003), was the first management plan for an AONB in Northern Ireland. Its preparation involved consultation with the local community, including farmers.

The whole of the AONB is covered by an Environmentally Sensitive Area (ESA) scheme that has had 100% take-up within the area. Works that have been done as part of this scheme include reinstating and managing existing landscape features, such as hedgerows, grassland habitats and traditional gateposts, with significant and positive results.

Prior to the management plan there was no detailed landscape character assessment for the AONB. Similarly, there was a lack of information on the landscape resource, including habitats outside existing designations, vernacular architecture, archaeological sites and historic landscape features such as field patterns and tracks. This meant that existing farm activity, and support through the ESA, was being undertaken without a thorough understanding of the resource and potential issues.

The preparation of the AONB management plan emphasised the following.

- Information about the landscape resource needs to be assimilated and made accessible, and used as a database for land managers and other advisors, in order to inform land management decisions.
- The need to address existing management conflicts and to provide the right financial support. For example, gaps in financial incentives for conservation work to vernacular buildings were revealed, and positive incentives for new rural buildings were found to exacerbate the loss of vernacular farm buildings. Similarly, due to tourism pressures, there was a demand for increased public access to farmland, but reluctance by landowners to permit this due to fears of liability and associated financial burdens.
- The importance of single-purpose schemes to bring about collaborative management between adjacent landowners. One particular scheme was the 'Return of the Chough' initiative which operated as a special project under the ESA with the support of the Royal Society for the Protection of Birds. It led to collaboration between adjacent landowners to create a sufficient area of appropriate habitat.
- The scope to re-establish closer links between the farmers and the general public, and the value of the process of preparing a management plan for this purpose.

Protected area management plans are just one type of strategic management plan. Arguably, there is no particular need to confine the preparation and use of strategic management plans to specific areas. They instead could have a wider relevance to landscapes, and perhaps be developed on a regional basis. There are early signs of the use of strategic land management plans to help focus minds on integrated management outside protected areas, where there is need and opportunity for this to be the case, such as in the development of urban fringe action plans (Entec UK Limited, 2003). This enables a more detailed understanding of landscape, so that funds can be targeted more successfully, based on more integrated policies and data. It is still early days, but this approach has the potential to be useful in interpreting or re-interpreting national and regional policy down to a more local level, and will offer greater clarity of understanding for farmers and for the preparation of more locally based initiatives, such as local farm/area plans.

Strategic management plans are, ultimately, just like any other management tool: they are only as 'good' as the people involved in their preparation and use. This emphasises the need to work with the grain of land managers when seeking to improve on the advice and support available. To do this, there is a need, amongst other things, for better communication, information exchange, education, innovation, commitment and partnership.

15.5 Landscape management guidelines and biodiversity

Landscape management guidelines are useful in disseminating advice on landscape management and are derived from the landscape character assessment (LCA) process (Countryside Agency & Scottish Natural Heritage, 2002).

Recent research into habitat networks known as 'Lifescapes' (Porter & Preston, 2001) and Biodiversity Habitat Networks (Forestry Commission Scotland & Scottish Natural Heritage, 2003) help us to understand ecology at a landscape scale. They present an opportunity, therefore, to incorporate this information into the LCA process and assist the development of management guidelines. Biodiversity habitat networks interpret the needs of species and habitats to survive and increase, in number and area, across a landscape, beyond existing ecological designations. The assessment therefore enables 'optimum' habitat patterns at a landscape scale to be determined.

Implicit in this is the creation, or re-creation, of habitats and therefore a potential change to the current landscape character of an area. The lack of existing examples of landscape assessments, which are integrated with an understanding and aspiration to create sustainable habitat networks, means that potential conflicts are not currently clearly understood. Kent County Council and English Nature are developing a methodology for addressing these issues as part of the Lifescape project (ADAS, 2003).

15.5.1 Case Study Two: The Kent Lifescape Programme

This study has used information on landscape character as the framework for establishing biodiversity habitat networks. It has involved the development of habitat capability maps by overlaying digital ecological information and developing a set of ecological rules for creating sustainable habitats. Information on cultural heritage, access and landscape character was used to identify constraints to habitat creation. This approach, therefore, considers not just optimum habitat creation but also tempers this with constraints and opportunities presented by cultural heritage, access and landscape character. It integrates different data and aspirations for the landscape and assists in the targeting of resources to achieve Biodiversity Action Plan (BAP) targets.

Kent County Council recognises that there is still further work to be done to refine the 'judgemental' aspect of this technique which relates to establishing appropriate ecological rules to maximise habitats and understanding constraints. Whatever the eventual assessment methods, there is a significant opportunity to improve the integration of biodiversity habitat networks into landscape assessment, which could result in the refinement and increasing sophistication of resulting landscape management guidelines with clearer, integrated priorities of benefit to farmers.

Land management guidelines derived from such studies should be able to provide more integrated advice at both regional and local levels. However, they are again unlikely to resolve all management and practical issues at a local farm level. There will continue to be a need for specialist one-to-one advice provided by organisations such as the Farming and Wildlife Advisory Group (FWAG) on such matters. There will also be a need for the right financial incentives and an understanding of the implications of advice on farm economics. This is explored more in the next case study.

15.6 Farm economics

Whether advice is from policy makers, landscape professionals or farming advisory groups, the advice given to farmers needs to be sensitive to the economic realities of running a farm or estate business. Similarly, integrated management ultimately requires financial incentives to be tailored in ways that achieve landscape management objectives. The following case study explores the inter-relationship of farm economics with the delivery of environmental improvements.

15.6.1 Case Study Three: East Hampshire Farm Research

This research, undertaken on behalf of Hampshire County Council (Landscape Design Associates, 1998), looked at a 1,000 acre (400 ha) mixed farm within the East Hampshire AONB. The research examined the links between international trends in agriculture, agricultural economics and the landscape. A number of different environmental and financial scenarios were developed, based on costs and benefits of existing, as well as alternative, land management approaches. The three scenarios were:

1. land farmed for maximum economic return;
2. reasonable care taken to farm with regard to the environment; and
3. a conservation farming-orientated system, balancing economic return with enhancing high landscape and biodiversity interest.

In each scenario, consideration was given to the impacts on landscape character, biodiversity and recreational use, and existing integrated management guidelines were tested against the profitability of the farm business unit. The results revealed that the approach in Scenario 3 reduced farm profitability by almost a half, and that relevant financial incentives available at the time were not adequate. This study also showed that recommendations and advice on landscape management would be more useful and realistic if they were informed by a real understanding of agricultural economics.

Current evolving changes in the funding available for environmental management of the landscape, such as the Common Agricultural Policy mid-term review (Edwards, 2004) make this case study an interesting area for further research. Further work could, for example, test to see if multiple objectives for the landscape are now more financially achievable than previously. This may also have implications for the development of 'economic modelling packages' for practitioners and land managers and the development of demonstration farms, and Land Management Contracts that it is hoped will be implemented in Scotland as part of the delivery of CAP reform (Newcombe, this volume).

All three case studies reveal that the preparation of plans and strategies, modelling packages and land management contracts can help to provide vision and structured advice, but they are dependent on commitment at grass-roots level. Also, Section 15.3 in this chapter highlighted farmers' concerns regarding increased bureaucracy and the loss of individual choice. There is a need, therefore, for land managers to be involved in developing solutions and exercising entrepreneurialism when it comes to delivery.

In response to these issues a number of schemes, which seek to harness the initiative of farmers, have been developed and are explored below.

15.7 Self-help for farmers

Maintaining individual choice and gaining access to relevant information have been recognised by the creation of 'one stop shops' using technology to disseminate information. An example of this is Environmental Management for Agriculture (EMA), which has been developed by the Agriculture & Environment Research Unit in the Science and Technology Research Institute of the University of Hertfordshire. This is an award-winning computer software package that provides a comprehensive suite of tools, information and assessment routines designed to help the farming industry improve its environmental performance.

Another initiative, the Land Management Information System (LaMIS), looks at the use of computer technology to enhance the accessibility and flow of information. It seeks to provide an inherently flexible system to develop whole farm plans and to deliver integrated management.

15.7.1 Case Study Four: Land Management Information System (LaMIS)

This is a research project currently being run as a pilot study in Hampshire, but with the expectation of repeating it across the South Downs and the North York Moors. The basis of the work is to improve communication in order to achieve more integrated planning and actions. It came about as a result of the debate on the future of the English countryside after the Foot and Mouth crisis.

The system consists of three components: a website (www.lamis.gov.uk) enabling access to a wide range of environmental data sets; software tools to help in the preparation of whole farm plans and other types of land management planning; and a portal for information exchange between farmers and others. The significance of LaMIS is that it is designed with the land manager in mind, rather than for the convenience of administrators.

The system is still being developed and the pilot study is likely to highlight the need for further refinement. However, it is hoped that the system will deliver the following advantages.

- It should provide a single portal for information. This should satisfy demand from all sectors of the land management community, including the growing number of 'new' members of the farming community who may have less practical farming experience or access to advice. It should therefore provide a one-stop-shop of improved information presented in a user-friendly way.
- It should inform and support farmers to achieve integrated and tailor-made approaches to land management as it fosters individual choice and entrepreneurialism by providing a diverse range of data rather than uniform prescriptions. Where specialist advice from organisations such as FWAG may be required, e-mail and web page links are given, enabling ready access to specialist technical and professional support.
- It will provide a template for the development and delivery of whole farm plans and/or land management contracts. It could also provide other templates to integrate biodiversity action across farms, entry into agri-environment schemes and could also potentially provide economic modelling to assist in predicting the implications of environmentally-based initiatives on farm economics.
- It will optimise knowledge transfer within the land management community and wider advisory bodies and policy makers. This will be invaluable in encouraging not just farm-based initiatives but also area-based initiatives, which are likely to be more useful in achieving biodiversity targets. This also assists in the sharing of best practice examples, and has the potential to create a stronger link between farmers and the general public.

Significantly, LaMIS has the potential to provide raw data sets, as well as their interpretation in the form of integrated data, from sources such as strategic management plans, management guidelines based on visual character and habitat network objectives and local access strategies. This information can also be linked to local targets such as Local Biodiversity Action Plans (LBAPs). Informing and supporting farmers with this information should enable a logistical problem of farm management plan production to be overcome, and it should in turn provide more inherent flexibility and individual choice.

LaMIS is, therefore, an important initiative providing for co-operative action and knowledge transfer. However, there are also a number of issues that have been identified and are being addressed.

The data held on LaMIS need to be up-to-date. Ways of tackling this will be investigated, whilst accepting that no system is perfect. Farmers need to have access to computer technology. Research undertaken in Hampshire and Oxfordshire (Dreweatt Neate & Giles Wheeler-Bennett, 2002) has demonstrated that 59% of farmers currently have such access and that this proportion is increasing. In the interim, all information available on LaMIS will also be made available in paper form.

LaMIS is a tool and will not solve all land management conflicts and decisions. It should instead increase awareness of the issues and assets pertaining to a particular area of land, and should help to highlight where professional and specialist advice should be sought. One-to-one personal advice will still be required.

15.8 Future research

The above areas of landscape work and supporting case studies suggest a number of improvements, which are being made to meet the advisory needs of farmers and to encourage integrated land management. These include:

- the provision of quality advice across a wide range of expertise and easy access to a wide variety of baseline data;
- advice which is based upon, and reflects, land economics;
- guidelines which are integrated and embrace the visual, ecological and historical dimensions of the environment;
- the encouragement of collaborative management to implement change at a landscape scale through co-ordinating action to achieve biodiversity gains;
- assistance in the preparation of tailor-made management plans at the local level to reflect individual choice;
- advice that is simple and in a consistent format, and set against common goals and clear priorities; and
- the need for people involved on all sides to both listen and learn in order to work together better.

These areas can be tackled by the way in which policies and targets are developed, assessments undertaken and guidelines presented. But equally, they are addressed by the level of information easily accessible to the farmer and the desire to use this information to deliver change on the ground at the whole farm and area level. Of particular interest is the engagement of people in the process of improving advice, information and action on the ground. In a people-based industry, the need for improved communication and guidance as well as individual choice is essential to achieving our aspirations for integrated land management. A key area for future research, therefore, is how best to encourage participation.

Arguably, integrated land management is likely to be most successful through a combination of a top-down approach (integrated policy and guidance) and a bottom-up approach, where individual integration of information and decision-making occurs at the grass roots level. A tricky balance, therefore, is needed between clear priorities at a strategic level and flexibility in delivery on the ground. Future landscape research into the use of technology, the development of land management contracts and the preparation of area management plans that connect between strategic and local scales are other key areas to take forward in our quest for integrated land management which includes landscape.

References

ADAS (2003). Development of the Kent Lifescapes Information System. Unpublished report. Kent County Council and English Nature, Peterborough.

Countryside Agency (2001). *Areas of Outstanding Natural Beauty Management Plans: A Guide (CA23).* Countryside Agency, Cheltenham.

Countryside Agency & Scottish Natural Heritage (2002). *Landscape Character Assessment: Guidance for England and Scotland (CAX 84F).* Countryside Agency, Cheltenham and Scottish Natural Heritage, Perth.

Dreweatt Neate & Giles Wheeler-Bennett (2002). LaMIS (Land Management Information System): Review of Farmer and Land Manager Information Needs and Attitudes in the Context of Local Land Management Plans. Unpublished Report. Hampshire County Council, Winchester.

Edwards, T. (2004). *CAP Reform: Implementation in Scotland. SPICe Briefing 04/12.* Scottish Parliament, Edinburgh.

Entec UK Limited (2003). Urban Fringe Action Plans: their role, scope and mechanisms for implementation. Unpublished report. Countryside Agency, Cheltenham.

Environment & Heritage Service (2003). *Causeway Coast Area of Outstanding Natural Beauty Summary Management Plan.* Department of the Environment, Belfast.

Forestry Commission Scotland & Scottish Natural Heritage (2003). *Habitat Network for Wildlife and People – the Creation of Sustainable Forest Habitats.* Forestry Commission Scotland, Edinburgh and Scottish Natural Heritage, Perth.

Landscape Design Associates (1998). East Hampshire AONB, Integrated Management Guidelines, Case Study. Unpublished Report. Hampshire County Council, Winchester.

Porter, K. & Preston, S. (2001). *Lifescapes: Progress Report. English Nature General Committee of Council GC POI 72 [Council paper].* English Nature, Peterborough.

Scottish Executive (2002). *Custodians of Change.* Report of the Agriculture & Environment Working Group, Scottish Executive, Edinburgh.

15 Farming, Forestry and the Natural Heritage: Towards a More Integrated Future

PART 5:
Moving Towards Integration: Some Case Studies

Farm manager and FWAG officer, near Blairgowrie © Lorne Gill, Scottish Natural Heritage

PART 5:
Moving Towards Integration: Some Case Studies

Some of the building blocks for achieving better integration of farming, forestry and the natural heritage already exist, most notably in terms of our knowledge of the impacts of farming and forestry on the natural heritage and the publication of new national strategies for agriculture, forestry and biodiversity. This part of the book looks at a number of case studies which, collectively, show that progress towards integration is now being made on the ground.

In Chapter 16, Angelstam *et al.* look at the challenges of moving from action at an individual farm level to co-operative action across a number of farms or at a landscape-scale. The authors focus on the use of modelling approaches to help plan future land use based on multiple objectives. The next two chapters look in more detail at the challenge of overcoming diffuse pollution. In Chapter 17, Macgregor & Warren look at the motivations and practices of farmers in relation to this issue. They conclude that because many farmers believed that they were not responsible for the pollution in the Eden catchment in Fife, a key challenge is to overcome the belief that change is not needed. Audsley (Chapter 18), however, notes that if advice is kept simple and is based on simple actions which can reduce the likelihood of problems getting bigger – and thereby save the farmer time and money – then acceptance of such advice amongst farmers is good.

Gibb (Chapter 19) reports on the work of the Royal Highland Education Trust, which is helping to educate children about farming and its role in looking after the natural heritage. Bartlett (Chapter 20) provides a good case study of how major new facilities for mountain biking are being successfully integrated with forestry in southern Scotland. Together, Gibb and Bartlett show that real progress is being made in taking a more positive approach to access issues.

Two case studies look in more detail at the encouragement of wild birds through changes in land management. It is clear from Tharme (Chapter 21) and Grant *et al.* (Chapter 22) that the abundance of wild birds is related to the structure and composition of vegetation. The use of predictive models to find the right balance of vegetation restoration to encourage wild birds and monitoring long-term trends is seen as vital.

Anderson (Chapter 23), Morgan-Davies *et al.* (Chapter 24) and Jenkins (Chapter 25) report on several initiatives where considerable progress has been made towards integration on the ground. These case studies confirm that involving farmers and other land managers in deciding on the balance of land use and on how best to integrate farming and forestry with the natural heritage and its enjoyment by people is essential. Involving other members of the local community is also very important. Jenkins in particular shows what can be achieved if a group of farmers is committed to improving the local natural heritage.

Land management contracts are seen by many as a key tool in supporting the integration of farming, forestry and the natural heritage. Newcombe (Chapter 26) looks at some of the issues involved in developing land management contracts from the fresh perspective of the new Cairngorms National Park Authority. Agreeing the outcome – the vision – is an important part of the process, as is co-ordinating action across a number of the contracts hence facilitating planning at a larger scale than may have been possible in the past. This issue is picked up in more detail in Part 6. Finally, in Chapter 27, Blaney & de la Torre report on the work of the Land Use Policy Group, which provides a mechanism for examining the effectiveness of European Union rural policies. This chapter looks at the effectiveness of the Rural Development Regulation and concludes that there remains considerable room for adding environmental objectives and for better co-ordination across a range of Government policies in the EU member states.

16 From forest patches to functional habitat networks: the need for holistic understanding of ecological systems at the landscape scale

P. Angelstam, J. Törnblom, E. Degerman, L. Henrikson, L. Jougda, M. Lazdinis, J.C. Malmgren & L. Myhrman

Summary

1. Maintaining biodiversity is a major challenge within the goal of sustainable development.
2. In both terrestrial and aquatic environments, loss of biodiversity is usually related to long-term or intensive management reducing the number of species and natural habitat structures (e.g. dead wood). Management also affects important biotic processes such as browsing by deer through impacts on large carnivore populations, and abiotic processes such as altered fluvial dynamics within watersheds. Management of species, habitats and ecosystem functions are closely intertwined with each other, as well as with society's institutions.
3. To assess the status of biodiversity, monitoring of state indicators over time should be combined with performance targets allowing assessment of the degree to which ecological sustainability has been achieved. Gap analysis and habitat models are two important tools for the strategic and tactical, respectively, assessment of the functionality of both terrestrial and aquatic habitat networks. However, unless the results of assessments can be communicated to different institutions and networks, appropriate management action may not take place.
4. To improve the implementation of biodiversity policies on the ground, it is argued that natural and social sciences must be integrated in case studies with proactive bodies and businesses managing the social-ecological systems in which biodiversity is found. Ideally, whole landscapes or watersheds should be used as 'landscape laboratories' for research, development and management.

16.1 Introduction

This chapter provides a review of the development of current approaches to sustainability, biodiversity and ecosystems, and argues in favour of an integrated and cross-disciplinary approach to the research or development work needed to maintain biodiversity in the landscapes of Europe's boreal forest region, and to introduce a toolbox suitable for this task. Secondly, we discuss research and development needs for the long-term maintenance of biodiversity, using case studies from both the natural and social sciences. Finally, the chapter suggests that landscape units such as watersheds may be an appropriate unit for management, research, development, education and training (Andersson *et al.*, 2000; Bissonette & Storch, 2003; Angelstam & Törnblom, 2004a,b).

16.2 A review of approaches to sustainability, biodiversity and ecosystems

The principles of sustainable development and the maintenance of biological diversity are two commonly used concepts in the development of policies and practices for landscape management. However, although these terms are often used in general contexts, it is uncommon to see them applied in operational management (Lee, 1993; Bissonette & Storch, 2003; Norton, 2003). There is, in other words, an implementation deficit.

A number of barriers need to be bridged to maintain biodiversity on the ground (Gunderson *et al.*, 1995; Clark, 2002; Angelstam *et al.*, 2004b). First, the fact that many concepts may have several definitions (Kaennel, 1998) makes them difficult to use. Second, there may be limitations in the communication between policy and science on the one hand, and between science and practice on the other (Clark, 2002). Third, in the absence of a single 'right' scale of management of ecosystems, there is a need to consider a hierarchy of scales, the sizes of which are dependent on the problem (Petersen & Parker, 1998). Fourth, even when concepts are defined unambiguously, and communication is good, there are often gaps in knowledge about how to maintain biodiversity in practice (Angelstam *et al.*, 2003a; Bissonette & Storch, 2003; Connelly & Smith, 2003). Fifth, both the natural sciences, such as biology and chemistry, and social sciences, such as political science and psychology, must be involved in implementing sustainability (Heberlein, 1988; Penn, 2003). Finally, management on the ground needs time to act so that the land develops and eventually reaches the desired goal of the policy. In reality, the policy life cycle is repeated over and over again as values and landscapes change.

Nature conservation has a long tradition of protecting resources from unwanted disturbances. Guruswamy & McNeely (1997) distinguished several phases in the protection of biological resources. One is single-species protection. This started with the regulation of hunting of particular species for food and fur, and continued with the protection of areas with unique vegetation in the form of reserves and national parks. By the early 1980s, it had become evident to policymakers that species diversity needed to be complemented by genetic and population variation in different habitats, leading to the establishment of the concept of biodiversity (Wilson, 1988). However, it was soon realised that these compositional (species) and structural (habitats) aspects of life were dependent also on the function of ecosystems including a wide range of abiotic, biotic and anthropogenic processes. To make this more explicit, concepts such as ecosystem health, ecosystem integrity and ecosystem management appeared (Karr, 2000; Bissonette & Storch, 2003). The ecosystem approach of the Convention on Biological Diversity (Convention on Biological Diversity 1998, 2003) can be viewed as a general guiding principle of integrated management of socio-ecological systems, such as Rametsteiner & Mayer's (2004) Sustainable Forest Management (SFM) approach for forests and woodland.

At the scientific level, there is some discussion as to whether or not processes – and thus ecosystem function – should be included in the biodiversity concept (Gaston, 2000). Here, we follow Noss (1990) and Larsson (2001) in viewing biodiversity as including species, habitats and different ecosystem processes. These then need to be considered at different scales (Wiens *et al.*, 2002) in policy formulation:

1. the presence of species within remnant patches of suitable habitat set aside as conservation areas;
2. ensuring population viability over the long-term including, if required, the renewal of areas as conditions deteriorate naturally by natural succession;
3. the integrity and health of ecosystems, including the processes affecting ecosystem function (Norton, 2003); and
4. ecological resilience, or the ability of systems to withstand temporary disturbances (Gunderson & Pritchard, 2002), and the role of society and government in this (Clark, 2002).

While the first level is a traditional criterion for setting areas aside, the others deal with the functionality of single patches in the context of similar patches elsewhere and the surrounding matrix (Lindenmayer & Franklin, 2002). The maintenance of biodiversity, therefore, requires a toolbox which allows for the planning, management and restoration of protected or specially managed areas. The knowledge needed to make these tools work comes from areas such as conservation biology and landscape ecology, especially when they are integrated with social and policy sciences (Angelstam et al., 2003a; Bissonette & Storch, 2003; Clark, 2002; Penn, 2003; Rydén et al., 2003). Before using these tools, several important questions need to be answered, including:

1. which variables are considered important for levels (1)-(4) of ambition listed above;
2. the degree to which parameter values for these variables differ between benchmark and managed areas;
3. which values and policies define what is a desirable state (Duelli & Obrist, 2003), such as naturalness (Peterken, 1996) or cultural landscapes (Kirby & Watkins, 1998); and
4. whether or not these differences are relevant to the values.

In Europe there is a long tradition of establishing protected areas to meet nature conservation goals. Satisfying the ecological dimension of sustainable development is, however, more than traditional nature conservation. To maintain a favourable conservation status, protected areas usually need to be considered as parts of a network of habitats of different types with different composition and structure (Anon., 2003). Typically, different groups of species need different habitat structures (Nilsson et al., 2001; Angelstam et al., 2004c). Just as economic and social elements of sustainable development require their own networks, the environmental dimension (Noss, 1990) requires a functional network of habitat patches, an ecological 'green' infrastructure, in the landscape. As nature conservation has become more complex, ecosystem management has developed as a scientific approach to maintaining biodiversity (Berkes et al., 2003).

16.3 A toolbox for assessing the functionality of habitat networks

Like sustainable wood production, biodiversity management requires planning at various levels. For forestry, this is usually based on forest management units, which can then be aggregated to coarser spatial scales, such as administrative regions or ecoregions. Forestry planning is usually divided into three sub-processes: strategic, tactical and operational. The role of strategic planning is to decide on long-term goals covering an entire rotation. Tactical planning focuses on the strategic goals, but on a shorter time horizon and smaller areas. Operational management deals with the actual operations within a year or a season. The same approach can be used to build a toolbox of analytic tools for the conservation, management and restoration of biodiversity (Angelstam, 1998, 2003; Angelstam *et al.*, 2003a, 2004a; Lazdinis & Angelstam, 2004).

16.3.1 Strategic planning using gap analysis

Gap analysis may mean several things. In this chapter, we focus on identifying the gaps in the different representative habitats without which viable populations of naturally occurring specialised species cannot be maintained in the long term. Such species, with their narrow ecological niche, cannot survive in the regularly managed landscape.

Extending the analysis of gaps in the representation of different habitats (Scott *et al.*, 1996), Angelstam & Andersson (2001) made a quantitative estimation of area gaps in the present network of protected forest areas within a region which would need to be filled to maintain viable populations of the most demanding specialised species (c.f. Olson *et al.*, 2002). They estimated the need for protected forest areas for different broad ecoregions in Sweden using a procedure which included the following steps (Angelstam & Andersson, 2001; Angelstam *et al.*, 2003a; Table 16.1).

Table 16.1 The 'ABC' of regional gap analysis for strategic conservation planning.

Code	Explanation
A	Reference condition
B	The present situation
C	Science-based policy target
B/A	Representation
A*C	Long-term goal for the amount of habitat needed
B-(A*C)	Area gap (if the value is negative)

The first step is to estimate the extent of different forest vegetation types in the past (A). This can be done from historical records, from reference areas and from the distribution of different soil and site types. The second step is to estimate the current amount (B) of the representative forest vegetation types defined in A. By comparing A with B, one can describe differences in the representation of different forest vegetation types among different ecoregions. To quantify the occurrence of any gaps in the amount of habitat to maintain viable populations, knowledge is needed about the proportion of representative forest types required in order to maintain viable populations of the most demanding species (C) (Fahrig, 2002). Finally, a quantitative gap analysis can be made by estimating the difference between B and A*C, where a negative value implies a gap in habitat area and a need for habitat rehabilitation and/or re-creation. This approach has also been used for the counties of Dalarna and Gävleborg in Sweden (Angelstam *et al.*, 2003d), for Estonia (Lõhmus *et al.*, 2004) and Latvia (R. Bergmanis, pers. comm.). These studies also identified the need for more detailed quantitative knowledge of reference conditions especially for regions with a long land use history; of the ways of describing present landscapes with a sufficient thematic resolution; and of the thresholds for both habitat loss and important processes.

There are two broad visions regarding the historic range of variability in terrestrial reference landscapes: naturally dynamic ecosystems and pre-industrial cultural woodland. In Sweden, natural ecosystems dominate in the north and the latter in the south. Estonia is a mixture of both (Kurlavicius *et al.*, 2004). Angelstam & Kuuluvainen (2004) reviewed the current knowledge about the 'naturalness vision' as seen in different natural disturbance regimes, which can still be seen in a few remaining large intact boreal forests in Russia. Similarly, Kirby & Watkins (1998) discuss the 'culturalness vision', for which reference landscapes can be found in economically remote regions in Central Europe (Angelstam *et al.*, 2003b). For riparian landscapes the challenge is similar (Rabeni & Sowa, 2002).

16.3.2 Tactical planning using habitat suitability modelling

When a gap analysis has been done for a particular ecoregion, identified gaps in the ecosystem types need to be evaluated for their suitability for species dependent on this habitat. Such evaluations usually result in the need to improve the selected 'ecological infrastructure' by acquiring additional protected areas, as well as the management and rehabilitation of existing reserves and forest areas in the matrix surrounding them. Because there is usually a range of vegetation types with different dynamics in a management unit, the network of each representative habitat (made up by one or several land cover types) must, as a rule, be analysed and managed separately.

Habitat suitability modelling evaluates species, distributions and occurrence by probability mapping, based on quantitative information (Scott *et al.*, 2002). Ideally, this should be combined with the focal or umbrella species approach, which is based on the idea that management for specialised and area-demanding species can contribute to the protection of many other species with similar requirements (Lambeck, 1997; Roberge & Angelstam, 2004). Combining empirical data for species' requirements with relevant land cover data can be used to produce habitat suitability maps. Information from habitat suitability modelling can then be used to evaluate the effects of different conservation strategies and forest management scenarios.

Habitat models require three components. First, knowledge about the habitat and the requirements of a species at individual and population level (Guisan & Zimmermann, 2000; Angelstam *et al.*, 2004c). Secondly, habitat data with the right sort of resolution. Finally, techniques such as Geographical Information Systems are used to integrate the data (Store & Jokimäki, 2003). The result is a step-wise process that selects the combinations of factors that describe habitats, then sufficiently large patches, and finally suitable patches that link together to allow populations to exist (Figure 16.1).

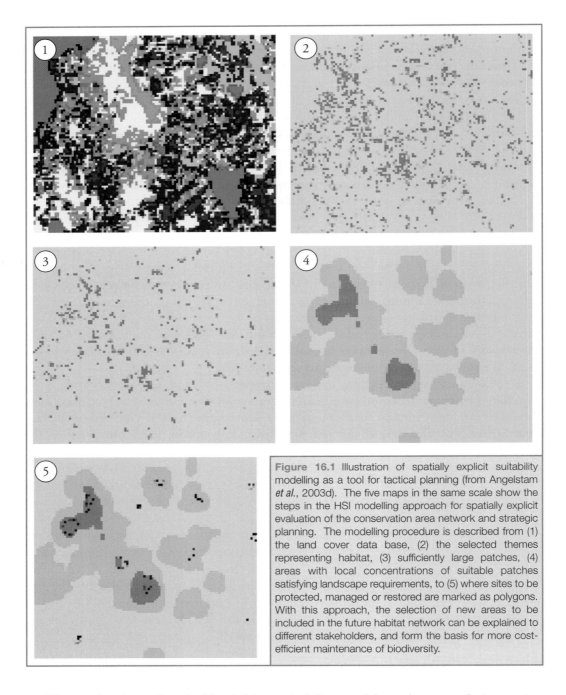

Figure 16.1 Illustration of spatially explicit suitability modelling as a tool for tactical planning (from Angelstam *et al.*, 2003d). The five maps in the same scale show the steps in the HSI modelling approach for spatially explicit evaluation of the conservation area network and strategic planning. The modelling procedure is described from (1) the land cover data base, (2) the selected themes representing habitat, (3) sufficiently large patches, (4) areas with local concentrations of suitable patches satisfying landscape requirements, to (5) where sites to be protected, managed or restored are marked as polygons. With this approach, the selection of new areas to be included in the future habitat network can be explained to different stakeholders, and form the basis for more cost-efficient maintenance of biodiversity.

The application of tools like habitat suitability models and gap analysis requires continuous development and critical evaluation. The following issues need to be considered when using them and evaluating their results. Firstly, with the exception of birds (Angelstam *et al.*, 2004c), quantitative knowledge about suitable suites of species is limited. Secondly, both gap analyses and habitat suitability models depend on the relevance and accuracy of data on land cover and habitat characteristics. Finally, there is a need to estimate how large areas of functional habitat need to be to maintain a viable population. Another important issue is whether or not managers know the species and their needs and characteristics (Uliczka *et al.*, 2004). This requires social science approaches (Bryman, 2001).

16.4 Research and development needs

16.4.1 Benchmarks and thresholds – the need for travelling in time and space

The maintenance of biodiversity requires good knowledge about the ecosystems in which species have evolved. This is essential for planning patterns of forest and woodland in a landscape. A fundamental issue, though, is to judge which benchmark of the forested landscape should be used. For example, how do altered forest landscapes, such as in NW Västerbotten County or the old mining district of Bergslagen in Sweden (Table 16.2) relate to other parts of the boreal region that have not been subject to such extensive human alteration? Another issue is how the pattern of land ownership affects the management of forests (Angelstam & Törnblom, 2004a,b).

With the town Vilhelmina (64.6°N, 16.6°E) as the centre, the former area encompasses the gradient from managed boreal forest to subalpine spruce (*Picea abies*) and birch (*Betula tortuosa*) forest, as well as mountains above the tree-line. Saami people have been practising reindeer herding for centuries: the area was colonised by Swedes establishing small-scale agriculture from the 18th century. Large-scale logging started in the late 19th century. The name of the Bergslagen area (~60°N, ~15°E) originates from the Swedish word 'Bergslag' alluding to the landowners combining agriculture, forest use and mining from the early Medieval period. During the 16th century the Crown established iron industries which later were sold to noble persons and businessmen. From the end of the 19th century, the mining and iron industry declined, and ceased completely during the 1990s.

Table 16.2 Issues of concern in two Swedish case studies in Vilhelmina in NW Sweden, and Bergslagen - an ancient mining district in south-central Sweden (see Svensson *et al.*, 2004).

Vilhelmina – at the Swedish timber frontier	Bergslagen - the old Swedish mining district
• Landscape-level biodiversity, mimicking natural disturbance regimes	• Competition among users of both wood and non-wood resources including tourism
• Influence of forest management methods, managing uneven aged forests for multiple values	• Landscape-scale restoration of biodiversity and cultural values
• Aquatic and wetland habitats in the landscape, watershed management	• Managing interactions between forest damage, abundant moose, returning wolves and people
• Involvement of indigenous people, combining forest management and reindeer husbandry	• Negative effects of anthropogenic pollution including acid compounds and nitrogen
• Local participation, education for forest-owners, professionals and the public	• Development of co-management and participatory planning where there are many land owners
	• Urban and social forestry issues related to the vicinity to large urban centres

Research to understand the historical range of variation of landscapes can be carried out using historical documents and maps, ecological records (such as pollen, charcoal and macrofossils in sediments, and fire scars) and by studying remnants of natural or culturally intact areas, as well as by modelling. All of these approaches have their advantages and disadvantages (Angelstam & Kuuluvainen, 2004). For example, in the remnant forest areas of Finland and Sweden, the distribution of more natural remnants is generally different and unrepresentative of managed landscapes. Additionally, remnants are often very small and have not been subjected to natural disturbances such as fire (Linder *et al.*, 1997) or flooding (Poff *et al.*, 1997) for decades. Finally, forest remnants may be affected by browsing by large herbivores altering the tree species composition (Angelstam *et al.*, 2000) or by other predators living in surrounding areas (Kurki *et al.*, 2000). As a result, combining different methods is recommended (Niklasson & Granström, 2000; Pennanen, 2002).

16.4.2 Encouraging a holistic perspective

There are both constraints and opportunities in taking this work forward though the challenges differ between regions. Table 16.3 shows how the pattern of forest use and development varies across Northern Europe.

Table16.3 The approximate temporal progression of different forest history phases from the perspective of biodiversity in boreal and hemiboreal forest (Angelstam *et al*., 2004d). Benchmark conditions are defined as a landscape with only local human use of resources. Selective harvest is defined as high-grading of certain tree species or dimensions of trees, and exploitation means that all dimensions are used. Sustainable yield timber production is defined as the use of harvesting, regeneration and stand treatment methods ensuring maximum sustainable yield.

Case study	Benchmark	Selective harvest	Exploitation	Sustained yield	Rehabilitation and re-creation
Scotland	Pre-Roman	Medieval	18th century	20th century	Present
Bergslagen	Pre-Medieval	Medieval	~1750	Present	Present
Latvia	Late 1700s	1800s	Early 1900s	Present	
Vilhelmina	Mid-1800s	Early 1900s	1950s/Present	Present	
Russia	Present	Present	Present		

In Scotland, the cover of natural forest is now very low (Summers *et al.*, 1999). Large plantation areas are approaching felling age, herbivore impact is large, and there is an ongoing discussion about access for recreation purposes. The relationship between forest, forestry and the aquatic habitat has been explored through the effects of afforestation on stream water quality (Ribbens, 2002). There are thus several challenges to be dealt with in order to restore different aspects of biodiversity and to satisfy social and economic values (Usher, 2003).

In Sweden, there is large regional variation in both the state of, and the visions for, the maintenance of biodiversity (Angelstam, 1997). In the south, a naturalness vision is competing with a cultural landscape vision, for which benchmark conditions are difficult to define quantitatively (Bengtsson *et al.*, 2000). In the boreal, a naturalness vision prevails.

In the Baltic States, the forest landscape is changing rapidly. In Estonia, for example, the final felling rates are four or five times as high as in Sweden and Finland (Kurlavicius *et al.*, 2004). However, the pool of species remains complete, and focal species such as the middle-spotted woodpecker (*Dendrocopos medius*) and the white stork (*Ciconia ciconia*) are even showing increasing numbers (M. Strazds, pers. comm.). However, this may change rapidly unless integrated landscape-scale management approaches are employed, covering large watersheds such as the North Vidzeme Biosphere Reserve at the River Salaca in Latvia, and the Lahemaa National Park covering the watershed of the River Loobu in Estonia (Ranke *et al.*, 1999). Although these examples are all different, we believe that by exchanging knowledge and experience from countries and regions with different experiences, all parties can gain.

While we can certainly learn much more about the elements of biodiversity, poor communication and insufficient dissemination of existing knowledge to policy-makers and land managers appears to be the major barrier to the implementation of ecological sustainability (Norton, 2003). Habitat thresholds for local viable populations provide a means of assessing the extent to which costly investments in conservation networks might be effective (Angelstam *et al.*, 2003c; Bütler *et al.*, 2004).

Establishing successful showcases could also spread insights about the need to build functional networks of habitat patches, whether protected, managed or restored, in order to conserve species. A prime example common to boreal and mountain forest in Europe is the capercaillie (*Tetrao urogallus*). Having gone extinct in Scotland in the late 1700s, it was re-introduced and thrived for a long time due to its importance as a game species. However, after a recent population decline, the present situation looks rather grave (Moss, 2001; Usher, 2003). In Germany considerable efforts are made to avoid regional extinction by applying spatial planning approaches (Suchant & Braunisch, 2004). In contrast, in boreal Sweden capercaillie are thriving, with numbers varying according to the available amount of habitat (Angelstam, 2004).

Brown trout (*Salmo trutta*) and salmon (*Salmo salar*) are important and are potential focal species in forest rivers and streams. As noted earlier, determining sufficient habitat size is difficult when building habitat models for terrestrial species, and perhaps even more so for aquatic ones. The brown trout is able to complete its whole life cycle within riffle-run areas of larger and deeper streams. In small forest streams, however, mature females remain small in size, and individuals are often forced to move to larger pools to be able to grow to mature size (Näslund *et al.*, 1998), but where they have become exposed to predators, shelter is then needed for larger individuals in the riffle areas and this, to a large extent, is regulated by the amount of woody debris. Thus the existence and growth of trout in forest streams is a product of interactions with the forest, the connections between different habitats, and other species, and these can be modelled if relevant data are available.

The Atlantic salmon is affected by forests and forestry. The decline of several Scottish salmon stocks has been linked with afforestation (Egglishaw *et al.*, 1986), probably due to acidification in stands of dense Sitka spruce (*Picea sitchensis*) (Ribbens, 2002). Ribbens (2002) also noted the negative effect of poorly designed bridge aprons and culverts associated with forest tracks on spawning habitat.

In Northern Europe, the biodiversity of riparian habitats varies depending on their use and management (Table 16.4). Amphibians use areas adjacent to streams in forest or woodland landscapes and are often found in areas rich in mossy rocks, decaying logs or moss and leaf litter, or in temporary or permanent standing water without predatory fish. Ponds or pools, especially tarns, bogs, beaver dams or oxbows are key habitats within the forest landscape. A focal species is the great crested newt (*Triturus cristatus*) which is widely distributed across the western Palaearctic (Thiesmeier & Kupfer, 2000). Due to habitat loss in cultivated areas (Langton *et al.*, 2001), it is now considered threatened throughout its distribution. However, in Scandinavia, and quite possibly in many other areas, it occurs also in boreal forests – often in small or medium-sized dystrophic lakes. Whether these habitats represent an original preference by the species, or a late successional adaptation, is unknown. The great crested newt has several life-history traits that make it particularly dependent on good quality terrestrial habitats (Malmgren, 2001), which also are important in migration (Malmgren, 2002; Schabetsberger *et al.*, 2004). Living in the interface between productive ponds and rich mature forest stands, the presence of great crested newts provides an important indicator of biodiversity and ecosystems functions.

Table 16.4 Comparison of how elements of the biodiversity of riparian landscapes differ in the boreal forest.

	Kola peninsula, Russia	**Southern Fennoscandia**
Species	• Intact composition	• Some species eliminated due to physical and chemical impact by humans
Habitat structures	• Large amount of dead wood • Intact riparian zones • Intact river channel • Intact watershed • Intact water quality	• Deficit in dead wood due to forestry • Missing or structurally affected by forestry • Decreased heterogeneity due to cleaning for timber floating • Altered due to land use and infrastructure • Eutrophication or acidification
Processes	• Intact water and ice dynamics • Functional riparian zones (shading, input of leaves, sediment trapping) • Natural fragmentation (connectivity)	• Regulated water table and flow • Non-functional riparian zones due to forestry • Fragmented due to dams (decreased connectivity)

16.4.3 Towards multiple case studies

New problems can often not be solved with old tools (Clark, 2002; Norton, 2003). Using a series of landscape-scale management units would allow a 'multiple case study' approach involving interdisciplinary research methods (Bryman, 2001; Jakobsen *et al.*, 2004). A sufficient range of units, encompassing both desired and undesired conditions, would allow more complete analyses of the relationships among species, habitats and ecosystem processes (Angelstam *et al,*. 2004d; Degerman *et al,*. 2004), and of the influence and effect of the institutions managing them (Figure 16.2).

Organisations wanting to adopt a holistic approach will need to modify management continuously, with the aim of promoting institutional learning based on an extended data collection and the use of new analytical tools (Lee, 1993). The concept of 'Adaptive Management Experiment Teams' (Boutin *et al.*, 2002; Angelstam *et al.*, 2004d), which includes active long-term collaboration between scientists and managers representing different elements of sustainability, and the comparison of multiple management

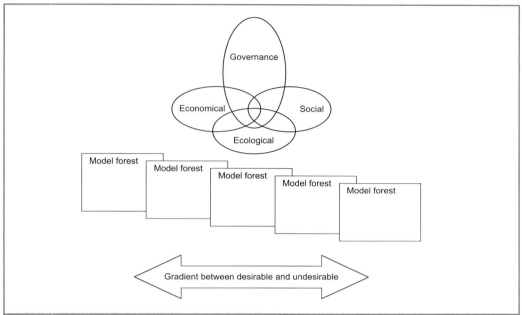

Figure 16.2 Idea for organising research, development and training to encourage the integration of disciplines, and policy, science and practice, in a multiple case study design using a network of model forests.

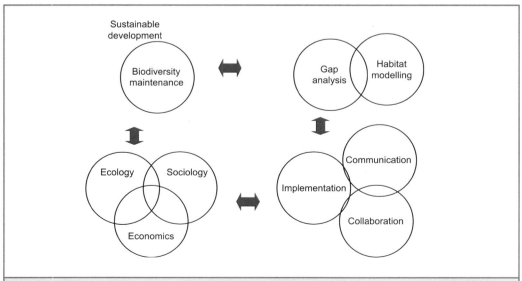

Figure 16.3 Conceptual description of integration among different policy areas, strategic and tactical planning, the implementation process and different scientific disciplines.

alternatives using experiments and simulation tools, is an attractive approach (Figure 16.3). Case studies can be both a research tool (Jasinski & Angelstam, 2002; Angelstam & Dönz-Breuss, 2004) and act as a way of demonstrating how to overcome problems with implementation. A holistic perspective requires that we understand and address forest and aquatic ecosystems together (Wiens, 2002; Schneider *et al.*, 2002). Viewing a stream in its geographic context is an important step for aquatic managers who have generally lacked the perspective of space and time long held by other disciplines (Swanson *et al.*, 1988).

To promote the ideas presented in this chapter we suggest that an international network of case studies should be established. One potential approach is the international model forest network initiated in Canada, which together forms a partnership between individuals and organisations sharing the common goal of sustainable forest management. In early 2004, Sweden officially embarked on the model forest concept with case studies representing long and short histories of forest use (Svensson *et al.*, 2004). In these case studies, it is important to use the social sciences before applying holistic approaches on the ground (Angelstam & Törnblom, 2004a,b).

16.5 Conclusions

A range of national and international case studies in regions with different conditions and different solutions to common problems would be an effective approach to promote ecosystem-based management principles in practice. Such a network of management units based on real landscapes, or ideally whole watersheds, can be used to spread good examples and to carry out research of wider relevance. The integration of natural and social sciences on the one hand and policy, science and practice on the other remains a major challenge to overcome.

Acknowledgements

The senior author thanks Scottish Natural Heritage for being invited to the stimulating meeting in Pitlochry, and in particular Helen Forster for organising all the practical details of the stay. We thank the editors of this volume, and two anonymous reviewers for constructive comments. Funding for this work was granted from Ludvika municipality and from the Swedish Environmental Protection Agency.

References

Andersson, F.O., Feger, K.-H., Hüttl, R.F., Kräuchi, N., Mattson, L., Sallnäs, O. & Sjöberg, K. (2000). Forest ecosystem research – priorities for Europe. *Forest Ecology and Management*, **132**, 111-119.

Angelstam, P. (1997). Landscape analysis as a tool for the scientific management of biodiversity. *Ecological Bulletin*, **46**, 140-170.

Angelstam, P. (1998). Maintaining and restoring biodiversity by developing natural disturbance regimes in European boreal forest. *Journal of Vegetation Science*, **9**, 593-602.

Angelstam, P. (2003). Reconciling the linkages of land management with natural disturbance regimes to maintain forest biodiversity in Europe. In *Landscape Ecology and Resource Management: Linking Theory with Practice*, ed. by J.A. Bissonette & I. Storch. Island Press, Covelo CA and Washington, DC. pp. 193-226.

Angelstam, P. (2004). Habitat thresholds and effects of forest landscape change on the distribution and abundance of black grouse and capercaillie. *Ecological Bulletin*, **51**, 173-187.

Angelstam, P. & Andersson, L. (2001). Estimates of the needs for nature reserves in Sweden. *Scandinavian Journal of Forestry*, Supplement, **3**, 38-51.

Angelstam, P. & Dönz-Breuss, M. (2004). Measuring forest biodiversity at the stand scale – an evaluation of indicators in European forest history gradients. *Ecological Bulletin*, **51**, 305-332.

Angelstam, P. & Kuuluvainen, T. (2004). Boreal forest disturbance regimes, successional dynamics and landscape structures – a European perspective. *Ecological Bulletin*, **51**, 117-136.

Angelstam, P. & Törnblom, J. (2004a). Towards targets and tools for the maintenance of forest biodiversity in actual landscapes. *Visnyk Lviv University, Ser. Geogr.*, **31**, 43-55.

Angelstam, P. & Törnblom, J. (2004b). Maintaining forest biodiversity in actual landscapes – European gradients in history and governance systems as a "landscape lab". In *Monitoring and Indicators of Forest Biodiversity in Europe – from Ideas to Operationality*, ed. by M. Marchetti. European Forest Institute, Joennsu. pp. 299-313.

Angelstam, P., Wikberg, P.E., Danilov, P., Faber, W.E. & Nygrén, K. (2000). Effects of moose density on timber quality and biodiversity restoration in Sweden, Finland and Russian Karelia. *Alces*, **36**, 133-145.

Angelstam, P., Mikusinski, G., Rönnbäck, B.-I., Östman, A., Lazdinis, M., Roberge, J.-M., Arnberg, W. & Olsson, J. (2003a). Two-dimensional gap analysis: a tool for efficient conservation planning and biodiversity policy implementation. *Ambio*, **33**, 527-534.

Angelstam, P., Boresjö-Bronge, L., Mikusinski, G., Sporrong, U. & Wästfelt, A. (2003b). Assessing village authenticity with satellite images – a method to identify intact cultural landscapes in Europe. *Ambio*, **33**, 594-604.

Angelstam, P., Bütler, R., Lazdinis, M., Mikusinski, G. & Roberge, J.M. (2003c). Habitat thresholds for focal species at multiple scales and forest biodiversity conservation – dead wood as an example. *Annales Zoologici Fennici*, **40**, 473-482.

Angelstam, P., Mikusinski, G., Eriksson, J.A., Jaxgård, P., Kellner, O., Koffman, A., Ranneby, B., Roberge, J.-M., Rosengren, M., Rönnbäck, B.-I., Rystedt, S. & Seibert, J. (2003d). *Gap analysis and planning of habitat networks for the maintenance of boreal forest biodiversity – a technical report from the wRESEx case study in Sweden*. Department of Natural Sciences, Örebro University, Örebro.

Angelstam, P., Dönz-Breuss, M. & Roberge, J.-M. (Eds) (2004a). Targets and tools for the maintenance of forest biodiversity. *Ecological Bulletin*, **51**, 510 pp.

Angelstam, P., Persson, R. & Schlaepfer, R. (2004b). The sustainable forest management vision and biodiversity – barriers and bridges for implementation in actual landscapes. *Ecological Bulletin*, **51**, 29-49.

Angelstam, P., Roberge, J.-M., Lõhmus, A., Bergmanis, M., Brazaitis, G., Breuss, M., Edenius, L., Kosinski, Z., Kurlavicius, P., Lārmanis, V., Lūkins, M., Mikusinski, G., Račinskis, E., Strazds, M. & Tryjanowski, P. (2004c). Habitat modelling as a tool for landscape-scale conservation – a review of parameters for focal forest birds. *Ecological Bulletin*, **51**, 427-453.

Angelstam, P., Boutin, S., Schmiegelow, F., Villard, M.-A., Drapeau, P., Host, G., Innes, J., Isachenko, G., Kuuluvainen, M., Mönkkönen, M., Niemelä, J., Niemi, G., Roberge, J.-M., Spence, J. & Stone, D. (2004d). Targets for boreal forest biodiversity conservation – a rationale for macroecological research and adaptive management. *Ecological Bulletin*, **51**, 487-509.

Anonymous (2003). *Natura 2000 and Forests – Challenges and Opportunities - an Interpretation Guide*. Office for Official Publications of the European Communities, Luxembourg.

Bengtsson, J., Nilsson, S.G., Franc, A. & Menozzi, P. (2000). Biodiversity, disturbances, ecosystem function and management of European forests. *Forest Ecology and Management*, **132**, 39-50.

Berkes, F., Colding, J. & Folke, C. (2003). *Navigating Social-Ecological Systems*. Cambridge University Press, Cambridge.

Bissonette, J.A. & Storch, I. (Eds) (2003). *Landscape Ecology and Resource Management: Linking Theory with Practice.* Island Press, Covelo CA and Washington DC.

Boutin, S., Dzus, E., Carlson, M., Boyce, M., Creasey, R., Cumming, S., Farr, D., Foote, F., Kurz, W., Schmiegelow, F., Schneider, R., Stelfox, B., Sullivan, M. & Wasel, S. (2002). The active adaptive management experimental team: a collaborative approach to sustainable forest management. In *Advances in Forest Management: from Knowledge to Practise,* ed. by T.S. Veeman, B. Duinker, A. Macnab, K. Coyne, G. Veeman, G. Binstead & D. Korber. University of Alberta, Edmonton. pp. 11-16.

Bryman, A. (2001). *Social Research Methods.* Oxford University Press, Oxford.

Bütler, R., Angelstam, P., Ekelund, P. & Schlaepfer, R. (2004). Dead wood threshold values for the Three-toed woodpecker in boreal and sub-alpine forest. *Biological Conservation,* **119**, 305-318.

Clark, T.W. (2002). *The Policy Process. A Practical Guide for Natural Resource Professionals.* Yale University Press, New Haven and London.

Convention on Biological Diversity (1998). *Report of the Workshop on the Ecosystem Approach, Lilongwe, Malawi.* Submission by the Governments of the Netherlands and Malawi. UNEP/CBD/COP/4/Inf.9. www.biodiv.org/doc/meetings/cop/cop-04/information/cop-04-inf-09-en.pdf.

Convention on Biological Diversity (2003). *Review of the principles of the ecosystem approach and suggestions for refinement: a framework for discussion.* Expert meeting on the ecosystem approach. UNEP/CBD/EM-MA/1/3. www.biodiv.org/doc/meetings/esa/ecosys-01/official/ecosys-01-03-en.pdf.

Connelly, J. & Smith, G. (2003). *Politics and the Environment: from Theory to Practice.* Routledge, London and New York.

Degerman, E., Sers, B., Törnblom, J. & Angelstam, P. (2004). Large woody debris and brown trout in small forest streams – towards targets for assessment and management of riparian landscapes. *Ecological Bulletin,* **51**, 233-239.

Duelli, P. & Obrist, M.K. (2003). Biodiversity indicators: the choice of values and measures. *Agriculture, Ecosystems and Environment,* **98**, 87-98.

Egglishaw, H.J., Gardiner, R. & Foster, J. (1986). Salmon catch decline and forestry in Scotland. *Scottish Geographic Magazine,* **102**, 57-61.

Fahrig, L. (2002). Effect of habitat fragmentation on the extinction threshold: a synthesis. *Ecological Applications,* **12**, 346-353.

Gaston, K.J. (2000). Biodiversity. In *Conservation Science and Action.* ed. by W.J. Sutherland. Blackwell Science, Oxford. pp. 1-19.

Guisan, A. & Zimmermann, N.E. (2000). Predictive habitat distribution models in ecology. *Ecological Modelling,* **135**, 147-186.

Gunderson, L.H., Holling, C.S. & Light, S.S. (Eds) (1995). *Barriers and Bridges to the Renewal of Ecosystems and Institutions.* Columbia University Press, New York.

Gunderson, L.H. & Pritchard Jr., L. (Eds) (2002). *Resilience and the Behavior of Large-Scale Systems.* Island Press, Washington DC.

Guruswamy, L. & McNeely, J. (Eds) (1997). *Protecting Global Biodiversity: the Converging Perspective of Science, Politics, Economics, Philosophy and Law.* Duke University Press, Durham, NC.

Heberlein, T.A. (1988). Improving interdisciplinary research: integrating the social and natural sciences. *Society and Natural Resources,* **1**, 5-16.

Jakobsen, C.H., Hels, T. & McLaughlin, W.J. (2004). Barriers and facilitators to integration among scientists in transdisciplinary landscape analyses: a cross-country comparison. *Forest Policy and Economics,* **6**, 15-31.

Jasinski, K. & Angelstam, P. (2002). Long-term differences in the dynamics within a natural forest landscape – consequences for management. *Forest Ecology and Management,* **161**, 1-11.

Kaennel, M. (1998). Biodiversity: a diversity in definition. In *Assessment of Biodiversity for Improved Forest Planning*, ed. by P. Bachmann, M. Köhl & R. Päivinen. Kluwer Academic Publishers, Dordrecht. pp. 71-81.

Karr, J.R. (2000). Health, integrity and biological assessment: the importance of measuring whole things. In *Ecological Integrity*, ed. by D. Pimentel, L. Westra, & R.F. Noss. Island Press, Washington, DC. pp. 209-226.

Kirby, K.J. & Watkins, C. (1998). *The Ecological History of European Forests*. CAB International, Wallingford.

Kurki, S., Nikula, A., Helle, P. & Linden, H. (2000). Landscape fragmentation and forest composition effects on grouse breeding success in boreal forests. *Ecology*, **81**, 1985-1997.

Kurlavicius, P., Kuuba, R., Lukins, M., Mozgeris, G., Tolvanen, P., Karjalainen, H., Angelstam, P. & Walsh, M. (2004). Identifying high conservation value forests in the Baltic States from forest databases. *Ecological Bulletin*, **51**, 351-366.

Lambeck, R.J. (1997). Focal species: a multi-species umbrella for nature conservation. *Conservation Biology*, **11**, 849-856.

Langton, T., Beckett, C.L. & Foster, J.P. (2001). *Great Crested Newt Conservation Handbook*. FrogLife, Halesworth.

Larsson, T.-B. (Ed) (2001). *Biodiversity Evaluation Tools for European Forest - Ecological Bulletin 50*. Blackwell Science, Oxford.

Lazdinis, M. & Angelstam, P. (2004). Connecting social and ecological systems: an integrated toolbox for hierarchical evaluation of biodiversity policy implementation. *Ecological Bulletin*, **51**, 385-400.

Lee, K.N. (1993). *Compass and Gyroscope*. Island Press, Washington DC.

Lindenmayer, D.B. & Franklin, J.F. (2002). *Conserving Forest Biodiversity. A Comprehensive Multiscaled Approach*. Island Press, Washington DC.

Linder, P., Elfving, B. & Zackrisson, O. (1997). Stand structure and successional trends in virgin boreal forest reserves in Sweden. *Forest Ecology and Management*, **98**, 17-33.

Lõhmus, A., Kohv, K., Palo, A. & Viilma, K. (2004). Loss of old-growth, and the minimum need for strictly protected forests in Estonia. *Ecological Bulletin*, **51**, 401-411.

Malmgren, J.C. (2001). *Evolutionary ecology of newts*. PhD thesis. Örebro Studies in Biology 1, Örebro University, Sweden.

Malmgren, J.C. (2002). How does a newt find its way from a pond? Migration patterns after breeding and metamorphosis in great crested newts (*Triturus cristatus*) and smooth newts (*T. vulgaris*). *Herpetological Journal*, **12**, 29-35.

Moss, R. (2001). Second extinction of capercaillie (*Tetrao urogallus*) in Scotland? *Biological Conservation*, **101**, 255–257.

Näslund, I., Degerman, E. & Nordwall, F. (1998). Effects of biotic interactions on brown trout habitat use and life history in streams. *Canadian Journal of Fish and Aquaculture Science*, **55**, 1034-1042.

Niklasson, M. & Granström, A. (2000). Numbers and sizes of fires: long-term spatially explicit fire history in a Swedish boreal landscape. *Ecology*, **81**, 1484-1499.

Nilsson, S.G., Hedin, J. & Niklasson, M. (2001). Biodiversity and its assessment in boreal and nemoral forest. *Scandinavian Journal of Forest Research*, Supplement, **3**, 10-26.

Norton, B.G. (2003). *Searching for Sustainability. Interdisciplinary Essays in the Philosophy of Conservation Biology*. Cambridge University Press, Cambridge.

Noss, R.F. (1990). Indicators for monitoring biodiversity: a hierarchical approach. *Conservation Biology*, **4**, 355-364.

Olson, D.M., Dinerstein, E., Powell, G.V.N. & Wikramanyake, E.D. (2002). Conservation biology for the biodiversity crisis. *Biological Conservation*, **16**, 1-3.

Penn, D.J. (2003). The evolutionary roots of our environmental problems: toward a Darwinian ecology. *The Quarterly Review of Biology*, **78**, 275-301.

Pennanen, J. (2002). Forest age distribution under mixed-severity fire regimes - a simulation-based analysis for middle boreal Fennoscandia. *Silva Fennica*, **36**, 213-231.

Peterken, G. (1996). *Natural Woodland Ecology and Conservation in Northern Temperate Regions*. Cambridge University Press, Cambridge.

Petersen, D.L. & Parker, V.T. (Eds) (1998). *Ecological Scale: Theory and Applications*. Columbia University Press, New York.

Poff, N.L., Allen, J.D., Bain, M.B., Karr, J.R., Prestegaard, K.L., Richter, B.D., Sparks, R.E. & Stromberg, J.C. (1997). The natural flow regime: a paradigm for river conservation and restoration. *BioScience*, **47**, 769-784.

Rabeni, C.F. & Sowa, S.P. (2002). A landscape approach to maintaining the biota of streams. In *Integrating Landscape Ecology into Natural Resource Management*, ed. by J. Liu & W.W. Taylor. Cambridge University Press, Cambridge. pp. 114-142.

Rametsteiner, E. & Mayer, P. (2004). Sustainable forest management and Pan-European forest policy. *Ecological Bulletin*, **51**, 51-57.

Ranke, W., Rappe, C., Soler, T., Funegård, P., Karlsson, L. & Thorell, L. (1999). *Baltic Salmon Rivers – Status in the Late 1990s as Reported by the Countries in the Baltic Region*. National Board of Fisheries, Gothenburg.

Ribbens, J. (2002). Overview of the forestry-aquatic biodiversity issue and future trends (UK). *Kungl. Skogs- och Lantbruksakademiens Tidskrift*, **7**, 69-75.

Rydén, L., Migula, P. & Andersson, M. (Eds) (2003). *Environmental Science*. The Baltic Press University, Uppsala.

Roberge, J.-M. & Angelstam, P. (2004). Usefulness of the umbrella species concept as a conservation tool. *Conservation Biology*, **18**, 76-85.

Schabetsberger, R., Jehle, R., Maletzky, A., Pesta, J. & Sztatecsny, M. (2004). Delineation of terrestrial reserves for amphibians: post-breeding migrations of Italian crested newts (*Triturus c. carnifex*) at high altitude. *Biological Conservation*, **117**, 95-104.

Schneider, R.L., Mills, E.L. & Josephson, D.C. (2002). Aquatic-terrestrial linkages and implications for landscape management. In *Integrating Landscape Ecology into Natural Resource Management*, ed. by J. Liu & W.W. Taylor. Cambridge University Press, Cambridge. pp. 241-262.

Scott, J.M., Tear, T.H. & Davis, F.W. (Eds) (1996). *Gap Analysis: a Landscape Approach to Biodiversity Planning*. American Society for Photogrammetry and Remote Sensing, Bethesda, MD.

Scott, J.M., Heglund, P.J., Morrison, M., Haufler, J.B., Raphael, M.G., Wall, W.A. & Samson, F.B. (Eds) (2002). *Predicting Species Occurrences: Issues of Scale and Accuracy*. Island Press, Covelo, CA.

Store, R. & Jokimäki, J. (2003). A GIS-based multi-scale approach to habitat suitability modelling. *Ecological Modelling*, **169**, 1-15.

Suchant, R. & Braunisch, V. (2004). Multidimensional habitat modelling in forest management – a case study using capercaillie from the Black Forest, Germany. *Ecological Bulletin*, **51**, 455-469.

Summers, R.W., Mavor, R.A. MacLennan, A.M. & Rebecca G.W. (1999). The structure of ancient native pinewoods and other woodlands in the Highlands of Scotland. *Forest Ecology and Management*, **119**, 231-245.

Swanson, F.J., Kratz, T.K., Caine, N. & Woodmansee, R.G. (1988). Landform effects on ecosystem patterns and processes. *BioScience*, **38**, 92-98.

Svensson, J., Fries, C. & Jougda, L. (2004). Synthesis of the model forest concept and its application to Vilhelmina model forest and Barents model forest network. *Skogstyrelsen Rapport* 6. National Board of Forestry, Jönköping.

Thiesmeier, B. & Kupfer, A. (2000). *Der Kamm-molch.b Ein Wasserdräche in Gefahr.* Laurenti Verlag, Bochum.

Uliczka, H., Angelstam, P. & Roberge, J.-M. (2004). Indicator species and biodiversity monitoring systems for non-industrial private forest owners – is there a communication problem? *Ecological Bulletin*, **51**, 379-384.

Usher, M.B. (2003). *Towards a Strategy for Scotland's Biodiversity: the Resource and its Trends.* Scottish Executive Environment and Rural Affairs Department, Edinburgh.

Wiens, J. (2002). Riverine landscapes: taking landscape ecology into the water. *Freshwater Biology*, **47**, 501-515.

Wiens, J.A., Van Horne, B. & Noon, B.R. (2002). Integrating landscape structure and scale into natural resource management. In *Integrating Landscape Ecology into Natural Resource Management*, ed. by J. Liu & W.W. Taylor. Cambridge University Press, Cambridge. pp. 143-175.

Wilson, E.O. (1988). *Biodiversity*, National Academy Press, Washington DC.

16 Farming, Forestry and the Natural Heritage: Towards a More Integrated Future

17 Perceptions and realities: the motivations and practices of farmers within Nitrate Vulnerable Zones

Colin J. Macgregor & Charles R. Warren

Summary

1. Coastal pollution as a result of nutrient over-enrichment (eutrophication) with nitrogen and phosphorous is a significant problem in European coastal ecosystems. Agriculture is blamed for much of the pollution and, consequently, the European Union's Nitrates Directive and its associated Nitrate Vulnerable Zone (NVZ) legislation was extended in Scotland in 2003, requiring more farmers to adopt sustainable management practices.
2. The motivations and management practices of 30 farmers representing 36% of the Eden catchment lying inside the Strathmore and Fife NVZ were investigated just prior to the legislation taking effect using interviews and farm-gate nutrient budgeting (NB).
3. The vast majority of the farmers did not believe that they were responsible for pollution, either on or off the farm. Nearly half thought that NVZ status would affect their management practices in some way, the perceived impacts being greater for livestock farmers than for cereal producers.
4. Nearly all of the farmers interviewed obtain income support from the Common Agricultural Policy but most seemed apathetic and even contemptuous towards other government funding support aimed at environmental stewardship initiatives.
5. The NBs indicate that the typical farm in the Eden catchment loses approximately 98 kg of nitrogen, 10 kg of phosphorous and 21 kg of potassium per hectare to the wider environment each year. The dairy farms and intensive livestock producers generate the most diffuse pollution. A large majority of farmers were unaware of the extent of their nutrient losses.
6. The NBs also revealed that the vast majority of farmers will not have to reduce the amount of fertilisers or farmyard manure they are using in order to comply with NVZ limits, suggesting that the legislation will have a minimal effect in reducing diffuse pollution.

17.1 Introduction

Coastal pollution as a result of nutrient over-enrichment (eutrophication) with nitrogen (N) and phosphorous (P) is a significant threat to coastal ecosystems (Nedwell *et al.*, 2001). This research is part of two European Union (EU) funded research projects which are adopting an innovative and inter-disciplinary approach to the issue of European coastal zone pollution. The approach combines detailed investigation of the marine biology of estuaries with a study of the land use practices which influence the water quality of rivers flowing into those estuaries. Here we report the initial results from a series of interviews that explored the environmental attitudes of farmers in a Scottish east coast catchment. More specifically, the research examined the motivations and management practices of 30 farmers located within the Eden catchment of the Strathmore and Fife Nitrate Vulnerable Zone (NVZ) (see Figure 17.1). The study was carried out between July 2002 and October 2003 and the farming area in question represented 36% (8,832 ha) of the catchment.

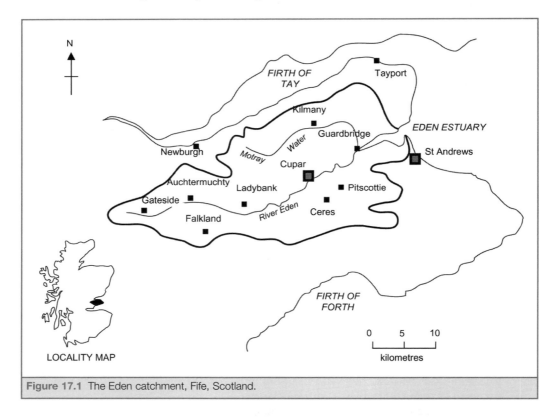

Figure 17.1 The Eden catchment, Fife, Scotland.

The Eden catchment is a useful case study because the area was designated a NVZ in 2003 and because farming is the predominant land use (76% of the land surface). The catchment is drained by two main rivers, the River Eden and the Motray Water. Its estuary, which includes the Eden Estuary Local Nature Reserve, is regarded as an important over-wintering site for wildfowl and waders (Environmental Change Network, 2002).

The project focused on four questions.

- How well-informed are farmers about the environmental impacts of their agricultural practices, and how accurate are their perceptions of nutrient output from their farms?
- What types of farms, and what land use/management practices, are most likely to cause pollution?

- Which farmers will bear most of the costs associated with adhering to environmental legislation?
- How do farmers regard environmental legislation, and what motivates them to adopt environmentally sustainable practices?

17.2 Research methods

Farmers were selected using a stratified sampling procedure that took account of the location, type and size of farms. Two research approaches were used. The first was a qualitative interview-based approach (using the general inductive method (Thomas, 2003)), an approach that ensures that interviews remain centred on the principal research but which also allows sufficient flexibility to discuss unanticipated issues. The interviews, conducted between July and October 2002, explored three themes: knowledge of farming's wider environmental impacts; farmers' attitudes to environmental legislation; and the management practices most likely to cause problems for the environment.

The second approach used farm gate nutrient budgets (NB). Data for these were provided by the farmers between August and October 2003, but the year under consideration was November 2001 to October 2002. The NB is a management tool that takes account of the flow of nutrients (N, P and potassium (K)) through the farm system (Sheaves, 1999). Nutrients enter the farm system in three ways; two are natural (deposition and mineralization) but the third, which raises environmental concerns, comprises agricultural fertilisers, manure, roughage and concentrated feedstuffs. All of the inputs and outputs (crop yields and/or livestock) must be accounted for and quantified. Only one of the selected farmers had completed a NB prior to this study, but nearly all of the farmers conduct regular soil testing. However, the frequency of tests varied widely, with less than a third testing within recommended frequencies (every 2–3 years).

17.3 Findings from the study

Most of the 30 farmers denied responsibility for any pollution, even though the NBs revealed a total surplus of 910,000 kg N and 86,000 kg P for the year. A simple extrapolation suggests that 2,505 tonnes of N and 236 tonnes of P per annum are being lost from all of the farms in the catchment.

Water quality evidence (Tay River Purification Board, 1994; Scottish Environment Protection Agency, 2000) suggests that the River Eden suffers more from N and P pollution than the Motray Water. This, combined with the spatial distribution of the farms, supports the NB findings, with the highest N concentrations found in the upper River Eden where dairy and livestock production is concentrated. The NBs also suggest that higher farm nutrient surpluses, for both N and P, are associated with dairy and livestock farms. Four farms that can

Table 17.1 Median nitrogen surpluses and efficiencies for 30 Eden catchment farms.

Farm type	Surplus Nitrogen (kg N/ha)	Nitrogen Efficiency (%)
Arable (crops) only (n = 10)	75.4	61.5
Open mixed farming (n = 16)	113.7	42.5
Dairy/intensive livestock (n = 4)	343.2	18.5

be described as dairy/intensive livestock producers contributed more than 21% of the total N from all 30 farms (Table 17.1). By contrast, the ten arable farms contributed less than 14%.

The principal findings from the interviews can be summarised as follows.

- All farmers were aware that the Eden would be designated a NVZ in 2003, although the details of what this meant in terms of management were not well understood by most. Their biggest concerns were associated with the storage of silage and slurry because many believed that they would have to invest in up-grading infrastructure.
- Half of the farmers had established riparian buffer strips within their farms, the most important motivation being to avoid spray regulations associated with LERAPS (Local Environment Risk Assessment for Pesticides).
- Most farmers displayed strong feelings against government-associated environmental initiatives. For example, antipathy towards bureaucracy is the main reason why most farmers do not access environmental stewardship-type funding support.
- Most of the farmers were not favourable towards the proposed changes to the European Union's Common Agricultural Policy (CAP) (which includes the removal of direct payments per farm based on products) because they believed that they might be geographically disadvantaged, especially under an expanded EU.

17.4 Conclusions

Most of the farmers interviewed were surprised at the amount of nutrient loss from their farms, although they were correct in assuming that the dairy/livestock producers are more likely to be responsible for diffuse pollution. This suggests that the links between farm management and off-site pollution (fluvial, coastal) are not well established in the minds of farmers.

There is also uncertainty as to whether NVZ legislation (as it currently stands) will go far enough to bring about improvements in water quality. All of the farms in this study – even the most efficient arable-only farms – were producing surpluses of N and P, despite the fact that the vast majority appear to be fertilising within prescribed limits.

More widespread use of NBs could help to inform farmers of the impacts of agriculture on the wider environment. Government agencies may need to consider more aggressive promotion of the potential benefits of NBs to dairy and livestock producers in particular, such as cost savings associated with reduced fertiliser use.

Since most farmers appear antagonistic towards environmental funding support, either the bureaucratic hurdles need to be reduced (less red tape) and/or further legislative controls may be required to ensure adoption of sustainable practices. The responses of farmers to LERAPS regulations demonstrate how restrictions can produce positive results. Importantly, few seemed particularly troubled by loss of productive land when they established riparian buffer strips.

Counter-productive bureaucracy is not, of course, a problem that is unique to the United Kingdom or EU. For example, evidence from Australia confirms that government bureaucracy can be a major factor impeding farmers' access to state-funded environmental stewardship-type support (Macgregor & Pilgrim, 1998). The current efforts to reform the CAP, and to steer it in a more environmentally sustainable direction, should take account of this most fundamental reality.

Acknowledgements

The authors acknowledge the European Union for funding support through TIDE (Tidal Inlets Dynamics and Environments) and HIMOM (Hierarchical Monitoring Methods). This study was only made possible with the generous help of the farmers in the Eden catchment.

References

Environmental Change Network (2002). *ECN Freshwater Sites – River Eden, Fife Region, Scotland.* www.ecn.ac.uk/sites/eden2.html.

Macgregor, C. & Pilgrim, A. (1998). Is landcare funding hitting the target? *Australian Journal of Natural Resource Management*, **1**, 4-8.

Nedwell, D.B., Dong, L.F., Sage, A. & Underwood, G.J.C. (2001). Variations of the nutrient loads to the mainland UK estuaries: correlation with catchment areas, urbanization and coastal eutrophication. *Estuarine, Coastal and Shelf Science*, **56**, 951-970.

Scottish Environment Protection Agency (2000). *SEPA Annual Report 1999/2000*. Scottish Environment Protection Agency, Stirling.

Sheaves, J. (1999). FWAG's Nutrient Budget Handbook. Unpublished. Farming and Wildlife Advisory Group, Dorchester.

Tay River Purification Board (1994). *A Catchment Study of the River Eden, Fife. Technical Report TRPB 1/94.* Tay River Purification Board, Fife.

Thomas, D.R. (2003). *A General Inductive Approach for Qualitative Data Analysis.* University of Auckland, New Zealand. www.health.auckland.ac.nz/hrmas/resources/qualdatanalysis.html.

17 Farming, Forestry and the Natural Heritage: Towards a More Integrated Future

18 The 4 Point Plan: helping livestock farmers to reduce diffuse pollution risk

Rebecca Audsley

Summary

1. The 4 Point Plan is a guidance document targeted at livestock farmers in Scotland to raise awareness of diffuse pollution and to suggest practical mitigation strategies to reduce pollution risk from livestock and operations involving slurry and manure.
2. The guidance has been well received by the agricultural sector as easy to read and informative. Implementing the 4 Point Plan could lead to benefits for farm businesses and assist compliance with current and forthcoming environmental legislation.

18.1 Introduction

Diffuse pollution is being highlighted across Europe as a priority for action. The European Union Water Framework Directive, incorporated into Scottish law as the Water Environment and Water Services (Scotland) Act 2003, aims to bring about the effective co-ordination of water environment policy and regulation to enhance and protect water quality across Europe (Scottish Environment Protection Agency, 2004). This will affect all water users and increase the focus on any input that could lead to a reduction in the ecological status of a waterbody.

Reducing pollution from agricultural sources has been stated as a priority environmental issue facing Scottish agriculture over the next 5 to 10 years (Scottish Executive, 2002). Estimates suggest that diffuse pollution from agricultural sources could have a major impact on water quality by 2010 (Scottish Environment Protection Agency, 1999).

This chapter reviews the diffuse pollution risks from agricultural sources and describes the 4 Point Plan (4PP), which is designated to raise awareness of these risks and of actions that can be taken to reduce them.

18.2 Diffuse pollution risks from agricultural sources

Diffuse pollution from agriculture can arise from a wide range of small sources, and can travel across land before reaching water, making it potentially difficult to monitor and control. Individually, impacts from small, diffuse inputs can be unimportant. However, when considered on a catchment scale they may result in significantly elevated nutrient levels, excess sediment deposition and an increased concentration of faecal bacteria with potential long-term implications for species diversity, drinking water quality and amenity value. For example, excess sediment entering a river could smother the riverbed, clogging pore spaces between gravels, suffocating fish eggs and reducing habitat for smaller invertebrate species. Invertebrates are an important food source for many species of bird and fish such as dippers (*Cinclus cinclus*) and brown trout (*Salmo trutta*). Common sources of diffuse pollution from agriculture include manure and slurry runoff from fields after spreading, and from livestock with direct access to burns and rivers.

18.3 Guidance for farmers

Guidance to reduce pollution risk from all agricultural sources is available in the Code of Good Practice for the Prevention of Environmental Pollution from Agricultural Activities (the PEPFAA Code) (Scottish Executive, 2005). A Scottish Executive report regarding the impact of agricultural practices on Ayrshire bathing waters (Aitken *et al.*, 2001) recommended a range of actions that could further reduce the risk of faecal contamination to surrounding waters. These recommendations formed the basis for the guidance contained in the 4PP.

18.4 The 4 Point Plan (4PP)

The 4PP provides straightforward suggestions for livestock farmers to minimise pollution risks from routine practices. Many of the suggestions have the potential to benefit the farm business through savings in time or otherwise hidden costs. Working through the 4PP could help farmers to:

- identify the risk of diffuse pollution on their farm;
- provide ideas and suggestions to reduce pollution risk; and
- cut hidden costs or save time due to a change or improvement in current practices.

Farmers who follow the 4PP will qualify for two points towards the Rural Stewardship Scheme (RSS), a competitive agri-environment scheme that can help farmers enhance and protect natural habitats around the farm. For example, measures under the scheme could fund the fencing of water margins leading to the development of natural vegetation around watercourses, benefiting both water quality and riparian habitat. The 4PP is split into four sections.

18.4.1 Managing dirty water around the steading

Farmers should collect and correctly dispose of any dirty water in the farm steading that is contaminated with slurry or manure. When clean water, for example rainwater runoff from roofed buildings, mixes with dung and urine on contaminated yards, it can significantly add to the amount of dirty water a farmer has to collect, store, handle and spread. A few simple actions can help to reduce the volume of clean water finding its way on to yards, reducing pressure on existing storage and cutting down the amount of dirty water with which the farmer has to deal. Keeping clean and dirty water separate, maintaining gutters, downpipes and drains, and changing hosing practices to reduce water use are all ideas within the plan for farmers to consider.

18.4.2 Better nutrient use

The second section of the plan gives a series of simple calculations allowing the farmer to estimate the nutrient content of slurry and manure produced by housed livestock. Taking account of nutrients in slurry and manure from housed livestock can make the most of the 'free fertiliser' in manure and lead to cost savings through targeted nutrient application, complementing planned inorganic fertiliser applications. Nutrient application above crop demand is not only wasted but may pollute the surrounding environment.

18.4.3 RAMS – a Risk Assessment for Manure and Slurry

The risk assessment for manure and slurry (RAMS) section of the guidance gives the farmer a step–by–step guide to completing a RAMS land plan, providing a reminder of no-spread zones or areas of high risk when spreading slurry or manure. The RAMS land plan acts as a guide for staff or contractors who may not be familiar with the location of ditches, burns, springs or wells on the farm. Spreading slurry or manure in close proximity to these areas increases the risk of contaminated material seeping into surrounding waters.

18.4.4 Managing water margins

Vegetated water margins act as a buffer strip between in-field practices and watercourses. Fencing off watercourses to prevent livestock access reduces the risk of livestock dunging directly into the water, reduces trampling at the water's edge, and can allow the development of vegetation through a reduction in grazing pressure.

18.5 Feedback from the agricultural sector

Overall, the guidance has been widely accepted as straightforward and common sense, and as benefiting the farm business and assisting compliance with environmental legislation. In turn, farmers' actions to reduce diffuse pollution can benefit the surrounding environment and help to protect water quality. Comments from farmers have included:

- "the plan is full of useful information";
- "it shows there are cheaper things most farmers can do to minimise pollution without massive investment"; and
- "most of the guidance seems to be relevant to my farm".

18.6 Conclusions

The 4PP has proved to be a useful document, helping livestock farmers to identify and reduce diffuse pollution risk and achieve compliance with environmental legislation. Individual actions could also benefit the natural heritage, for example through the creation of vegetated buffer strips intercepting polluting runoff and providing food and cover for a range of insect, bird and mammal species. Farmers can obtain a copy of the 4PP from their local agricultural consultant. Alternatively, a copy of the plan can be downloaded from the Scottish Agricultural College's website at www.sac.ac.uk/4pp.

Acknowledgements

As Project Officer in Water Resource Management, based with the Scottish Agricultural College, and responsible for the 4 Point Plan, I would like to acknowledge the funding and support provided by Scottish Agricultural College, Scottish Executive Environment and Rural Affairs Department, Scottish Natural Heritage, WWF (Scotland) (funded through The BOC Foundation), Farming and Wildlife Advisory Group Scotland, Scottish Environment Protection Agency and NFU Scotland.

References

Aitken, M., Merrilees, D.W. & Duncan, A. (2001). *Impact of Agricultural Practices and Catchment Characteristics on Ayrshire Bathing Waters.* Scottish Executive Central Research Unit, Edinburgh.

Scottish Environment Protection Agency (1999). *The State of the Environment Report – Water.* Scottish Environment Protection Agency, Stirling.

Scottish Environment Protection Agency (2004). *Introduction to the Water Framework Directive.* www.sepa.org.uk/wfd/introduction.htm.

Scottish Executive (2002). *Custodians of Change.* Report of the Agriculture and Environment Working Group, Scottish Executive, Edinburgh.

Scottish Executive (2005). *The Prevention of Environmental Pollution from Agricultural Activity (PEPFAA) – Code of Good Practice.* Scottish Executive, Edinburgh.

Water Environment and Water Services (Scotland) Act 2003. www.opsi.gov.uk/legislation/scotland/acts2003/2003003.htm.

19 The Royal Highland Education Trust

Lindsey Gibb

Summary

1. The Royal Highland Education Trust is an educational charity which aims to create opportunities for children to learn about the countryside.
2. Training is available for both teachers and farmers throughout Scotland.
3. Learning resources are made available to school teachers, together with initiatives to increase awareness of farming, food production and the working countryside amongst children.
4. The annual Royal Highland Show is used as an educational tool by many primary and secondary schools.

19.1 Introduction

The Royal Highland Education Trust (RHET) is an educational charity which aims to create opportunities for children to learn about the countryside. This chapter describes how the RHET arranges for schools to visit farms, provides independent information about the countryside, food and farming, and supplies resources to assist teachers and pupils in their understanding of the countryside. The main area of work is with primary schools and the RHET has recently started working in secondary and informal education sectors. The RHET also has links with tertiary education and teacher training colleges.

19.2 Training and placements

The RHET creates and delivers training for both teachers and farmers and contributes to in-service training throughout Scotland. Teacher training is delivered in partnership with local authorities in line with the requirements of their teachers. Workshops can focus on the curriculum or involve a farm visit to gain an understanding of the learning opportunities open to children through visiting a working farm. Along with the regular in-service training, there are placement opportunities for teachers to help develop resources, learn more about the organisation and get involved with the Royal Highland Show. In 2003/04, 15 teachers were involved in a variety of week-long placements which were managed and evaluated by Careers Scotland as part of their Enterprise in Education through Business Links programme. As a result of this close liaison and joint working with education staff and teachers, the RHET is highly regarded within the education system. In its publication *Health & Safety on Educational Excursions* (Scottish Executive, 2004) the Scottish Executive recommends that schools wanting to do farm visits should go through the RHET. Local authorities regularly request resources for all of their schools and are represented on six of the ten local Countryside Initiative groups. The RHET also provides training for farmers before they host farm visits or speak in schools. This includes presentation skills, guidance on working with children and health and safety issues.

19.3 Countryside initiatives

In addition to the core education team based in Edinburgh, there are a growing number of RHET Countryside Initiatives (CIs) around Scotland. There are currently ten CIs from Dumfries & Galloway and the Borders to the Royal Northern in Aberdeenshire. CIs are groups of volunteers, often farmers and landowners, who work to promote awareness and education about farming in their local areas. This is delivered mainly through school visits to farms or farmer talks to schools. The demand for these has been increasing steadily since they began in 1999. In 2003/04, there were over 200 farm visits and over 20 farmer talks. Eight of the CIs have paid Project Co-ordinators who are responsible for establishing and maintaining links with local schools, organising the volunteers, and undertaking preparation work (such as risk assessments) with the schools and farms.

19.3.1 Resources

To support the CIs and encourage farm visits, the RHET has produced various resources including the *Looking at the Farm* pack (Royal Highland Education Trust, undated), *Working Countryside* resource boxes and guidance for teachers on farm visits. The *Looking at the Farm* pack was developed in partnership with the Royal Northern Countryside Initiative (RNCI) and Aberdeen Environmental Education Centre and has sections on food production, safety on the farm, seasons, weather, the role of the farmer, wildlife and machinery.

RHET staff developed the *Working Countryside Box* (Royal Highland Education Trust, undated) with help from teachers on placement. It contains materials from other organisations, such as SNH biodiversity posters, the Linking Environment and Farming (LEAF) virtual farm walk CD-ROM, Health & Safety Executive *Stay Safe* booklets, and the forest trees guide from the Forestry Commission. The contents of the *Working Countryside Box* are always being updated and the RHET management committee evaluates new materials before they are included. Evaluation is from an educational and countryside perspective to ensure that materials are educationally valid and free of bias.

19.3.2 Countryside Classroom on Wheels

Many of the CIs produce their own activities and resources, notably the Royal Northern Countryside Initiative (RNCI) which has created the successful Countryside Classroom on Wheels (CCOW) project. The CCOW is a purpose-built, bio-secure trailer designed to bring livestock, crop and feed samples to the school playground. A 7 m long trailer enables local farmers to bring part of their farm to the school playground. There are hand-washing facilities on board and a portable ramp allows assisted wheelchair access. Tanks beneath the floor of the livestock pens collect the waste and slurry.

This pioneering project was created as part of the RNCIs efforts to increase awareness of farming, food production and the working countryside amongst north-east youngsters at a cost of £12,500. It was funded by donations from individual farmers, businesses, agricultural show societies and local authorities. It continues to be financed through donations received by the RNCI from benefactors and funding members.

The CCOW is an important resource in delivering information to school children about farming and the countryside. It is part of the RNCI's education package which, like the other CIs, includes farm visits and farmer talks. In 2003/04 the CCOW visited a large number of north-east schools, agricultural shows and various exhibitions and events.

19.4 The Royal Highland Show

The RHET also delivers the education programme at the Royal Highland Show every year. The show lasts for four days and the RHET has been involved since its inception in 1999. Schools receive a guided tour of the showground and visit the education centre where they take part in a wide range of activities from building tractors to grinding flour. In 2003, 222 schools visited the show and 5,500 children took part in the RHET schools programme.

A curriculum-linked education pack is available for teachers to use pre- and post-visit. The pack was developed and piloted by teachers on placement through Careers Scotland. The primary school material is cross-curricular as many come for an end-of-term trip, whereas the secondary school material is subject-specific. The pack is evaluated every year after the show.

19.5 Biodiversity

The initial focus for the RHET was on food production but for the last three years it has slowly expanded into other areas such biodiversity, healthy eating and access issues. *Biodiversity and the Farming Year* (Royal Highland Education Trust, undated) is a learning and teaching resource developed to enable teachers and pupils to learn about biodiversity on farms. The activities promote learning about species and habitats that can be found on Scottish farms. It is a partnership project between the RHET and East Lothian Council's Biodiversity Officer. After a successful pilot, it is being rolled out nationally with support from SNH.

19.6 Working with others

The RHET and SNH are also working together to create links with the new access education programme by supporting the creation of a 'good practice' model farm and helping to develop a section on access for the training provided to farmers.

Working with other organisations has been a key ingredient in the RHETs growth and has resulted in several projects and significant support over the years. For example, the Clydesdale Bank currently supports *Count and Grow* which is a new project which enables teachers and children to grow a crop of potatoes in school with the help of a local farmer. This project is based on a pilot project in Perth and Kinross. Lantra, the sector skills council for the environment and land based sector, supports the RHET by funding teacher training and farmer training aspects of their work, and the NFU Scotland offers practical advice and support, such as compiling a fact sheet on Common Agricultural Policy reform for secondary geography teachers.

The management committee reflects this partnership approach with a wide variety of organisations represented including SNH, NFU Scotland, Edinburgh University, the Scottish Executive and the Scottish Rural Property & Business Association.

19.7 Conclusion

Raising levels of awareness and understanding of farming amongst children is very important to the long-term future of farming. Working closely with farmers and teachers, the RHET has been successful with a number of good initiatives. The key to this success is the partnership approach adopted by the Trust.

References

Royal Highland Education Trust (undated). *Looking at the Farm.* Royal Highland Education Trust, Edinburgh.

Royal Highland Education Trust (undated). *Working Countryside Box.* Royal Highland Education Trust, Edinburgh.

Royal Highland Education Trust (undated). *Biodiversity and the Farming Year.* Royal Highland Education Trust, Edinburgh.

Scottish Executive (2004). *Health & Safety on Educational Excursions: A Good Practice Guide.* Scottish Executive, Edinburgh.

20 Integrating access and forestry: the 7stanes case study

Karl Bartlett

Summary

1. This chapter explores the management of access in a working forest through the case study of the 7stanes mountain bike project. The 7stanes Project is creating seven mountain bike centres in key Forestry Commission locations across the south of Scotland.
2. Key factors in the success of this project include planning to accommodate the needs of forestry and a range of different recreational users in the same area, trail design that includes consideration of environmental, safety and potential user conflict factors, and detailed consultation with landowners, managers and users.
3. The project has contributed to greatly increased visitor numbers and the trail design has won international acclaim. However, funding ongoing costs such as maintenance and marketing continues to be a challenge.
4. The project demonstrates that integrated access planning and management can yield huge benefits but that the time and effort needed to handle the complexity generated should not be underestimated.

20.1 Introduction

This chapter briefly explores the issue of managing access in a working forest environment through a case study of the 7stanes mountain bike project.

Before looking at the project in detail, it is important to note that forests are adaptable and accommodating to changes in the nature of access and recreational demands. Forests have always satisfied a recreational role in the community and it is often repeated, but worth reiterating, that woods and forests are able to give sufficient cover for Girl Guide picnics to take place in the same wood as a mountain bike race, without either being aware of the other.

Woods and forests offer tremendous cover: they can absorb noisier activities and still give visitors a sense of being alone or, at least, a feeling of privacy and seclusion; they can hide activities from each other more effectively than open hillsides – many inter-user tensions stem from the simple presence of other users and forests are able to allow near passage of different users without contact; and they can accommodate facilities such as tracks and trails that, on an open hill, would be unwelcome.

Forests are, above all else, adaptive because of the longer timescales used in growing and managing trees. Blocks of forestry often need little management input for long periods of time, unlike other land uses which can require annual, or more frequent, management. This allows foresters to take a longer-term view of access in forests.

20.2 Forest access policy

Despite current perceptions, the Forestry Commission does not have a long-standing tradition of access provision; some foresters can still remember the 'keep out' signs on forest gates. However, successive Government policy changes have steered the Forestry Commission away from its focus on tree production and there is now strong acceptance of the capacity of forests to deliver a wide range of multiple benefits. There has also been a general realisation that countryside recreation is now a larger industry than either agriculture or forestry.

As policy has changed, so have the trends within the recreational use of forests. A list of the countryside uses taking place at any one forest may be small or large, but across the country foresters are now faced with managing and providing for such diverse activities as commercial ropes courses and husky dog sled training. Demand for access has grown slowly, but the tendency for access to become increasingly specialised and inherently resource-demanding has increased more quickly, making a state of change the norm for forest managers.

20.3 Mountain biking and forestry

Mountain biking itself has a relatively recent pedigree, with the first bikes being produced by the late 1980s and the sport gaining wide popularity in the early 1990s. Initially, confined by the technology of the bikes and the availability of cycling opportunities, most mountain biking took place on forest roads. These new users put pressure on existing forest roads and trails, as well as increasingly impacting on other forest uses. The Forestry Commission was especially well placed to meet this new demand given its land holdings and recreational remit.

By the mid-1990s, mountain bike use was nearing its first peak and the success of the purpose-built off-road route at Coed Y Brenin in Snowdonia, Wales illustrated the popularity of mountain biking in forests. In southern Scotland, growth in mountain bike use was confined to local riders using local knowledge on which trails were best and this may have remained so but for the success of the Coed Y Brenin trails. These trails established mountain biking as a major attraction and highlighted the opportunities of developing specific forest trails and experiences as tourism products.

This mountain bike boom encouraged local Forestry Commission staff to look at developing the forests of southern Scotland in a conscious attempt to increase recreational forest use. The 7stanes Project grew from an assessment of each forest's individual potential, its location and possible matches with changing holiday market requirements. Southern Scotland was identified as an area well placed to become a destination for short break biking holidays, although there was not enough riding in the area or at individual sites to warrant investment. However, the Foot and Mouth outbreak focussed energy and minds and the 7stanes Project was conceived. It quickly generated enough support to become a reality.

20.4 The 7stanes Project

The aim of the 7stanes Project was to create seven new mountain bike centres in key Forestry Commission locations across the south of Scotland by 2004. A range of partners raised £1 million for the initiative and European structural funding brought the £2 million project to life. Figure 20.1 provides a location map of the project. The development of the 7stanes mountain biking centres will have a number of benefits:

- providing employment and training;
- increasing tourism revenue;
- providing a sustainable recreation resource;
- encouraging the health benefits associated with cycling;
- involving communities; and
- supporting local enterprises based on the success of the 7stanes.

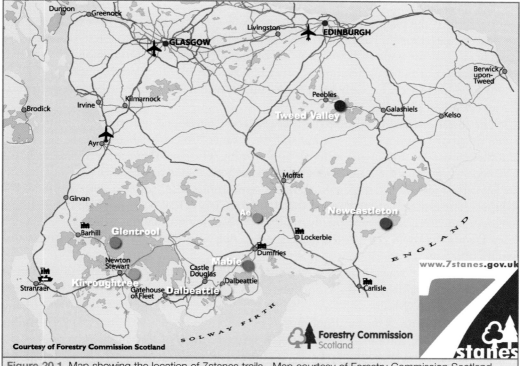

Figure 20.1 Map showing the location of 7stanes trails. Map courtesy of Forestry Commission Scotland. © Crown copyright. All rights reserved Forestry Commission Licence No. 100025498 2005.

But what makes a mountain bike centre a 'stane'? At the heart of the 7stanes is the concept of sustainability: in the construction of the trails; in their management; and in their ability to grow without consuming the very resource people are coming to enjoy. Although the idea is simple, the task remains substantial. The 7stanes has the same four objectives faced by any project, at any level, that wants to be sustainable (Department of Culture, Media and Sport, 1998):

- social progress which recognises the needs of everyone;
- effective protection of the environment;
- prudent use of natural resources; and
- maintenance of high and stable levels of economic growth and employment.

This has meant finding ways of working with land managers and users, understanding their needs and priorities, whilst creating the type of trails that mountain bikers will want not only to ride, but to ride time and again. The 7stanes Project has learned to accept changes to our ideas, whilst encouraging a willingness to accommodate new ideas amongst others. For instance, some sections of the early trail designs did not work in practice in local soil conditions. In redesigning these sections, the Project successfully sought the views of local mountain bikers and gave them considerable input into the final designs, and met the needs of the Forestry Commission.

20.4.1 Planning

A brief look at the planning process illustrates how access and forestry have been managed in the 7stanes Project. Planning the trails has involved foresters, forest planners and designers, and landscape architects, the use of forest design plans, production forecasts and coupe plans, and the consideration of other factors such as conservation, water quality and drainage. This process has only been successful because access became a part of the regular forest operation planning.

The forest districts have, over the past few years, developed systems to accommodate a wide range of considerations. Access and the 7stanes Project has been, on the one hand, another constraint, but on the other an opportunity. The constraints are obvious, especially the need to consider the trails as a new fixed asset, one that is treated with the same respect as any other feature in the forest. But the opportunity also exists to co-ordinate access, coupe felling and replanting dates at the planning stage. Thus, trails can help define future coupe boundaries. Environmentally, trails have become fixed habitat networks. Although they favour the more competitive edge species, they also break up blocks of conifer that would otherwise be continuous stands.

Mountain bike trails, being 15-20 km long, tend to penetrate deeper into the forests than other recreational facilities, including most forest walks which are often clustered around car parks. Mountain bikers, therefore, tend to see more of forestry operations and the impact of them. This has implications for land managers, both in terms of planning and day-to-day operations, and encourages foresters to think beyond production to the experience for users and wildlife.

At the early trail design stage, foresters have been able to flag up areas of particular concern or opportunity, listening to concerns, negotiating, resolving problems and attending joint on-site meetings. Although this can be time-consuming, it is essential if the trails are to become an asset rather than a liability. The mutual sharing of ideas and skills has been of great interest. The project, armed with a textbook full of sustainable trail designs, is being challenged to prove they work in the wet forests of South Scotland. Forestry civil engineers have shared ideas, and also learned from the project experience. The trails built by the 7stanes, mainly for mountain bike use, employ techniques that are now being applied to pedestrian use. In the longer term, every hour spent planning trails, before they are constructed on the ground, means less maintenance and liability for the land manager.

20.4.2 Managing potential conflict with other forest users

Apart from forest operations, the project has also had to consider other recreational users. This was implicit in the project planning before the project started, as the Forestry Commission has for a while practiced an open access policy. In Scotland, this came into sharp focus with the Land Reform (Scotland) Act 2003 which will mean that trails are now open to all forest users. This has been a major challenge.

On the one hand, the success of the trails has been their design for purpose – mountain biking; the flow of the trail, the singletrack experience and the technical challenges have been enjoyed by thousands of mountain bikers. Yet the project has noticed regular, and increased, use by walkers and some use by equestrians. This has posed problems, but by incorporating this into planning the project has been able to design-in features that are great for mountain bikers yet less attractive to other users, thereby discouraging other uses without banning them. For instance, singletrack, by its very nature, is narrow, less than 0.5 m wide. A well established trail can be as little as 0.3 m wide. This tends to militate against use by walkers who enjoy walking in groups side-by-side. The nature of the routes themselves, which often weave in and out of the trees and hug contours, can also be frustrating to other forest users, who tend to be more destination focused. The undulating trail surfaces are often rougher than other users enjoy, and the banked corners and jumps offer a great ride to mountain bikers but are less attractive to others.

Despite these design features, some other users will inevitably use mountain bike trails creating potential conflict and safety issues. The Project has tried to tackle this primarily at the design stage, but also at the information stage. At the design stage, the use of sustainable gentle grades, by their nature, control speed. Most accidents in forests involving bikes occur on forest roads, less so on the purpose built trails. Forest roads are broad, fast and sometimes steep, yet are attractive to novice cyclists. Mountain bike trails are narrow, technical and challenging. Rider speed can build up, but can be reduced at key points through appropriate trail design so that bikes do not cross paths, tracks and roads at speed. Blind corners are avoided and sight lines planned, all before the digger goes anywhere near the forest.

Information to all users is also important, though less so than trail design. At information points the forest code is displayed, as it is on leaflets. On the trails themselves the project has resisted turning the forests into waymarked and over-signed nightmares. A standard waymarking system has been employed and at particularly hazardous sections, or where other users might normally be expected, the project has offered the advice 'Mountain Bike preferred route'. So far, it seems to work.

The other potential source of conflict is that some mountain bikers will go onto walkers' paths and into other sensitive areas. Our experience is that purpose-built trails are better fun to ride, so concentrate usage. Walkers' paths, in contrast, are often dull and featureless for the biker. At the planning stage, it is important to consider who the target market for a trail is, and gear promotion and publicity to that market. Mountain bike users are also becoming used to trail grading systems, as used by skiers, and the 7stanes has been keen to promote an expectation of experience before riders even arrive on trail.

Purpose-built trails also facilitate access through sensitive areas. It is a fact that open access, responsible or not, will open up sensitive areas and purpose-built trails can help manage such sites more effectively than written codes. But it is inevitable that there will always be a few mountain bikers who are less responsible then others. As Sprung (1998) observed "The problem perhaps extends to all brands of recreationists. How many climbers can accept closure of some routes to protect bird nesting sites? Probably most, but not all. And how many hikers can accept the notion that some natural places should be *off-limits to everyone?"* So, despite an overall positive impact, trails and some of their users do continue to pose some problems.

20.4.3 Costs and potential income sources

For the 7stanes Project the biggest problem is one faced by most land managers: finance. Trail costs have been higher than anticipated in the original project proposal, and there is also the longer term issue of creating sustained income to fund maintenance and other ongoing costs such as marketing. The success of the trails can also encourage demand for facilities for other users, such as horse riders, creating further financial demands for the Forestry Commission.

There are some potential revenue sources. Car parks are one of the inevitable features of such centre-based developments. Mountain bike growth at the centres has come hand-in-hand with car use and the ability of riders to travel, sometimes not inconsiderable distances, to ride. Planning for cars is essential, but more than just a place to park, riders can and will reasonably expect to pay for additional, sometimes essential, parts of a good visit: bike wash; toilets; information points; showers; locker room; and a cup of coffee. Not all of these are essential, but the success of Glentress, one of the 7stanes, indicates that where such facilities are available, visitors are willing to spend – a potential source of funds for maintenance. However, car park planning also needs to be sensitive, considering other access points to the network, trying to avoid the unnecessary trend towards 'industrial tourism' as Abbey (1968) termed it; where the impact of the trails themselves is minimal compared to the loss of land to hard landscaping and additional facilities. The double bind is that developing revenue streams that do not require investment in such infrastructure seems to be very difficult.

To help forest managers to generate income, the Project has also been looking at sponsorship opportunities. By its very definition, sponsorship demands a commitment from us as trail builders and managers to secure mutually satisfactory deals from sales-focused marketing professionals. Looked at from a sponsor's point of view, mountain bike trails offer good penetration into a key target market, mostly higher social group males aged 30+ with high disposable incomes. For trail managers, sponsorship can offer a guaranteed income for, normally, a 3 year period. However, the 7stanes trails have been created through large public investment, beyond the means of most of the bike-related companies, and for the Forestry Commission tight public sector guidelines limit just how effective such deals could be. So sponsorship in the public sector, whilst useful, is not the answer to longer-term revenue generation.

20.5 The impact of the 7stanes Project

Projects such as the 7stanes can have a big impact. As of 2003, three of the seven centres were operational, 120+ km of trail had been created, 43 jobs had been either directly created or supported, and three major sponsorship deals had been struck. Significant media coverage was been generated both at home and abroad, and visitor numbers jumped considerably. For instance, Glentress has gone from below 100,000 visitors in 1999 to over 160,000 in 2003, with the growth attributed mainly to the mountain bike trails and the infrastructure on site.

This seemingly endless supply of high-spending mountain bikers has convinced local communities, councillors, foresters, land managers and funding bodies to support applications to develop mountain bike projects, but a word of caution at this point would be useful. Just consider how good it would be as a recreational user to see large numbers of facilities springing up across the country. The range of sites is potentially great, the merits of sites perhaps somewhat similar. So the recreational user instead of coming to your site monthly, may now come only once or twice a year. We are in the dangerous position of having a plethora of planned mountain bike trails all offering great riding experiences, but as yet the Project has not invested in the market, either by expanding the numbers taking up the sport at home or through targeted marketing further afield.

Now is a good time to take stock. We can envisage a network of properly planned and designed world-class centres of excellence, offering the biker great trails and all the accompanying facilities. This would be supported by national and international marketing, really establishing Scotland as one of the world's premier mountain bike destinations. Below this would be trails of regional importance, but without such large capital investment and at the grass roots level, via the core paths network, there would be access opportunities from the doorstep for all Scots to enjoy.

Mountain biking can thus be a potent force for rural investments, but it has to be developed in its national and international context. We certainly have some of the best riding in the world. The International Mountain Bicycling Association's (IMBA) latest score card gives Scotland an A- score, the top mark internationally, partly thanks to the work undertaken by the 7stanes Project, partly due to the Land Reform (Scotland) Act 2003 and partly due to the great landscapes we have for riders to enjoy.

20.6 Conclusions

In summary, managing access has become an essential tool in the forest manager's toolkit and purpose-built mountain bike trails have proved to be a valuable way to manage access. They can also bring considerable benefits to adjoining land owners and communities, bringing riders into areas often not visited by many tourists.

For nature conservation, properly planned trails offer both riders and wildlife the chance to co-exist. In addition, we are seeing riders' enjoyment of the trails invoking a sense of commitment that is expressed through volunteering for trail building and the growth in membership of the IMBA–UK. Far from the countryside being merely a testing ground for an individual's skill and stamina, there is a genuine interest developing amongst some riders in the forests they are enjoying, and in the work and livelihoods of those managing those assets.

The study of mountain bikers' behaviour on-site, rather than perceived behaviour by off-site managers or even the bikers themselves, offers important lessons for future trail planning and design. At a simple level, many bikers will avoid muddy and more difficult sections, despite a perceived desire for both, so if easier routes are provided they will be used. On a grander scale, therefore, challenging our sometimes negative perceptions about mountain bikers, and other recreational users, is an important start that can open up opportunities for local economies and nature conservation.

This case study echoes the results of a study of the potential for sustainable of mountain biking in the Cairngorms (Adams & Alexander, 2000). This study shows that managing access through an holistic approach, involving good communication and co-ordination, and effective management, will result in sustainable development in the long term.

Management is fundamental to successful access planning and development. However, managing access in practice is a far cry from the theory. It is complex and wide-reaching and often difficult to implement in practice. Integration is the back-bone of managing access. Yet the pressure on managers to deliver immediate solutions often precludes taking a wider view – let alone sharing good practice.

The forest environment is generally at the forefront of public changes with regard to access provision and a motivated and multi-disciplined approach is essential to deliver public expectations.

Successfully managing access is no accident. It has a context, a tradition and a code (Scottish Natural Heritage, 2005), but as the 7stanes is revealing, it requires us to challenge our own perceptions, planning to accommodate change, and listening to our own and others' common sense.

References

Abbey, E. (1968). *Desert Solitaire.* Ballantine Books, Random House, New York.

Adams, A. & Alexander, N. (2000). *Sustainable Tourism and Outdoor Recreation: Mountain-biking in the Aviemore-Cairngorm Massif Region: A Case Study.* www.ocio.deusto.es/formacion/ocio21/pdf/P11319.pdf.

Department for Culture, Media and Sport (1998). *Tourism - Towards Sustainability: a Consultation Paper on Sustainable Tourism in the UK.* Department for Culture, Media and Sport, London.

Land Reform (Scotland) Act 2003. www.opsi.gov.uk/legislation/scotland/acts2003/20030002.htm.

Scottish Natural Heritage (2005). *Scottish Outdoor Access Code.* Scottish Natural Heritage, Perth.

Sprung, G. (1998). *Recreation as an ally for environmental protection.* Presented at the conference 'Outdoor Recreation: Promise and Peril in the New West'. University of Colorado School of Law, Boulder. www.imba.com/resources/conservation/rec_ally.html.

21 Scottish Borders LBAP project: farming for partridges and sparrows

Andy Tharme

Summary

1. The Scottish Borders Local Biodiversity Action Plan partnership enlisted the support of farmers to grow wild bird cover on set-aside land in a core area of the Scottish Borders for two UK Biodiversity Action Plan priority species: the grey partridge (*Perdix perdix*) and the tree sparrow (*Passer montanus*).
2. By 2004, 72 farms had grown 76 ha (188 acres) of wild bird cover with targeted placement within farms to benefit both species.

21.1 Introduction

Grey partridge (*Perdix perdix*) and tree sparrow (*Passer montanus*) are both UK Biodiversity Action Plan (UKBAP) priority species. Between 1968 and 1999, they declined in the UK by 85% and 96% respectively (Baillie *et al.*, 2002). Key populations of both species are still found in the Scottish Borders and the Local Biodiversity Action Plan (LBAP) partnership wished to take action to help safeguard important core areas for both species. The Scottish Borders LBAP started in June 2001.

Research by The Game Conservancy Trust (GCT) has identified three main causes of population decline in grey partridge: reduced chick survival through loss of insect prey; habitat loss (especially hedgerows and grassy margins); and increased predation (Potts, 1986). Research by the British Trust for Ornithology has indicated that the decline in farmland passerines, including tree sparrow, has been driven by both reduced post-fledging and over-winter survival, probably linked to food supply (Siriwadena *et al.*, 1998).

The project aims to create more tailor-made habitat in the Scottish Borders for both grey partridge and tree sparrow, using the 'Wild Bird Cover' (WBC) option for growing areas of suitable plant species as part of a set-aside agreement. WBC provides a rich seed supply for both species over winter and also provides an ideal brood rearing habitat for grey partridge as it is rich in insects and provides cover from predators. WBC has also been shown to provide increased food resources for farmland songbirds (Henderson *et al.*, 2001; Parish & Sotherton, 2004). This project is a further development of the grey partridge conservation project initiated in East Lothian by the East Lothian LBAP in 2002. The following summarises progress to date and highlights some of the issues that we hope to develop during the course of the project.

21.2 Methods

A Project Steering Group, consisting of the Borders Farming and Wildlife Advisory Group (FWAG), Borders Forest Trust, GCT, NFU Scotland, Royal Society for the Protection of

Birds (RSPB), Scottish Agricultural College (SAC), Scottish Borders Rural Partnership, Scottish Rural Property & Business Association (SRPBA), Scottish Natural Heritage (SNH) and chaired and coordinated by Scottish Borders Council, identified a core area for grey partridge and tree sparrow in the Scottish Borders. Using GCT research data (Tapper, 1999), preferred areas for grey partridge were identified. The local Scottish Ornithologists' Club breeding bird atlas (Murray *et al.*, 1998) was used to identify the core areas for tree sparrows.

Three hundred and thirty farms were invited to participate, 25% of which wished to take part. Seventy-two farms were selected and, beginning in 2003, were given a choice of three WBC crop mixes (Table 21.1) and targeted advice on WBC placement to benefit grey partridge and tree sparrow (Table 21.2). Wild Bird Cover seed and conservation advice were given free of charge. Figure 21.1 shows the core areas and the participating farms.

Table 21.1 Wild Bird Cover crop mixes and their uptake.

Crop type	Sowing rate	Area grown (ha)
Kale & Triticale	5 kg ha^{-1} & 100 kg ha^{-1}	38.24
Kale & Quinoa	6 kg ha^{-1} & 1.25 kg ha^{-1}	19.63
Triticale & Linseed	125 kg ha^{-1} & 10 kg ha^{-1}	18.00
Total		75.87

Table 21.2 Placement of Wild Bird Cover crops within farms

Species to benefit	Location within farm
Grey partridge only	South-facing open farmland along fence or dyke, rank grassy margin, no backing woodland or tree line
Tree sparrow only	South-facing mature hedge, rank grassy margin, backed by mature broad-leaved woodland or tree line
Both species (preferred)	South-facing, along mature thick hedge, wide rank grassy margin. No backing woodland, though scattered tree line acceptable

Three project demonstration farms have been set up to host events to discuss Wild Bird Cover and other conservation management on farms, bird identification and monitoring and to share the ideas and experience of the farmers in the region.

21.3 Monitoring
21.3.1 Grey partridge monitoring
All farmers participating in the project are asked to record both spring counts and autumn (brood) counts of grey partridge under the GCT grey partridge UKBAP monitoring programme. Under the programme, all farmers receive an information pack from GCT with advice on monitoring and grey partridge management. By April 2004, 112 farmers were participating in the monitoring scheme in the Scottish Borders (72 under this project),

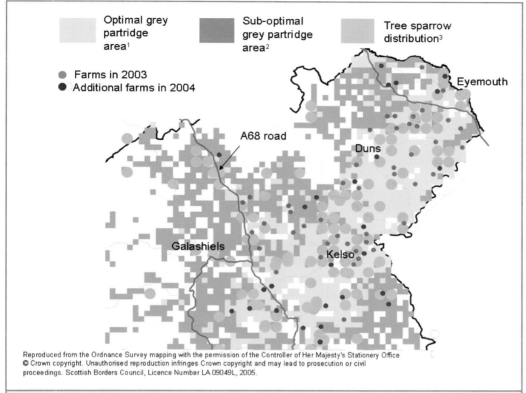

Figure 21.1 Core areas for grey partridge and tree sparrow in the Borders and farms in the project. Explanation of coloured key: 1) optimal = >50 ha km^{-2} tilled land, <10 ha woodland km^{-2} (Tapper, 1999); 2) sub-optimal = 10-49 ha km^{-2} tilled land, <10 ha woodland km^{-2} (Tapper, 1999); 3) tree sparrow distribution (Murray et al., 1998).

an increase from two farmers participating in 1999. The Scottish Borders now has the highest number of farmers in Scotland participating in the UK grey partridge BAP scheme and the second highest regional total in the UK (after Norfolk).

21.3.2 Breeding bird surveys

RSPB, Scottish Borders Council and a team of volunteers carried out breeding bird surveys on the three demonstration farms in 2003. This will provide a baseline index of all breeding birds including tree sparrow and red-listed passerines. Species not included are feral (*Columba livia*) and wood pigeon (*C. palumbus*), carrion crow (*Corvus corone*), rook (*Corvus frugilegus*), jackdaw (*Corvus monedula*), gulls (Laridae) and pheasant (*Phasianus colchicus*). An average of 10.22 ±9.22 (SD) grey partridge registrations per visit and 12.17 ±14.99 (SD) tree sparrow registrations per visit were recorded on the three demonstration farms.

21.3.3 Winter passerine monitoring

RSPB are co-ordinating winter monitoring of tree sparrow, red-listed passerines and the use of the Wild Bird Cover plots by other birds. In order to establish a winter baseline, 42 WBC plots on 30 farms were monitored over the winter of 2003-2004. Three visits were made to each site, to give a baseline index of the subsequent effectiveness of the management.

Ongoing RSPB research is investigating the link between winter numbers of tree sparrow and any local changes in breeding populations. Initial results indicated that tree sparrows were recorded on 31% of plots and overall were the fourth most abundant passerine recorded. Grey partridge occurred on 24% of plots. A large number of other red-listed passerines were also recorded.

21.4 Community involvement

Nest boxes for tree sparrows were made and put up by local school children at three farms and a golf course within the core project area. The project aims to develop further schools and community initiatives.

21.5 Links to existing schemes

21.5.1 Links to forestry

Under the Scottish Forestry Grant Scheme (SFGS) (Forestry Commission, 2003) there is a specific element targeted to improve the diversity of the farmed landscape. Whilst any increase in native woodland in farmland is, in principle, to be welcomed, there may be implications for open farmland species such as grey partridge, which is a UKBAP priority species. GCT research indicates that preferred areas for grey partridge have less than 10 ha of woodland per km^2 (100 ha), and that even small increases above this area of woodland can have significant negative effects on the grey partridge population (Game Conservancy Trust, unpublished).

With this in mind, the Forestry Commission, in liaison with GCT and SBC, has agreed to monitor annually the area of new woodland created under the SFGS in the optimal and sub-optimal grey partridge areas in the Scottish Borders. Results from the study will be important for designing integrated landscape support mechanisms in the future.

21.5.2 Links to other schemes

The Scottish Executive Environment and Rural Affairs Department (SEERAD) initially approved the project as an environmental scheme, so participating farmers who submitted an application to the Rural Stewardship Scheme (RSS) (Scottish Executive, 2003) were able to include this as part of other packages. The environmental scheme element of RSS applications has now been dropped by SEERAD.

Participation in the project is also one of the selection criteria used in assessing applications to the Scottish Borders Council's 'Living Landscapes' scheme, for planting hedgerows and small woodlands. The project is also proving to be an ideal means of implementing management on farms that have Whole Farm Plans for grey partridge produced by Borders FWAG.

21.6 Future developments

It is hoped to run the project for several years if sufficient funding can be secured. For some farmers, this project represents their first participation in a conservation scheme. It is hoped that the LBAP partnership can build on the relationship established with farmers in the region, so they can assist and support them with any future environmental management following the Common Agricultural Policy reforms. Provision of Wild Bird Cover through further roll out of RSS (Unharvested Crops prescription), or perhaps through an option available under a future Land Management Contract for farmers, will help provide key habitat for grey partridge, tree sparrow and other priority species. The strong partnership approach adopted by the Scottish Borders LBAP is taking important steps to safeguard the populations of national priority species within the region.

Acknowledgements

Funding for the project was provided by Scottish Natural Heritage and Scottish Borders Council. The project was co-ordinated by Dr Andy Tharme, Scottish Borders Council. The project partners are Derek Robeson (Borders FWAG), Hugo Straker (GCT), Chris Walton (NFU Scotland), Peter Gordon (RSPB Scotland), David Kerr and Moira Gallagher (SAC), Keith Robeson (Scottish Borders Council), Graeme Wilson (Scottish Borders Rural Partnership), Hilary Dunlop (SRPBA), Sarah Eno and Steve Hunt (SNH), the demonstration farms of Bruce and Rob Cowe, George Farr and Chris Walton, and the farmers of the Scottish Borders.

References

Baillie, S.R., Crick, H.Q.P., Balmer, D.E., Beaven, L.P., Downie, I.S., Freeman, S.N., Leech, D.I., Marchant, J.H., Noble, D.G., Raven, M.J., Simpkin, A.P., Thewlis, R.M. & Wernham, C.V. (2002). *Breeding Birds in the Wider Countryside: their conservation status 2001. BTO Research Report No.* 278. British Trust for Ornithology, Thetford.

Forestry Commission Scotland (2003). *Scottish Forestry Grant Scheme*. Forestry Commission Scotland, Edinburgh.

Henderson, I.G., Vickery, J.A. & Carter, N. (2001). *The Relative Abundance of Birds on Farmland in Relation to Game-cover and Winter Bird Crops. BTO Research Report No.* 275. British Trust for Ornithology, Thetford.

Murray, R.D., Holling, M., Dott, H.E.M. & Vandome, P. (1998). *The Breeding Birds of South-East Scotland: A Tetrad Atlas 1988-94.* The Scottish Ornithologists' Club, Edinburgh.

Parish, D.M.B & Sotherton, N.W. (2004). Game crops and threatened farmland songbirds in Scotland: a step towards halting population declines? *Bird Study*, **51**, 107-112.

Potts, G.R. (1986). *The Partridge: Pesticides, Predation and Conservation*. Collins, London.

Scottish Executive (2003). *Rural Stewardship Scheme*. Scottish Executive, Edinburgh.

Siriwardena, G.M., Baillie, S.R., Buckland, S.T., Fewster, R.M., Marchant, J.H. & Wilson, J.D. (1998). Trends in the abundance of farmland birds: a quantitative comparison of smoothed Common Birds Census indices. *Journal of Applied Ecology*, **35**, 24-43.

Tapper, S. (Ed) (1999). *A Question of Balance: Game Animals and Their Role in the British Countryside*. The Game Conservancy Trust, Fordingbridge.

21 Farming, Forestry and the Natural Heritage: Towards a More Integrated Future

22 Determining the effects of grazing on moorland birds: a summary of work underway at the RSPB

Murray Grant, James Pearce-Higgins, Graeme Buchanan & Mark O'Brien

Summary

1. Few quantitative data exist to determine the effects of grazing on moorland birds, although high grazing pressure is considered a likely cause of decline in their populations.
2. Research that seeks to address this issue is currently underway, and several different projects that examine aspects of the relationships between moorland grazing and birds are described.
3. The results from broad-scale correlative studies in southern Scotland describe relationships between moorland bird abundance and vegetation composition and structure. These relationships are interpreted in terms of likely effects of grazing on vegetation, and indicate how certain species (e.g. red grouse (*Lagopus lagopus scoticus*) and stonechat (*Saxicola torquata*)) may be susceptible to high grazing pressure, whilst others (e.g. golden plover (*Pluvialis apricaria*)) may benefit.
4. Following from the initial broad-scale studies, a number of other projects are underway that seek to build upon this work and to investigate the effects of the relationships between moorland grazing and birds in greater detail. The aims and approaches of these various projects are described.

22.1 Introduction

UK moorlands are of international conservation importance, in part because of the unique breeding bird assemblage they support (Thompson *et al.*, 1995). Increasingly, evidence suggests that several of the species in this assemblage are declining (e.g. Sim *et al.*, 2005), with higher grazing levels (from increasing sheep (*Ovis ovis*) and red deer (*Cervus elaphus*) numbers) often considered as a likely cause (Fuller & Gough, 1999). While it is established that sufficiently intense grazing on moorland generally causes a shift from dwarf shrubs to graminoids, along with a reduction in vegetation height and density (Thompson *et al.*, 1995), few quantitative data exist to assess how variation in grazing affects moorland birds, despite the fundamental nature of the associated habitat changes.

To address this issue, a number of projects concerned with moorland birds and grazing have been initiated and are currently underway. These include two consortium-run projects that consider wider issues associated with moorland grazing (see Table 22.1). The aim of this short chapter is to provide an overview of this work, focusing on the rationale and development of the work. Thus, some of the findings from this research to date are summarised briefly, whilst the aims and approaches being adopted in the different projects are outlined.

Table 22.1 Details of projects described in this chapter. Abbreviations used: RSPB – Royal Society for the Protection of Birds; SEERAD – Scottish Executive Environment and Rural Affairs Department; Defra – Department of Environment, Food and Rural Affairs; CCW – Countryside Council for Wales; ADAS – ADAS Consulting Ltd.

Project	Operative dates	Funding organisations	Lead organisation	Main RSPB contact
Moorland grazing and birds study	1999-2001	RSPB	RSPB	James Pearce-Higgins
Effects of reducing livestock numbers on moorland bird abundance at Geltsdale	1999-2006*	RSPB	RSPB	Murray Grant
Grazing and Upland Birds	2002–2004*	SEERAD	Macaulay Institute+	Murray Grant
Determining environmentally sustainable and economically viable grazing systems for the restoration and maintenance of heather moorland in England and Wales	2002–2007	Defra, English Nature, CCW	ADAS+	Murray Grant
Determining effects of the 2001 outbreak of foot and mouth disease on upland breeding waders	2002–2004	English Nature, RSPB	RSPB	James Pearce-Higgins

* Refers to initial phases of projects only.
+ Indicates consortium projects involving several organizations.

22.2 Relationships between bird abundance and vegetation composition and structure

Initial investigations of moorland grazing and birds examined how the breeding densities of nine species varied with vegetation composition and structure, using data from 85 2 km² plots (mainly in southern Scotland). On each of these plots, bird abundance was measured by standard methods (O'Brien & Smith, 1992; Brown & Shepherd, 1993; Thirgood et al., 1995), whilst detailed measures of vegetation composition and structure were made (Pearce-Higgins & Grant, 2002). Additionally, a wide range of physical and management variables were also recorded for each plot (e.g. topography, climate, soil type, gamekeeper density, proximity to forestry). The approach adopted uses vegetation condition as an indicator of historical grazing pressure, as opposed to current stocking density, which may not reflect past grazing and can produce varying effects on vegetation according to physical factors, livestock type and habitat (Marrs & Welch, 1991). Generalized linear models were used to relate bird abundance to the vegetation measures, conducting analyses so that the effects of the potentially confounding physical and management variables were controlled first, thereby reducing the risk of detecting spurious vegetation effects.

Using heather (*Calluna vulgaris*) cover and vegetation density as examples of compositional and structural features that are sensitive to grazing, and will be reduced by high levels, relationships suggest that red grouse (*Lagopus lagopus scoticus*) and stonechat (*Saxicola torquata*) are most susceptible to high grazing pressure, as both species are most abundant where there is extensive heather cover. Other species, such as skylark (*Alauda arvensis*) and golden plover (*Pluvialis apricaria*), are likely to benefit on many moors from relatively high grazing pressure, as they are most abundant on short, open, swards (Figure 22.1). Although whinchat (*Saxicola rubetra*) abundance increased with vegetation density, in a similar way to that of stonechat, the relationship for this species was linked to an association with bracken (*Pteridium aquilinum*) cover (Pearce-Higgins & Grant, in press). Therefore, high grazing levels could still create suitable habitat conditions where grazing is important in maintaining or increasing bracken cover by reducing the competitive ability of heather (Marrs & Pakeman, 1995; Ninnes, 1995). Meadow pipits (*Anthus pratensis*) tended to be most abundant on moors with a grass-heather mix, but with grasses predominating. The densities of curlew (*Numenius arquata*), snipe (*Gallinago gallinago*) and wheatear (*Oenanthe oenanthe*) were unaffected by variation in either heather cover or vegetation density. Other relationships indicate the importance of wetland vegetation, such as sedge and rush communities, and of compositional and structural heterogeneity in determining bird abundance (Table 22.2). Thus, for several species increased grazing may be detrimental if it causes a loss of such heterogeneity, although heterogeneity could also increase with grazing levels.

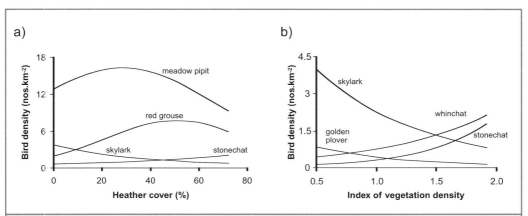

Figure 22.1 Predicted relationships between density and heather cover (a) and vegetation density (b) for species where statistically significant effects of these vegetation variables were detected using generalized linear models. Relationships are shown having first accounted for statistically significant physical and management effects (see text). Note, meadow pipit and skylark densities are divided by 10 for presentation.

Table 22.2 Simplified summary of the direction of relationships between abundance and broad vegetation measures for several moorland breeding bird species, in southern Scotland. The summary is derived from relationships, using generalized linear models, determined for a wider range of more detailed compositional and structural measures, in analyses where vegetation effects are considered only after accounting for statistically significant physical and management effects. The relationships presented are summarised from statistically significant vegetation effects only (details provided in Pearce-Higgins & Grant, in press). Cover refers to percentage cover.

Bird Species	Height/density of vegetation	Variation in vegetation height	Dwarf shrub cover	Dwarf shrub/graminoid heterogeneity	Wetland plant cover
Red grouse			+	+	
Golden plover	–				+
Curlew		+			+
Snipe		+			+
Skylark	–		–		+
Meadow pipit				+	+
Whinchat	+			+	+
Stonechat	+		+	+	

22.3 Moorland restoration and bird populations

Current agri-environment schemes in England and Wales can halt dwarf shrub loss on moorland, but have had limited success in achieving dwarf shrub recovery (ADAS, 1998). Therefore, a five-year consortium project was initiated in 2002 to determine grazing systems, including both cattle and sheep and different breeds, that will achieve restoration of heather moorland (Table 22.1). Impacts on farm economics, vegetation and biodiversity are assessed, with the unified approach to addressing environmental and economic sustainability being a key feature. Effects on bird populations are assessed using bird-habitat models as outlined above, but with the inclusion of data from additional regions (North Wales, North Pennines and South Pennines) to determine their generality across moorland areas. The bird-habitat models are then linked to data on the vegetation response to different grazing systems to predict bird response, determining the impact of grazing systems on vegetation through a combination of modelling and field trials. While this work is still in its early stages, preliminary analyses suggest that, although bird abundance may differ between regions for a given habitat condition, the direction of the response to variation in vegetation composition and structure is consistent. Thus, generalized predictions on the response of bird abundance to grazing systems may be possible in many cases.

22.4 Verifying spatial models and determining mechanisms

Monitoring temporal trends in bird abundance following changes to grazing systems provides a means of verifying predictions from the above spatially based models. This is undertaken for meadow pipit and skylark within the grazing trials of the moorland restoration project, and for a wider range of species in projects investigating the impact of foot and mouth disease and longer-term effects of stock reductions at RSPB's Geltsdale reserve (Table 22.1). Associated changes in vegetation are also examined in these studies, enabling close linkages to be made with the bird-habitat models.

Greater understanding of underlying mechanisms should emerge from the Grazing and Upland Birds (GRUB) project (see Table 22.1). Involving several institutes, a fully replicated experiment, at Glen Finglas in Stirlingshire, has been used to determine how different grazing regimes affect meadow pipit breeding ecology, via changes in habitat structure and arthropod abundance (Dennis, 2003). The applicability of the findings from this experiment to other moorland types and bird species is determined by linking bird distributions to arthropod abundance and habitat on an extensive scale using a series of plots in southern Scotland. Initial fieldwork will be completed in 2004, with findings to be reported soon after.

Acknowledgements

Thanks to RSPB staff (past and present) who have collected data for these projects, and helped in other ways, and to many landowners who allowed access. ADAS, Institute of Grassland and Environmental Research, Centre for Ecology and Hydrology (CEH), Scottish Agricultural College (SAC) and Newcastle University are partners on the moorland restoration project, whilst the Macaulay Institute, CEH, SAC, Stirling University, Aberdeen University and BioSS are partners on GRUB. The Woodland Trust provides access to their land for the purposes of the GRUB experiment. Work has been funded by Defra, SEERAD, English Nature, Countryside Council for Wales and RSPB.

References

ADAS (1998). Effect of stocking rates and vegetation management practices on the regeneration of *Calluna vulgaris* and dwarf shrub heath communities. Unpublished report. Countryside Management Division, MAFF, London.

Brown, A.F. & Shepherd, K.B. (1993). A method for censusing upland breeding waders. *Bird Study*, **40**, 189-195.

Dennis, P. (Ed) (2003). *Effects of Grazing Management on Upland Bird Populations: Disentangling Habitat Structure and Arthropod Food Supply at Appropriate Spatial Scales (GRUB).* 2nd year interim report to SEERAD.

Fuller, R.J. & Gough, S.J. (1999). Changes in sheep numbers in Britain: implications for bird populations. *Biological Conservation*, **91**, 73-89.

Marrs, R.H. & Pakeman, R.J. (1995). Bracken invasion: lessons from the past and prospects for the future. In *Heaths and Moorland: Cultural Landscapes*. ed. by D.B.A. Thompson, A.J. Hester & M.B. Usher. HMSO, Edinburgh. pp. 180-193.

Marrs, R.H. & Welch, D. (1991). *Moorland Wilderness: The Potential Effects of Removing Domestic Livestock, Particularly Sheep.* ITE Report to Department of Environment, Project T09052e1.

Ninnes, R.B. (1995). Bracken, heath and burning on the Quantock Hills, England. In *Heaths and Moorland: Cultural Landscapes*. ed. by D.B.A. Thompson, A.J. Hester & M.B. Usher. HMSO, Edinburgh. pp. 194-199.

O'Brien, M.G. & Smith, K.W. (1992). Changes in the status of waders breeding on lowland wet grasslands in England and Wales between 1982 and 1989. *Bird Study*, **89**, 165-176.

Pearce-Higgins, J.W. & Grant, M.C. (2002). The effects of grazing-related variation in habitat on the distribution of moorland skylarks *Alauda arvensis* and meadow pipits *Anthus pratensis*. *Aspects of Applied Biology*, **67**, 155-163.

Pearce-Higgins, J.W. & Grant, M.C. (in press). Relationships between bird abundance and the composition and structure of moorland vegetation. *Bird Study*.

Sim, I.M.W., Gregory, R.D., Hancock, M.H. & Brown, A.F. (2005). Recent changes in the abundance of British upland breeding birds: Capsule Breeding wader populations have more often shown declines than passerine populations during the last 10–20 years. *Bird Study*, **52**, 261-275.

Thirgood, S.J., Leckie, F.M. & Redpath, S.M. (1995). Diurnal and seasonal variation in line-transect counts of moorland passerines. *Bird Study*, **42**, 257-259.

Thompson, D.B.A., MacDonald, A.J., Marsden, J.H. & Galbraith, C.A. (1995). Upland heather moorland in Great Britain: a review of international importance, vegetation change and some objectives for nature conservation. *Biological Conservation*, **71**, 163-178.

23 The Breadalbane Initiative for Farm Forestry

Ruth Anderson

Summary

1. This chapter describes the Breadalbane Initiative for Farm Forestry, which is a farmer-led initiative involving several partner organisations. The Initiative aims to safeguard the future viability of farming in the area and stimulate local economic development by diversifying into forestry.
2. The focus of work is on awareness-raising, timber utilisation projects and policy development.
3. Current funding schemes do not provide effective support to the creation and management of farm woodlands, and the chapter suggests several ways of improving this support.

23.1 Introduction

The Breadalbane Initiative for Farm Forestry (BIFF) is a farmer-led initiative, working within the Breadalbane Environmentally Sensitive Area (ESA) in Perthshire. Its aim is to encourage integration of forestry with farming, in order to diversify and strengthen the status of farming in the area, to support and generate additional woodland- or timber-based activity on farms and in the local economy, and to help revitalise local woodland culture and industry.

The Initiative is run by a group of local farmers, with support from the Scottish Agricultural College, the Worldwide Fund for Nature Scotland, and Scottish Native Woods. It employs a part-time coordinator, funded by Perth & Kinross Council, Forestry Commission Scotland, and the Scottish Executive's Environment and Rural Affairs Department (SEERAD).

The Breadalbane ESA equates roughly with Highland Perthshire (see Figure 23.1). Farm woodlands are a significant component of the landscape in this area, and are part of the basis on which the ESA was designated. Uptake of the (voluntary) ESA scheme by Breadalbane farmers has been high (90%) and some 1,800 ha of woodland are now under ESA management. The area also has a high proportion of commercial conifer plantations, both private and public. Forest cover around Aberfeldy is about 30%, well above the national average of 17% and the Scottish Executive's target of 25%.

It is the extent and diversity of the forest resource, in the context of wide-reaching changes in the agricultural sector, which encouraged local farmers to consider forestry as a further development of their businesses and as a means of stimulating local economic development. The challenge is how to make this wood resource economically relevant to farmers under conditions of very low timber prices.

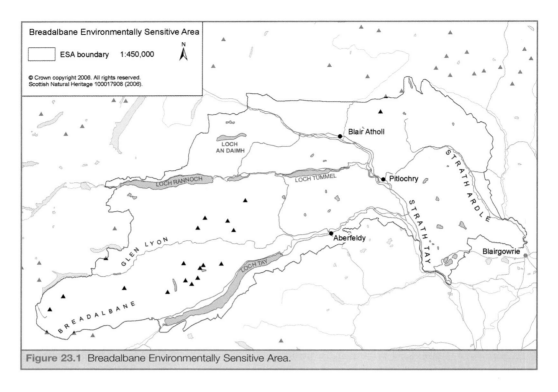

Figure 23.1 Breadalbane Environmentally Sensitive Area.

23.2 Farm forestry – a special case

By their nature, farm woodlands are well-placed to provide many of the environmental and social benefits now expected from forestry. They tend to be smaller rather than larger, diverse in form and species composition, and are an important component of the landscape. Appropriate management can safeguard a range of habitats and species and can also support the farm enterprise through provision of shelter for stock, cover for game, firewood, fencing materials and other products.

From a policy perspective, farm woodlands are a key route to deliver strategic objectives of the Scottish Executive, such as land use integration, rural development, and woodland expansion. However, farm woodlands are not well supported by current funding schemes. SEERAD's Rural Stewardship Scheme (Scottish Executive, 2000), now subsuming the ESA scheme, supports only limited woodland management, in a limited range of situations. The definitions and eligibility criteria used in the Scottish Forestry Grant Scheme (SFGS) (Forestry Commission Scotland, 2003) effectively debar many farm woodlands on grounds of stocking densities, percentage of open ground or shrub species, or grazing. As a result, most farm woodlands remain unmanaged. Part of BIFF's work has been to encourage better integration of these schemes and new policy initiatives are now being worked up which recognise this need.

23.3 BIFFs approach

The Initiative is seeking to develop a local economy where farm woodlands make an effective contribution which complements, rather than competes with, agricultural activity. Work focuses on three approaches.

23.3.1 Raising awareness, building capacity and bridging skills gaps

This is achieved through provision of advice, and demonstration and training events. The Initiative has held a number of open days looking at issues of woodland management and planning. The most recent of these was to learn about the uses of woodchips on farms, which can include corrals for wintering cattle and a fuel source. This event was very well-attended, which illustrates the level of interest from the farm management perspective and in woodchips as a saleable product from farm woodlands.

23.3.2 Developing local projects which stimulate and support timber utilisation and marketing opportunities

The main focus here is on a renewable energy project designed to meet the heating requirements of the proposed new school in Aberfeldy, with woodfuel supplied from local sources. This technology is widely used in Europe and is of current interest to many in rural Scotland. BIFF, and other stakeholders, commissioned a feasibility study (Henderson, 2003) to determine the type and cost of technology needed to meet the school's requirements, the potential level of supply of woodchips from local forests, the type and cost of infrastructure development needed to extract, process and supply that biomass, as well as sources of potential funding and an indication of the likely employment, economic and environmental impacts of the scheme. The study concluded that provision of a wood-fuelled heating system at the proposed new school was a practical alternative to reliance on fossil fuel systems.

Perth and Kinross Council is supportive of the project and Forestry Commission Scotland locally has offered to underwrite the supply. BIFFs aspiration is that a guaranteed proportion of the woodfuel required would be drawn from small woodland owners in the locality. An outline assessment of potential supply from local farms found more than adequate capacity within the immediate area. The next step for BIFF is to undertake a full feasibility study of the supply chain and develop a business plan.

BIFF farmers are also hoping to become involved in a local access and recreation project, as supplier of the timber for construction of gates, stiles, bridges and signposts throughout a local path network.

Finally, a link has been created with an education project at Breadalbane Academy, which offers vocational modules in agriculture, environmental conservation, forestry and game-keeping. This provides pupils aged 15 to 18 with a chance to acquire basic skills and qualifications in land-use activities, and is also an innovative way of encouraging deeper ties between young people, local resources and potential employment opportunities.

23.3.3 Policy development

As outlined above, farm woodlands are presently at a disadvantage in terms of grant support, relative to other farm habitats and indeed to non-farm woodlands. This situation reflects and perpetuates the historical separation of farming from forestry. However, as Forestry Commission Scotland and SEERAD begin to work more closely together, BIFF has been collaborating in the development of ideas on improving grants for farm woodlands. This dialogue has highlighted a number of principles:

- that farmers will be more likely to enter a woodland management scheme if there is a common entry point through the agri-environment programme, thus allowing farm woodlands to be treated comparably to other farm habitats;

- a broad definition of farm woodland needs to be applied, which recognises the value and eligibility of scattered or open woodlands, small copses, field margin trees, scrub, mixed plantations and shelter belts, in addition to current descriptions;
- grants should be objective-orientated, supporting a fuller range of objectives, including those which contribute to the agricultural business, such as shelter for stock, production of fencing materials, and/or conservation of other habitats (for example, woodland grazing to maintain diversity of ground flora);
- rates of grant should reflect real costs, recognising the implications of both income foregone and additional management; and
- training, cooperative working, and establishment of local businesses which process, market and use timber also require support.

In the wider context, the European Union's Common Agricultural Policy (CAP) continues to exert a major influence. Uptake of any new grants which may come onstream for farm woodlands will be in the still uncertain climate engendered by CAP reform. It remains to be seen how farmers will respond to the decoupling of support and production. In particular, the way in which rules on land use and single farm payment entitlement are developed will either free up or block land for woodland expansion. The creators of the SFGS will have to keep a close eye on grant rates and integration rules, so as to take advantage of new opportunities which may arise in a positive and structured fashion. In this respect, Habitat Action Plans and Forest Habitat Networks (Fowler & Stiven, 2003) may be used to guide expansion of tree cover on farms.

23.4 Conclusion

Given the continuing uncertainties in the agricultural sector, the current downturn in the forest industry and dearth of markets for forest products, BIFFs aspirations may seem ambitious. But if, through the work of BIFF and others, trees and timber can come to be seen as an integral part of the farm's natural resources, then there may be a very positive outcome for farm woodlands from CAP reform. Of course, for this to happen, we will all have to work hard to make sure that trees on farms are not ignored for another generation.

References

Forestry Commission Scotland (2003). *The Scottish Forestry Grant Scheme.* Forestry Commission Scotland, Edinburgh.

Fowler, J. & Stiven, R. (2003). *Habitat Networks for Wildlife and People: the Creation of Sustainable Forest Habitats.* Forestry Commission Scotland, Edinburgh and Scottish Natural Heritage, Perth.

Henderson, R. (2003). The potential for wood-fuelled heating at the proposed new Breadalbane Academy: pre-feasibility study for WWF Scotland and Perth & Kinross Council. Unpublished report. WWF Scotland, Aberfeldy and Perth & Kinross Council, Perth.

Scottish Executive (2000). *The Rural Stewardship Scheme.* The Stationery Office, Edinburgh.

24 Hill farming and environmental objectives: conflict or consensus?

C. Morgan-Davies, A. Waterhouse, K. Smyth & M.L. Pollock

Summary

1. This study investigated the attitudes of hill farmers and crofters, and environmental specialists, to environmental land management objectives.
2. A participative research approach was used to assess how both interest groups viewed the land management problems and solutions and how they succeeded or failed in reaching consensus on three topical environmental management issues: woodland regeneration; recreational access on the land; and livestock numbers.
3. Both groups agreed on the need to combine economic, biological, and environmental goals for land management, but they differed on how best to accomplish this.
4. The key results arising from the participatory approach are presented as well as comments on group dynamics, interactions and the means of achieving consensus.

24.1 Introduction

Although establishing high environmental goals is fundamental to Scottish hill farming systems (Scottish Executive, 2001), opinions differ on how to do it, rendering the process difficult and leading to conflict between those farming the land and those prescribing rules (McEachern, 1992; Marshall, 2004). However, considerable practical knowledge in both hill farming and conservation communities is available. Working together, there is potential to evaluate how management might be delivered and tailored to a local context, both to the underlying biodiversity and to land management practices, as demonstrated by Schusler & Decker (2002).

In order to understand how both environmental and economic sustainability in hill farming and crofting systems might be achieved, and to identify related trade-offs and conflicts, a programme of participative research involving both hill farmers and crofters and local environmental specialists was undertaken to determine whether these two groups could agree and provide a model for future land management planning.

24.2 Methods

Workshops were held in 2001 and 2002 with local farmers and nature conservation and policy specialists from non-governmental organisations and Government agencies. Three regions were considered, covering the range of hill farming systems in Scotland: central Highlands (held at Loch Lomond in February 2001), North Western Cairngorms (held in Aviemore in June 2002) and crofting counties and islands (held in Kyle of Lochalsh in October 2002).

Participants were invited from contact lists of local agricultural and conservation advisers. They were sent a short questionnaire with 20 topic areas, drawn up by the research team, and asked to choose three topics they would like to discuss. Topics were then ranked according to the number of times they were chosen. The three most popular topics, common to both local farmers and conservation specialists, were identified: recreational access to the land, livestock ratios on the land, and woodland regeneration on hill farms. Two of the topics were covered at each workshop.

Through group exercises, the participants identified and discussed key issues relating to the topics, then solutions, whilst considering trade-offs and assessing the environmental and economic impacts of the different approaches. In total, 38 people participated in the workshops.

24.3 Results

For each topic, a number of issues and problems, with their solutions, were identified and proposed by farmers and conservationists (see Tables 24.1, 24.2 and 24.3).

Table 24.1 Issues, problems and solutions regarding access, as identified in the Loch Lomond and Aviemore workshops (not in rank order). Number of attendees is given in brackets.

Issues and problems	
Farmers (12)	**Conservationists** (11)
• Livestock issues (lambing time interference, loss of income, extra management required)	• Visitor management and pressure
• Tourists/walkers issues (irresponsible parking, litter, crime and damage to flora, fauna and archaeological features)	• Lack of education about the countryside
• Problems arising from organised walking holidays and inaccurate guide books	• Biodiversity degradation (disturbance/crime to wildlife, habitat degradation, path erosion)
• Maintenance of the land	• Problem of litter, gates left open, fly-tipping
• Biosecurity issues, safety/liability	
Identified solutions	
Farmers (12)	**Conservationists** (11)
• More ranger services, available 24 hours	• Management of the visitors through rangers
• Establish official pathways and proper signage, provide more local consultation on legislation and walking routes	• More rural education towards farming, wildlife and field sports (in schools, countryside visits, hill phones, eco-tourism)
• Provide countryside education for the general public (education trusts, linking schools to farms, farmers talking in schools, etc.) and encourage visitors to take responsibility	• Proper maintenance of the land and access to it, at the local authority level
	• More information about wildlife, livestock and farming activities
	• Better funding of these activities for land managers

Table 24.2 Issues, problems and solutions regarding livestock ratios, as identified in the Loch Lomond and Lochalsh workshops (not in rank order). Number of attendees is given in brackets.

Issues and problems	
Farmers (17)	**Conservationists** (10)
• Grazing management – controversy over which species (sheep, cattle or deer) causes problems or is most appropriate for land management • Animal control issues (less shepherd/labour leads to uncontrolled grazing, feeding areas are problematic and damaging) • Local grazing issues (under-grazed hills, vegetation changes, local problem of ragwort infestation without grazing) • Local policies and monetary issues (locals versus visitors, issues of support payment, profitability, changing and inconsistent policies)	• Grazing issues (sheep are overgrazing, not enough cattle, lack of definition of inappropriate grazing) • Management issues (quantity versus quality, loss of local common sense, lack of flexibility in agri-environmental schemes) • Breed of animal should be considered • Problems of unburied livestock

Identified solutions	
Farmers (17)	**Conservationists** (10)
• Need for site specific add-ons and individual prescription for environmental schemes (Rural Stewardship Scheme is too competitive and inflexible) • Land managers (including grazing commons) should decide on land use, with viable choices, in consultation with local people • Area payments should be worked out on an individual farm basis and the viability of the farm must be considered	• Proper maintenance of the land and its access, at the local level, is needed • Individual farm management plans must be developed, with both farmers and conservationists • Payments should be based on whole unit management plan and subject to environmental audit, and solutions applied locally • Communication must be developed, with larger spectrum of people to decide on outcomes at a system level • Real costs of implementing policies must be recognised

Table 24.3 Issues, problems and solutions regarding woodland regeneration, as identified in the Aviemore and Lochalsh workshops (not in rank order). Number of attendees is given in brackets.

Issues and problems	
Farmers (19)	**Conservationists** (7)
• Woodland management and planning issues (fence lines and fencing, lack of use of local knowledge, need for Township plan in the crofting counties, lack of proper use of farm woodland)	• Management issues (need for a balanced strategy, more integration of land management, clear objectives, appropriate stock management systems)
• Livestock/grazing issues (removal of sheep leads to depopulation and the increase of impenetrable scrub areas; re-introduction of livestock should be an option, to keep people who can manage the land with local skills and knowledge)	• Legislation/framework (allowing controlled grazing in and beside woodlands, avoiding 'hard' woodland edges with conservation issues, allowing strips of woodlands rather than large blocks)
• Legislation issues (financial problems with grants and tenancy)	• Financial issues (is there economic benefit from timber, problem of tenant versus owner for benefits, need incentive payments, need to recognise the true costs)

Identified solutions (both by farmers and conservationists)

- Strategic plan is necessary, based on a Farm Management Plan
- Site specific planning: should be objective-led and include land managers in the decision-making
- Give ownership of woodlands to the Crofting Township (where appropriate)
- Site specific management, with flexible prescriptions and local management
- Adequate payments, with more funding within new schemes
- Better publicity for consultations on woodland planting
- More use of existing woodlands

Both groups recognised the others' viewpoints. Although presenting different arguments, and often approaching issues in a different way, there was broad consensus on land management requirements. Disagreement occurred when considering the finer details of how management plans should be implemented. There were differences of view on issues such as shifting from a production grant to a visitor management grant for recreational access, the problem of compensation for woodland regeneration (landlord versus tenant legislation), and livestock ratios and the issues relating to overgrazing. However, the degree of agreement between the groups was very high.

24.4 Discussion and conclusions

A number of benefits and opportunities for the participants and facilitators were derived from this exercise. Firstly, it gave a better understanding of the inter-related effects of farming management and objectives with those involved in biodiversity and landscape management. The participants said the meetings were beneficial, with many stating that it was an 'eye-opener'. Thompson *et al.* (1999) also recognised the effectiveness of small group discussions to improve knowledge and change of attitudes among farmers on certain issues.

It also gave an opportunity for the different groups to really express their views on the subject. Although both groups differed in their approaches, they had similar views regarding local issues. Both acknowledged the value of local discussion, local knowledge and the solving of local problems with locally derived solutions, a concept already highlighted by Berkes *et al.* (2001).

Both groups also stressed the need for area-based strategic management when considering land management objectives and agri-environment legislation. The need for broader schemes, as highlighted in England by the Policy Commission on the Future of Farming and Food (2002), which should review the whole farm rather than focusing on conservation only, appears to be being taken forward (Land Management Contract Steering Group, 2002) in a similar way to that already in place in Europe, e.g. France (ENGREF, 2000).

Finally, this approach provided considerably more consensus than conflict among the participants, proving that active local communication between all parties was a key to the success of local land management, along with flexible schemes and policies.

Acknowledgements

This research received financial support from the Scottish Executive Environment & Rural Affairs Department.

References

Berkes, F., Mathias, J., Kislalioglu, M. & Fast, H. (2001). The Canadian Arctic and the Oceans Act: the development of participatory environmental research and management. *Ocean & Coastal Management*, **44**, 451-459.

ENGREF (2000). *Les CTE dans le Haut Languedoc héraultais: de nouvelles relations possibles entre agriculture, forêt et société*. ENGREF, Montpellier.

Land Management Contract Steering Group (2002). *Land Management Contract Working Group, Report to Agriculture Strategy Implementation Group*. Scottish Executive Environment and Rural Affairs Department, Edinburgh.

Marshall, G.R. (2004). From words to deeds: enforcing farmers' conservation cost-sharing commitments. *Journal of Rural Studies*, **20**, 157-167.

McEachern, C. (1992). Farmers and conservation – conflict and accommodation in farming politics. *Journal of Rural Studies*, **8**, 159-171.

Policy Commission on the Future of Farming and Food (2002). *Farming and Food, a Sustainable Future*. Cabinet Office, London. Available from http://archive.cabinetoffice.gov.uk/farming/pdf/PC%20Report2.pdf.

Schusler, T.M. & Decker, D.J. (2002). Engaging local communities in wildlife, management area planning: an evaluation of the Lake Ontario Islands search conference. *Wildlife Society Bulletin*, **30**, 1226-1237.

Scottish Executive (2001). *A Forward Strategy for Scottish Agriculture*. Scottish Executive, Edinburgh.

Thompson, G.K., Larsen, J.W.A. & Vizard, A.L. (1999). Effectiveness of small workshops for improving farmers' knowledge about ovine footrot. *Australian Veterinary Journal*, **77**, 318-321.

24 Farming, Forestry and the Natural Heritage: Towards a More Integrated Future

25 The Pontbren Farmers' Group

David Jenkins

Summary

1. The Pontbren Group consists of ten neighbouring families who farm in the catchment of the Pontbren stream in mid-Wales.
2. Their response to rising costs and falling commodity prices was to develop a system of farming which would make their businesses more viable by reducing their use of feed, fertiliser and straw. This involved reducing livestock numbers and changing to hardier breeds. Hedgerows and shelter woodlands were created and restored to enable this change.
3. Changes in bird populations and soil structure were observed and these have triggered a programme of monitoring and research which is investigating the effect of native woodland planting within a farmed landscape.

25.1 Introduction

The Pontbren Group consists of ten neighbouring families who farm about a thousand hectares in the catchment of Pontbren Stream near Llanfair Caereinion in North Powys (Figure 25.1). They came together in 1997 as a group of three farmers, with the remainder joining in 2001. They have invested a great deal of time considering the future of their farming enterprises and have identified a number of changes which they can make, individually and collectively, to improve their lot.

The land they farm is productive beef and sheep land. It ranges from 200 m to 400 m above sea level in the rolling countryside of the old county of Montgomeryshire (Trefaldwyn). Most of the land has been ploughed and re-seeded since the 1970s. There is a small area of unimproved grassland and some interesting small wet areas. Woodland occupies 1.5% of the land

Figure 25.1 Map showing the approximate location of Pontbren, Wales.

25 Farming, Forestry and the Natural Heritage: Towards a More Integrated Future

and there is a complex network of hedgerows which are highly valued for stock shelter.

The same families have occupied the land for generations. Welsh is the first language of most of the adults and all of the children. Co-operative working was necessary in the past when farm work was labour intensive, but with greater mechanisation this has changed.

25.2 The work of the Group

When the original three families came together, their primary aim was to restore hedgerows and shelterbelts (some of which were common boundaries) with a view to keeping hardier breeds of sheep which could lamb outdoors. That process gathered momentum very quickly with some financial support from the local LEADER (Liaison Entre Actions de Development de l'Economie Rurale' (Links between actions for the development of the rural economy)) group.

Coed Cymru (an all-Wales initiative to promote the management of broadleaf woodlands and the use of locally grown hardwood timber in Wales) had worked with one of the original members since 1992 and a number of streamside plantings and broadleaf shelterbelts had been established. Our first step when the Group was formed was to provide large-scale maps to each farmer to identify their priorities in the first and second five-year periods. Their hand-drawn maps were then digitised, the plans costed and work started. At this stage there was no grant aid available for this work, but the sheep market was buoyant and the businesses bore the cost. Eventually the Group secured funding for two years from Scottish Power under their Rural Care Scheme.

As the landscape began to alter, a farm walk was arranged to explain the plans to immediate neighbours. Without exception, they asked to join and the group of three became ten. This all happened a few weeks before the foot and mouth (F&M) outbreak which badly affected this particular area. Despite this, the maps were quickly extended to include the new farms and the Group was formally constituted as a legal entity.

At this point there was already plenty to show for their efforts so it was relatively easy to grab the attention of the Welsh Assembly Government and its agencies. Having already done things on the ground it became much more than an idea on paper and this was crucial in securing political, and hence financial, support for the Group and their ideas. It has been a lengthy process but the Group is now formally recognised and funded as a unique project. Long-term funding for woodland work and de-stocking has been agreed with the Welsh Assembly Government and the Forestry Commission, and further funding for hedgerows, ponds and wetlands has been agreed through Enfys – a lottery funded programme administered by Wales Council for Voluntary Action (WCVA). See Figure 25.2 for a detailed representation of the work undertaken by the project.

It is important to remember that all of this development took place under the shadow of BSE (Bovine Spongiform Encephalopathy), F&M and unfavourable commodity markets. Weekly meetings began to explore ways of making the community more self-sustaining by maximising incomes and reducing external costs. With beef and lamb as their main products the choice of options to increase income was limited and the group now sells pre-packed meat at farmers' markets and the weekly market in Welshpool. Reducing costs offered greater potential. The first cost to be tackled was straw. Wales imports 300,000 tonnes of straw as bedding every year from as far away as East Anglia. Members of the group began to experiment with dry woodchip derived from woodland and hedgerow

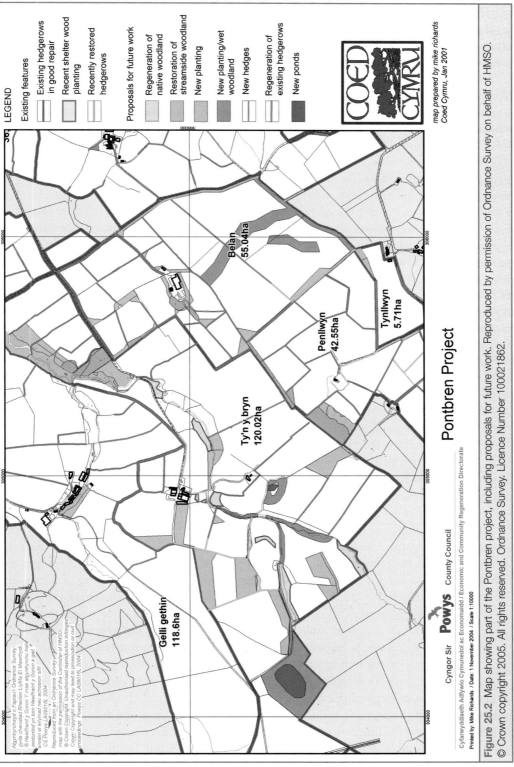

Figure 25.2 Map showing part of the Pontbren project, including proposals for future work. Reproduced by permission of Ordnance Survey on behalf of HMSO. © Crown copyright 2005. All rights reserved. Ordnance Survey. Licence Number 100021862.

restoration and slabwood from a local sawmill. Used indoors under sheep and cattle the results were encouraging. Labour inputs were low and the incidence of footrot in sheep was reduced. After a series of trials, the best chipping machine was purchased two years ago by the Group with help from the Welsh Development Agency.

Members then turned their attention to the cost and traceability of trees. With over 40 ha of new woodland and 30 km of hedgerow restoration in the first 5 year plan it was decided to establish a small tree nursery. This was partly to reduce costs, but also in response to problems they had experienced using imported stock, which was obviously unsuitable for wild Welsh hillsides. One member set up the nursery and has achieved notable success in the first two years, collecting seed from around the farms and producing some excellent planting stock. It will be a while before the Group is even self-sufficient, but there have already been a number of enquiries from potential customers outside the Group. Initially, a commercial peat-based compost was purchased for use in the nursery, but a small trial established that composted farmyard manure based on woodchip was equally good with species like birch and alder, and noticeably better with species like ash and sycamore. All production has now shifted to home-produced compost.

A composting facility was recently constructed on one of the farms and the first batch of material is nearing the end of the composting process. Its progress has been monitored by the Agriculture Development and Advisory Service. Analysis of samples has shown good moisture retention and high levels of plant nutrients. It is also eligible for use in organic agriculture and horticulture.

Animal feedstuff is a major expense on most Welsh hill farms, few of which now grow arable crops. In order to reduce expenditure on feedstuff, the Group has reduced ewe numbers and increased the proportion of hardy breeds in their flocks. The Welsh Assembly Government is supporting this experiment with compensatory payments and monitoring the financial consequences. Overall, the programme of capital works is ambitious – too ambitious to be undertaken by farm labour alone. The Group has taken the decision to use local labour and materials, wherever possible.

The Group's approach is unambiguous and now well established. Substantial progress has been made, particularly with hedgerow and woodland restoration and pond creation. One unexpected consequence was observed during heavy rain when surface water running off grassland was absorbed immediately it passed under the fence into the newly planted areas. A few simple experiments with spade and plastic pipe showed that remarkable changes in surface porosity had occurred. In order to quantify these changes the Countryside Council for Wales commissioned a survey by the Centre for Ecology and Hydrology and University of Wales, Bangor. The results of their preliminary survey have been published (Bird *et al.*, 2003). They confirm the field observations and demonstrate significant differences in surface soil compaction, rooting depth and water absorption. In consequence of this work, a consortium of public bodies is now considering proposals to fund a major piece of research on this site over the next three years.

As expected, there have been interesting changes in flora and fauna, such as with the return of green woodpeckers (*Picus viridis*) and barn owls (*Tyto alba*), and every opportunity has been taken to record information. The Group has been assisted by the Montgomeryshire Wildlife Trust and local volunteers. The first student project, supported by the University of Wales, Bangor, recently surveyed the carabid beetle populations.

25.3 Conclusions

The Pontbren approach is unconventional in many ways. Farming, since the Second World War, has been driven and coaxed with sticks and carrots and it has responded to the extent that the industry is now heavily dependent on public sector support. This Group has made a bold attempt to take control of their own destinies. Rather than undertaking work which attracts grant aid, they have set their own programme and sought funding that fits. In the case of the Enfys funding, they have taken a step further by administering the fund on behalf of WCVA. This involves setting their own standards and inspection procedures. The consequences could be far-reaching. Pontbren has become an agri-environment scheme, tailored to a particular catchment and combined with a co-operative marketing project. Its achievements are manifest and it has won support across the political spectrum. Perhaps its impact is best illustrated by the proposal to introduce a new higher tier of agri-environment support for groups of Welsh farmers willing to come together to manage whole landscapes. This proposal is based very closely on the Pontbren blueprint.

Reference

Bird, S.B., Emmet, B.A., Sinclair, F.L., Stevens, P.A., Reynolds, B., Nicholson, S. & Jones, T. (2003). *Pontbren: Effects of Tree Planting on Agricultural Soils and Their Functions.* Centre for Ecology & Hydrology, Bangor.

25 Farming, Forestry and the Natural Heritage: Towards a More Integrated Future

26 A brief review of Land Management Contracts and their possible use in the Cairngorms National Park

Fiona Newcombe

Summary

1. The development of Land Management Contracts in Scotland is described.
2. The key points identified by the Land Management Contracts Working Group, and their implications, are summarised.
3. The Cairngorms National Park Authority's recognition of the opportunity offered by Land Management Contracts as a way of delivering integrated land management is explained.

26.1 Introduction

Support for farmers and crofters in Scotland is characterised by a wide range of different schemes and regulations, each with their own administration arrangements and requirements. The requirements of these schemes can be contradictory and there has been no clear single process to help a land manager plan his or her business. It is widely recognised that this system has resulted in a heavy burden of administration and the development of systems that harvest subsidies rather than producing what the market or public want.

The Land Management Contract (LMC) approach provides an opportunity to deliver a wide range of environmental, economic and social objectives. This chapter explores the progress made so far in developing the approach and considers the ways in which it might be implemented in the Cairngorms National Park.

26.2 The development of Land Management Contracts

In its *A Forward Strategy for Scottish Agriculture* (Scottish Executive, 2001), the Scottish Executive proposed the development and introduction of land management contracts (LMCs) to pay farm businesses for the economic, social and environmental benefits needed by their area. The aim of LMCs is to:

- enable farmers and crofters to make market-orientated business decisions;
- reward farmers and crofters for the public goods they produce; and
- reduce paperwork and red tape.

The Scottish Executive set up a Working Group in 2002, which included a wide range of stakeholders, to develop the proposal for LMCs in more detail. In October 2002, the Working Group produced a report which summarised its discussions on the introduction of LMCs (Scottish Executive, 2002). This report developed ideas for LMCs, explored the contexts within which LMCs would be developed, regional involvement, set out a proposed forward strategy and put forward a work plan to oversee their development and implementation. In summary, the Working Group proposed a three-tier system for LMCs:

- Tier 1 – an annual base payment paid to all farmers;
- Tier 2 – a further annual payment available to all farmers; and
- Tier 3 – top-up payments.

The report identified two main issues that might limit the development of LMCs. First, at the time, the outcomes of the Common Agricultural Policy (CAP) Mid-Term Review negotiations were unknown. This was considered relevant to the development of annual base payments across Scotland to all producers and to good farming practice. Second, there is an issue about the availability of sufficient funding and its distribution between the tiers. The Working Group proposed that the Rural Development Regulation (RDR) offered an opportunity to fund Tiers 2 and 3. However, it was proposed that the scheme receiving the largest allocation of funding under the RDR, the Less Favoured Area Support Scheme, be maintained, leaving only the funding made through agri-environment and forestry schemes available for the LMCs. Given that Scotland is one of the lowest funders of agri-environment schemes in the European Union, the environmental stakeholders were concerned that the proposed transfer of funds from agri-environment and forestry schemes to LMCs (which have social and economic as well as environmental objectives) would result in a net loss of support for environmental objectives for farmers and crofters in Scotland.

The Working Group supported the regionalisation of LMCs to help with the strategic direction for the menus of options available for Tiers 2 and 3 of the LMC model. This would allow each region to design and administer its own tiers and schemes. The Working Group identified a number of considerations relating to the regionalisation of LMCs, including:

- administrative costs and arrangements;
- existing partnership arrangements for implementing existing schemes;
- differences in geography, environmental conditions and land management; and
- linkages with other policies and structures such as Local Biodiversity Action Plans.

The Working Group recognised that further work was needed to develop a more regional approach. The following regional groupings were provisionally agreed, based on partnership arrangements for implementing current schemes:

- Highland and Islands;
- Aberdeenshire and Moray;
- Perth & Kinross, Angus, Fife and the Lothians;
- Borders, Dumfries and Galloway; and
- Ayrshire and Clyde Valley.

Unfortunately, these regional groupings do not fit comfortably with the role of the National Park Authorities in Scotland, which is to ensure that the four aims of the National

Parks are achieved in a co-ordinated way. The proposed regional groupings, for example, would mean that the Cairngorms National Park might be divided into three LMC regions, making this integration very difficult.

26.3 The LMC modelling exercise

The Working Group agreed that it was important to ensure that proposals for LMCs were practical and applicable, and for ideas from farmers and crofters for Tiers 2 and 3 to be incorporated as early as possible in the development of the scheme. A modelling exercise was undertaken and reported on in March 2003 (Scottish Executive, 2003b). Its aim was to test the model developed by the Working Group, as described previously in this chapter. Twenty-one farmers, three crofters and an associated Common Grazing were selected across Scotland to give a good geographical spread as well as a range of agricultural types. Scottish Executive Environment and Rural Affairs Department (SEERAD) staff undertook the work.

During the first visit, the exercise was outlined, and a 'kitchen table' audit undertaken (a number of assumptions were made about the opportunities in the CAP Mid-Term Review settlement which had not yet been agreed). A questionnaire was used to guide discussion and identify possibilities for economic, social and environmental activities. The information gathered included proposals for measures under Tiers 2 and 3. Four regional meetings involving representatives of Local Authorities, Local Enterprise Companies, and tourism, environmental and farming bodies were held. The aim of these meetings was to discuss and refine the information for Tiers 2 and 3 gathered from farmers and crofters. The meetings focused on explaining the process of LMCs. Second visits were then made to farmers and crofters. This enabled the development of a complete model LMC for the farm or croft.

The report showed that the proposed tiered system generally worked well when used in the modelling exercise. It was a practical and integrated way of delivering agricultural policy. The process reflected the aim of the exercise and the limited resources available, rather than the ideal LMC development. However, the report identified a number of issues which needed further consideration in the future development of LMCs.

1. The main weighting for Tier 2 was for environmental measures. Consideration needs to be given to developing social and economic measures so that these are also available.
2. In drawing up environmental prescriptions, there are many benefits to developing a 'what is wanted' approach rather than a 'how to do it' approach. For example, the height of grass is important for many bird species. Traditional agri-environment schemes have prescribed set livestock per unit grazing levels, which do not recognise variations in weather, vegetation or landscape. Prescriptions based on vegetation height rather than on a preferred way of achieving this would give farmers and crofters more flexibility to achieve this desired outcome in a variety of ways. This also appeals to entrepreneurialism.
3. The role and membership of the regional groups needs further consideration. Participants in the process wish to contribute to the range of prescriptions available in their region and to the methods of administering LMCs (for example, to

encourage collaborative prescriptions between farmers and crofters to give a wider area of biodiversity management).
4. Payment rates will have to be carefully considered. For environmental objectives, the opportunity to reward farmers and crofters rather than to compensate them, while delivering value for money for the taxpayer, is a major challenge under CAP reform negotiations. For economic and social objectives, it is even more difficult to set payment rates.
5. It will also be important to review the LMC plan with the farmer and crofter to reflect new knowledge on economic, social and business opportunities, and to incorporate the practicalities of the farming system.
6. Regionalisation of LMCs needs to be considered further. Work is required to assess how to reflect regional differences and priorities.
7. The need to balance local needs with national and international priorities is essential. This could be done through careful targeting, information and adequate payment rates.
8. The interaction between agricultural and other land management schemes, such as the Scottish Forestry Grant Scheme, needs to be considered to enable the true integration of plans. The current LMC model utilises support currently paid through agri-environment and forestry RDR schemes. Support available through other agricultural schemes such as the Farm Business Diversification Scheme also needs to be integrated.

After the reporting of the modelling work, the uncertainty surrounding CAP reform negotiations meant that the development of LMCs was suspended. There was a further commitment in the Partnership for Scotland Agreement (Scottish Executive, 2003a) to "implement LMCs to deliver reformed CAP support which takes account of the diversity of Scottish agriculture and its economic, social and environmental impact". This agreement was reached between the Scottish Labour Party and the Scottish Liberal Democrats in forming the current government in the Scottish Parliament. The agreement was reached in 2003 and covers the Parliamentary session from 2003 to 2007.

26.4 The Cairngorms National Park Authority's interest in Land Management Contracts

The Cairngorms National Park, created in 2003 as Scotland's second national park, has four aims:

- to conserve and enhance the natural and cultural heritage of the area;
- to promote sustainable use of the natural resources of the area;
- to promote understanding and enjoyment (including enjoyment in the form of recreation) of the special qualities of the area by the public; and
- to promote sustainable economic and social development of the area's communities.

The Cairngorms National Park Authority (CNPA) believes that farmers and crofters across the National Park contribute greatly to all four aims and supports them in their efforts. For example, their management creates the landscape and helps support the wildlife that helps to make the area so special. The National Park contains the highest density of breeding wading birds such as lapwing (*Vanellus vanellus*) and redshank (*Tringa totanus*) on mainland Scotland (Royal Society for the Protection of Birds, 2000). This is a direct result of the farming systems found there.

Agriculture contributes to the third aim as many people visit and live in the area because of this special landscape and wildlife. Agriculture also plays a direct role in the economic and social development of the area's communities, providing employment and housing, and supporting many other industries such as fencing contractors. The contribution of agriculture to the special qualities of the area has been recognised in a variety of publications including the Spey Catchment Management Plan (Spey Catchment Steering Group, 2003) and the Local Biodiversity Action Plan (Cosgrove, 2002).

Agriculture in the Park has changed in recent years (Bayfield & Conroy, 1996). The area of cropped land has declined, and the permanent grassland has increased. More trees have been planted. While there are fewer cattle, the number of sheep has increased. There is a trend towards the amalgamation of units, and employment is declining. The CNPA is concerned about the threats facing agriculture in the National Park. These threats include:

- low prices;
- an ageing farming and crofting community;
- difficulties for young entrants to start farming;
- an overburden of rules, regulations and paperwork;
- national schemes that do not fit local requirements;
- a lack of funding for environmental schemes;
- a lack of training (for traditional agricultural skills); and
- a downturn in agricultural output.

The CNPA recognises that many of these issues reflect national trends, but the National Park has been specially designated to reflect its special qualities, many of which depend on agriculture. It is essential that support and efforts be directed towards overcoming these threats to agriculture in the National Park.

The CNPA Board has identified some immediate priorities to deliver their aims and to try to mitigate the concerns listed above. They propose the integration and simplification of land management schemes through a one-stop-shop approach to land management schemes, which will support land managers across the Park for their contribution towards the four aims, from all land management sectors. The Scottish Executive's consultation on CAP reform, and its planned implementation through LMCs, offers an early and important step towards delivering this vision.

The National Park Plan will be developed over the next two years. It will set out the policies for managing the National Park and co-ordinating the functions of all of the public bodies that influence it. Agricultural support has a major influence on the delivery of the National Park's aims and the CAP reform consultation offers the opportunity for the potential regionalisation of agricultural support through the development of LMCs. The implementation of the National Park as a LMC region would enable the development of a targeted delivery mechanism that supports farmers and crofters in their efforts to deliver all four of the National Park's aims.

26.5 How might LMCs be developed in the Cairngorms National Park?

The CNPA strongly supports the introduction of LMCs, via CAP reform, as a simplified and flexible mechanism for delivering economic, environmental and social objectives through agriculture. The Authority proposes that the National Park is a LMC region for a number of reasons:

- the CNPA could act as a co-ordinating body across the area;
- there is a strong record of collaboration and partnership through the Cairngorms Partnership to be built upon;
- the identification of economic, environment and social priorities has already started (for example, through the Cairngorms Local Biodiversity Action Plan, Communities Council Group and agricultural waste schemes);
- the integration of land management schemes is underway (for example, the Cairngorms Moorland Project which aims to integrate sporting and agricultural management);
- the Cairngorms is well-placed to deliver a wide range of public goods that attract thousands of visitors each year and provide excellent value for money; and
- it provides an opportunity to act as a pilot area to trial the new policy mechanisms.

The CNPA vision is for every farmer and crofter to have the opportunity to receive face-to-face advisory support to prepare a development plan. This plan would identify all agricultural (and off-farm) activity, and identify and integrate funding streams and other sources of help, such as those available under LMCs, Leader+ and other schemes. *The Whole Farm Review Pilot Scheme* (Scottish Executive, 2003c) could provide a useful model for developing such a scheme.

The CNPA suggest that the CAP reform proposals be implemented through the tiered LMC structure outlined by the LMC Working Group.

- Tier 1 would be the decoupled Pillar 1 payment, with agricultural and environmental conditions attached. The payment rate and conditions would meet local needs. The National Park would be a region for Tier 1.
- Tier 2 would include the Rural Development Plan/national envelope schemes. The National Park would be a region for Tier 2 and the options available (and their funding levels) would be agreed to help deliver the National Park Plan.
- Tier 3 would be more specialised Rural Development Plan/national envelope schemes. The National Park would be a region for Tier 3 and the options available (and their funding levels) would be agreed to help deliver the National Park Plan.

The National Parks provide opportunities to test different models for the design and administration of land management schemes. However, it is recognised that a model developed to suit the conditions of a National Park may not be easily replicated elsewhere. The geography of areas may differ, meaning that prescriptions are not applicable. Indeed, different prescriptions may be developed to suit different parts of the National Park. The framework for public policy differs too as the National Park Authority has a unique function to integrate public sector activity across its area. However, by testing such models in a National Park area there is an opportunity for general lessons to be learned and then applied to other areas.

26.6 Acknowledgements

The contents of this chapter reflect experiences developed through working at the Scottish Agricultural College, RSPB Scotland, and the Cairngorms National Park Authority. There are many individuals who have influenced the thinking and development of integrated land management schemes, and in particular I wish to thank members of Scottish Environment LINK's Agricultural Task Force, and staff and Board members of the Cairngorms National Park Authority.

The Scottish Executive is thanked for having the vision to develop this new method of developing integrated land management to achieve multiple objectives. The two anonymous referees are thanked for their comments and suggestions, though I take full responsibility for the contents of this chapter.

Editors' note

Since this chapter was written there have been several developments in land use policy in Scotland. In February 2004, the Scottish Executive announced proposals for implementing the CAP reforms. These included decoupling subsidies from production and introducing a single farm payment using a historic reference period, and increasing rates of modulation. It was announced that these decisions would help in the development of delivery systems using the LMC model.

In August 2004, the Scottish Executive announced a consultation on the Land Management Contract Menu Scheme. The consultation proposed the introduction of a Scotland-wide Tier 2 LMC in 2005. It noted that the Scottish Executive planned to work with stakeholders over the next two years to integrate LMC tiers more fully, with the longer term aim of having all support to farming, and possibly some to rural development more generally, within the LMC framework. Discussions have been held on the opportunity to regionalise the prescriptions and priority setting of the scheme from 2007 onwards. The Menu Scheme began during the first half of 2005.

References

Bayfield, N.G. & Conroy, J.W.H. (1996). *The Cairngorms Assets; a Cairngorms Partnership Working Paper.* Cairngorms Partnership, Grantown-on-Spey.

Cosgrove, P.J. (2002). *The Cairngorms Local Biodiversity Action Plan.* Cairngorms Partnership, Grantown-on-Spey.

Royal Society for the Protection of Birds (2000). Strathspey Breeding Farmland Wader Survey 2000. Unpublished report. Royal Society for the Protection of Birds, Edinburgh.

Scottish Executive (2001). *A Forward Strategy for Scottish Agriculture.* Scottish Executive, Edinburgh.

Scottish Executive (2002). Land Management Contract Working Group. Report to Agriculture Strategy Implementation Group. Unpublished report. Scottish Executive, Edinburgh. Available from www.scotland.gov.uk/library5/agri/lmcwg_report.pdf.

Scottish Executive (2003a). *A Partnership for a Better Scotland: Partnership Agreement.* Scottish Executive, Edinburgh.

Scottish Executive (2003b). Land Management Contracts: Modelling Exercise. Unpublished report. Scottish Executive, Edinburgh. Available from www.scotland.gov.uk/library5/development/lmcme.pdf.

Scottish Executive (2003c). *National Strategy for Farm Business Advice and Skills: Whole Farm Review Pilot Scheme Explanatory Booklet.* Scottish Executive, Edinburgh.

Spey Catchment Steering Group (2003). *River Spey Catchment Management Plan.* Spey Catchment Steering Group, Aviemore.

27 European rural development policies and the Land Use Policy Group

Ralph Blaney & Maria de la Torre

Summary

1. The Land Use Policy Group aims to advise on policy matters of common concern to agriculture, woodlands and other rural land uses.
2. An important area of Land Use Policy Group work has been to examine the effectiveness of European Union rural policies. To this end, a two-stage research project was commissioned, in partnership with WWF Europe, on Europe's Rural Futures.
3. The project found that the flexible, coherent, integrated and partnership-orientated approach that was originally envisaged has not been widely achieved in reality, and that the Rural Development Regulation requires a much more substantial funding base.
4. Publicising these findings helped start a debate on how European Union resources can be better used to meet the needs and opportunities for Europe's rural areas.

27.1 Introduction

Comprised of the British statutory conservation, countryside and environment agencies, the Land Use Policy Group (LUPG) aims to advise on policy matters of common concern to agriculture, woodlands and other rural land uses. It seeks to improve understanding of the pros and cons of policy mechanisms related to land use, particularly farming and forestry, to develop a common view of desirable reforms to existing policies and to promote these views. It does this through a joint programme of work which includes research into land use policy issues, input into European and other international policy-making, and advice to national agricultural and forestry departments where there is a joint interest.

An important area of LUPG work has been examining the effectiveness of European Union (EU) rural policies. To this end, research was commissioned, in partnership with WWF Europe, on Europe's Rural Futures. This was a major project on the nature of rural development in the EU and was carried out in two phases.

27.2 The scoping study

As a first step a scoping study was undertaken by the Institute for European Environmental Policy (IEEP) (Baldock *et al.*, 2001). The countries chosen included six EU Member States, three accession countries and Switzerland. The aim of this scoping study was to investigate actors, institutions and attitudes towards rural development in Europe in order to clarify key issues. Broad areas of interest were identified as the driving forces for rural change; the institutions involved, perspectives on rural development, the role of the environment in rural development policies, the resourcing of rural development policy, and accountability versus flexibility and innovation in policy delivery.

An expert team in each of the ten countries explored attitudes and institutional behaviour through a series of interviews with key officials, stakeholders and expert observers in the field. The findings were as follows.

- In all countries there were common economic and social threads underlying the way in which rural areas have evolved in recent decades.
- Agricultural employment has declined, and many rural areas face long term challenges posed by an ageing population. However, in more densely populated countries the decline of agriculture has resulted in a switch of rural employment into other areas and increased commuting to neighbouring urban areas, whilst in more sparsely populated countries commuting was less common and the depopulation of rural areas is a major problem.
- Where living standards were rising, there was a greater emphasis on amenity, quality of life and a healthy environment.
- There were common aspirations amongst all countries' rural development policies, and most countries shared environmental concerns.

In terms of approaches, it was found that countries were divided between those with a strong national agenda and institutional pattern and those more influenced by the driving force of EU policy. In every country the delivery of rural development involved a hierarchy from central governments to local authorities. Levels of involvement and different stages of the hierarchy vary and there appeared to be no 'ideal model' for organising the formulation and delivery of rural development policy. Countries reported that the Ministry of Agriculture, rather than the Ministry of Environment, took the lead in rural policy development, thus sustainability was still not a key objective of rural development policies. Finally, improvements that are needed in existing policies were highlighted and positive examples of structures and policy approaches, which suggest potential solutions, were identified. Work then proceeded to the main study.

27.3 The main study

The main study compared planning and implementation of the Rural Development Regulation (RDR) and the Special Accession Programme for Agriculture and Rural Development instruments across Europe (Dwyer *et al.*, 2002). Its main aims were to assess how far these instruments were likely to achieve the EU's objectives and promote sustainable rural development and biodiversity protection, identify good practice and learn lessons that will help to improve implementation and inform future reform of the Common Agricultural Policy and the RDR.

In-depth studies of six European Union Member States (France, Germany, Sweden, Spain, UK and Austria) and two Candidate countries (Hungary and Poland) provided the basis for the pan-European comparative study. Case studies in each country illustrated some of the strengths and weaknesses of current practice. This study evaluated the early implementation of the RDR against a broad set of goals for sustainable rural development.

The findings were as follows.

Firstly, there were striking national differences in the patterns of proposed expenditure on the RDR, which broadly reflect historical allocations to similar measures (see Figure 27.1). There was a marked difference between countries for whom the RDR is a tool to promote environmental land management and those for whom the RDR is a tool to modernise agriculture. Secondly, there was a lack of coherence between RDR funds and other EU policies. Thirdly, many RDR programmes were criticised by stakeholders for their continuing preoccupation with agriculture. Fourthly, innovation in Rural Development Plans occurred both at the level of programme design and at the level of individual projects. Fifthly, RDR programmes varied greatly between countries in their treatment of the environment. Overall, more apparent weight is given to environmental issues and measures in Sweden, Austria, France, the UK and Germany than in Spain, Poland and Hungary. In all plans there were very few objectives relating to environmental outputs, yet many concerns exist regarding the potential environmental effects of the Rural Development Plans. Sixthly, the success of programmes has been hampered by three sets of constraints: budgetary, institutional and practical. Overall, the flexible, coherent, integrated and

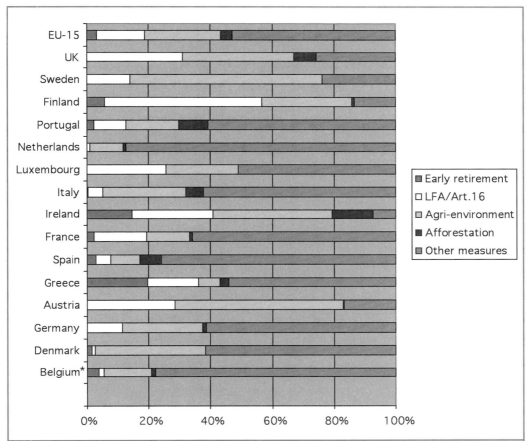

Figure 27.1 Planned allocation of RDR spending in Member States 2000-2006 (note that for Belgium (*) this does not include the full range of support measures which total € 1.7 million). Source: Dwyer et al. (2002).

partnership-oriented approach that was originally envisaged has not been widely achieved in reality. It appears that the RDR requires a much more substantial funding base. In addition, a European body to exchange information and promote best practice amongst Member States is also required.

27.4 Conclusion

As a follow-up to this work, the LUPG held a workshop on the research report findings and later a conference on 'Future Policies for Rural Europe 2006 and Beyond' (Land Use Policy Group, 2003). The conference was aimed at European, national and regional policy makers. It started a debate on how EU resources can be better used to help meet the needs and opportunities for Europe's rural areas. A third stage of research to help identify environmental priorities for the future Rural Development Plans (2007-13) is currently being taken forward. The LUPG will continue to play a role in informing European policy-makers about rural policy needs and how agriculture and the environment can best be integrated into sustainable rural development.

References

Baldock, D., Dwyer, J., Lowe, P., Petersen, J. & Ward, N. (2001). *The Nature of Rural Development: Towards a Sustainable Integrated Rural Policy in Europe. A Ten-Nation Scoping Study.* Institute for European Environmental Policy, London.

Dwyer, J., Baldock, D., Beaufoy, G., Bennett, H., Lowe, P. & Ward, N. (2002). *Europe's Rural Futures, The Nature of Rural Development II. Rural Development in an Enlarging European Union.* Institute for European Environmental Policy, London.

Land Use Policy Group (2003). *Future Policies for Rural Europe 2006 and Beyond - Long Term Support to Rural Areas in an Expanding Europe.*
www.lupg.org.uk/uploaded_photos/pubs_note_of_seminar_final_(PROCEEDINGS).pdf.

PART 6:
Looking to the Future: Achieving Better Integration

Uath Lochan, Inshraich, Cairngorms © Lorne Gill, Scottish Natural Heritage

Farming, Forestry and the Natural Heritage: Towards a More Integrated Future

PART 6:
Looking to the Future: Achieving Better Integration

So far, the book has looked in detail at the progress being made in moving towards the integration of farming, forestry and the natural heritage. Earlier chapters have indicated the importance of making sure that European funding delivers wider environmental objectives, of having a clear vision that different interests can work towards, of thinking about the needs and perceptions of farmers and other land managers, and on the importance of good research and advice. This part of the book builds on these themes and looks at ways of achieving better integration in the future.

Achieving better integration does require a European funding framework that helps rather than hinders progress. In Chapter 28, Baldock provides a timely review of the current state of play on European funding and warns of the need to ensure that there are sufficient incentives for farmers to manage their land in an integrated way. Although land management contracts may be strongly dependent on European funding for their implementation, they also provide an opportunity to link the ways in which land is managed with a wider range of funding mechanisms. However, land management contracts will only be effective at doing this if all interests can agree on a common vision for a multi-benefit countryside. This theme is picked up by Thomson (Chapter 29), who concludes that whilst some good progress has been made with the publication of the Scottish Forestry Strategy and the Scottish Biodiversity Strategy, there remains the key challenge of incorporating landscape and access into an overall vision for the countryside.

The human dimension remains critical in achieving integration on the ground. Balfour (Chapter 30) echoes many of the points raised in other chapters that the needs of farmers and other land managers must be respected if better integration is to be achieved.

In the final chapter, Galbraith & Davison pull together many of the threads running through the book. This chapter concludes that whilst progress towards better integration is mixed we are now moving in the right direction. Some parts of a shared vision are in place but there remains a need to make sure that existing strategies are co-ordinated and for landscape and access to be included. They recommend that more research into predicting the likely outcomes of different policies and funding regimes is needed, and that this will then help to reinforce the delivery of simple, effective and well-targeted advice and support. The overall message is that whilst some good progress is being made there is now a unique opportunity to take a decisive step towards better integration on the ground and to achieve a countryside for everyone.

Farming, Forestry and the Natural Heritage: Towards a More Integrated Future

28 The European policy and funding horizon and its implications for integration

David Baldock

Summary

1. Policies for agriculture, forestry and the natural heritage in Scotland are shaped in part by a European framework, which is in a process of active evolution.
2. The policies constituting this framework are being shaped not only by sectoral concerns, but also by the forthcoming enlargement of the European Union and other political, institutional and constitutional developments.
3. The relatively radical revision of the Common Agricultural Policy, arising from the 'Mid-Term Review' of June 2003, will be significant in several respects, and will come into effect from 2005 onwards.
4. Further changes to the 'Second Pillar' – rural development support – of the Common Agricultural Policy are due to be agreed before 2006, with implications for a range of rural activities including forestry and nature conservation.
5. European environmental policy will continue to be a force influencing the national agenda, but the level of European Union funding available for nature conservation purposes in the future is uncertain.
6. In this and other respects, the revised regulations and budget for the European Union Structural Funds will have a bearing on the integration debate.
7. As a new European Union framework emerges over the next three years, there will be a need to identify and develop the options which best suit local conditions. Greater integration on the ground is unlikely to flow automatically from the introduction of new European Union funding and procedures.

28.1 Introduction

The influence of the European framework on policy and the availability of funding for the management of farms, forests and the natural environment in Scotland has been considerable for many years. Many of the keystones of European Union (EU) policy are now shifting. The period to 2007 will see substantial revisions to a raft of policies concerned with rural Europe. In some areas the direction of change is apparent, in others the formation of new policies is at an earlier stage. Despite these uncertainties, the process of preparing for implementation decisions in Scotland is already underway, and it is not too soon to be considering ways of pursuing a more integrated approach, tying together natural heritage, agriculture and forestry objectives.

Events on the near horizon are inseparable from the more fundamental realignment now occurring in Europe. The EU is pacing itself for a major enlargement, with ten new

Member States due to join in May 2004. These are Cyprus, the Czech Republic, Estonia, Hungary, Latvia, Lithuania, Malta, Poland, Slovakia and Slovenia. This will mean that the EU will extend eastwards further than ever before, with a whole raft of new neighbours. Beyond this there will be a further shift in the Eastern borders, with Bulgaria, Romania and then perhaps Croatia next in line. We can anticipate a rolling programme of enlargement which, in due course, is likely to include Turkey. The process makes the *status quo* untenable at several levels.

The proposed new Constitution for Europe is partly a response to governance issues, which will be exacerbated within a larger Union. The pressure to agree significant reforms to the Common Agricultural Policy (CAP) during 2003 stemmed partly from the enlargement timetable, and from a sense that changes might well be more difficult to agree when 25 agriculture ministers are seated at the table for the Agriculture Council. A redistribution of the EU budget beyond 2007 will be necessary. There is little doubt that Structural Fund allocations for the existing EU regions with Objective One status, parts of Scotland amongst them, will be reduced in favour of poorer areas to the East. Few would venture to forecast all of the implications of enlargement but its significance as a policy driver should not be underestimated.

28.2 Looking forward to 2004

A series of institutional changes is scheduled to occur in 2004. As well as the enlargement of the EU in May 2004, this year will also see European Parliament elections and changes to the composition of the Commission, taking place in two stages.

There will be a change in the representation of each individual Member State in the European Commission and Parliament, with some existing Member States (including the UK) having fewer Commissioners after November and fewer Members of the European Parliament after the European Parliament elections. This is in order to allow a substantial influx of new representatives to these two institutions without prejudicing their work with an unwieldy number of people. A significant change in representation will take place in the Council as voting rights are altered to accommodate the new countries. The changes mean that the balance of power within the Council in relation to qualified majority votes will alter, with the new entrants being potentially important allies for existing Member States, for example if they are trying to block the progress of a particular policy item. Enlargement could also affect the speed of the policy process, after the new institutions to be bedded in with changes to voting in the Council and a stronger role for the European Parliament in several areas of policy.

While their longer term implications are debatable, these events will certainly have an impact in the short term and are likely to slow the process of policy development. The Commission's work programme has been cut down in length in order to produce a 'programme that is as realistic as possible, both in terms of what it (the Commission) can deliver and the other EU institutions can absorb'.

The anticipated reduction in institutional momentum will take place as a series of strategic issues come onto the agenda. These include the beginning of discussions on the review of financial and funding matters for the Union post-2006; the emergence of the first of the Sixth European Action Programme's (6EAP) Thematic Strategies with the pesticide, soil protection and waste prevention and recycling strategies due in September 2004; the delivery of the first action plan under the environment and health strategy; and progress on the controversial proposal on the Registration, Evaluation and Assessment for Chemicals, known as REACH (European Union, 2003a). In addition, 2004 will see the first extended impact assessments – looking at the social, economic and environmental costs and benefits of proposals – being introduced for all the Commission's major initiatives. Potentially this is a tool for bringing sustainability criteria to bear on new policy proposals and increasing the role for stakeholders in policy formation but it will be a considerable challenge to make it work in practice with limited resources.

Underlying the tangle of policy initiatives lie the EU Treaties where further change is afoot. It is probable that the current debate on the proposed Constitution for Europe will still be live during 2004. Despite the Italian Presidency's hope that discussions at the Intergovernmental Conference will be completed before the end of the year, leading to a new Treaty of Rome, it is still not clear if this is practicable and even if it is completed, there may be some further negotiations and loose ends to tidy up in 2004.

28.3 EU Structural Funds

The Structural Funds are a key source of finance for regional development, especially in the less affluent parts of Europe. The current Structural Fund Regulations and programmes expire at the end of 2006, and work within the Commission to develop a new framework for the 2007-2013 period has been gathering momentum in recent months. Within Directorate General (DG) Regions, drafting has proceeded on the Third Cohesion Report (European Union, 2004a) which provides a report on recent progress and a clearer indication of the Commission's thinking for the period ahead. The Commission's 2004 Work Programme promises new draft Regulations by May 2004, together with a draft EU budget, known as the Financial Perspective for the post-2007 period. However, both are likely to be delayed since the ten acceding Member States will want to have their say at an early stage, and negotiations are likely to be protracted.

Meanwhile, however, some of the main features of the new post-2007 Structural Funds (European Union, 2004b) landscape have become clearer. These are set out below.

- Assistance for Objective 1 regions (targeted at the poorest Member States) will be focused mainly on the new Member States, so that regions in many existing Member States (including the UK) will lose their Objective 1 eligibility. There will be transitional arrangements, however, which will provide assistance for a number of years.
- Despite calls from some Member States, including the UK and Germany, to abolish Objective 2, the Commission is proposing its continuation, with an allocation of between 20-30% of the total Structural Funds budget. The new Objective 2 would encompass the current Objectives 2 and 3, and would be re-orientated to focus on support for the priorities set out in the Lisbon agenda – i.e. improving innovation and competitiveness.
- Member States would have considerable flexibility to identify their own Objective 2 regions or themes, choosing from a menu established at EU level. Environmental priorities are likely to include support for the management of the Natura 2000 network. One encouraging sign is that the recently revised Commission guidelines

for the mid-term review and revision of current Structural Funds programmes list the following as potentially eligible for support in the remaining 2004-2006 period:
- the Natura 2000 network;
- integrated river-basin management under the Water Framework Directive;
- rural development (although there are questions about this);
- renewable energy and greenhouse gas reduction programmes; and
- sustainable urban transport.

- It is not yet clear whether current 'horizontal' environmental requirements relating to the *ex ante* environmental assessment of draft programmes and the role of environmental authorities in their development and management would be required for new programmes, which are likely to take the form of 'tri-partite agreements' between the Commission, a Member State and one of its regions.

In practice, final details of the size, distribution and future management of the Structural Funds will have to await agreement on the new EU Financial Perspective, expected to set the budget for the period from 2007 to 2013. This must be agreed by July 2006 at the latest, and Commission proposals are not expected before the new Commission assumes office in November 2004.

28.4 LIFE

The EU's only budget line dedicated solely to environmental objectives, known as 'LIFE', is relatively small but has played a useful role in some areas, such as supporting the implementation of Natura 2000. Following a mid-term evaluation of the LIFE fund, the European Commission has proposed a two-year extension to the current programme – but there are no guarantees that it will continue after 2006.

The current phase of the programme – LIFE III – was launched in 2000 and expires on 31 December 2004 (European Union, 2004c). Because the new multi-annual Financial Perspective will not come into effect until 1 January 2007, the Commission has proposed as an interim measure to extend the current programme from 1 January 2005 to 31 December 2006, with a budget of €317.2 million.

A number of technical changes and shifts in emphasis are proposed, several of which reflect the findings of the mid-term evaluation of the programme. They include the following.

- The priority areas are to reflect more closely those in the Sixth Environmental Action Programme and the forthcoming EU Environmental Technology Action Plan.
- Greater emphasis will be given to using LIFE as a means to improve policy implementation.
- In LIFE-Environment, a better definition of priority fields of intervention, will be adopted to avoid dispersing limited resources.
- A new committee procedure is proposed to conform to a recent ruling of the European Court.

The longer term future of LIFE and the assistance it provides for selected nature conservation projects is less certain. There is an argument that nature conservation funding should be integrated into the mainstream funding instruments, such as the Structural Funds, reducing or limiting the need for this element of LIFE. Whilst integration is attractive in principle, it may be less appealing in practice, unless these are strong incentives for Member States to give sufficient priority to nature conservation. The track record in this area is not entirely encouraging. In this uncertain climate, it would be unwise to bank on a significant stream of European funding for Natura 2000, despite the conclusions of the Markland report (European Union, 2002) that substantial expenditure is needed for appropriate site management in Europe.

28.5 Nature conservation

No major new EU measures on nature conservation are expected in 2004. However, DG Environment is preparing for a year during which the EC Biodiversity Strategy and associated Action Plans are to be 'relaunched', the 25 year-old Birds Directive celebrated, and the national lists of Sites of Community Importance within Natura 2000 are, in principle, to be finalised.

The review of the Biodiversity Strategy is underway, structured around a set of workshops considering the individual Biodiversity Action Plans. The results are to be synthesised in the spring, in time for a major focus on this neglected area of policy at the May Biodiversity Hearing in Dublin under the Irish Presidency. It is hoped that the conference conclusions will be taken up by the Environment Council in October 2004 and will enthuse the new Environment Commissioner coming into post in the autumn.

The debate on the scale and shape of EU co-funding for Natura 2000 sites will run in parallel to the Biodiversity Strategy. Progress on the Commission's communication on the subject has been disappointing. Initially due in late summer 2003, the document has only recently reached the Environment Commissioner and final adoption by the whole Commission is scheduled for early 2004. It is expected to comment on both the need for funding at a European level and on the sources of funding available; notably the structural and Rural Development funds. It is unlikely to propose a major new European fund.

28.6 CAP reform

The July 2003 agreement on CAP reform was less radical than originally proposed by the Commission but perhaps still qualifies as a watershed, given the generous use of this term in European policy. Agriculture ministers have now approved the final text for a regulation establishing common rules for the new direct support schemes under the CAP, establishing how decoupling will work in practice. The pivotal regulation (1782/2003) was published in the Official Journal on 21st October 2003 (European Union, 2003b). It is a critical part of the overall CAP reform package, determining the rules for payments to farmers and cross-compliance, setting out modulation rates and the requirements for a farm advisory system and establishing the system of 'national envelopes'. The regulation applies to the current 15 Member States but will be adapted over the next few months to make it applicable to all ten new Member States.

A further package of reforms to 'Mediterranean' agricultural products lies ahead. At the end of September 2003 the Commission proposed reforms for the olive oil, tobacco and cotton sectors within the CAP and tabled a number of controversial options for the sugar regime. The principles of decoupling and more direct aid to producers run through the proposals, including a number of environmental elements. They received a mixed reception from agriculture ministers, some of whom will seek to minimise changes to the regimes, whilst aware of the scrutiny they will receive within the current round of World Trade Organisation negotiations.

The Mid-Term Review agreed in June 2003 will affect agriculture from January 2005 onwards. It brings about major changes to both the market support element of the CAP, now known as Pillar 1, and the rural development strand, the much smaller Pillar 2. There will be some transfer of funding from Pillar 1 to Pillar 2 from 2005 onwards through 'modulation', which will in time transfer €1.166 billion per year. For the UK, however, the effect of modulation at a European level is not expected to be large. Voluntary modulation, a transfer from Pillar 1 to Pillar 2 within the UK, is likely to be more significant up to 2006. In the longer run, however, it is difficult to forecast how much funding will be devoted to rural development measures in the UK. This will depend on the outcome of the EU budget debate, the share of CAP funding devoted to Pillar 2 after 2007, the proportion of this budget allocated to the UK and the effects of both compulsory and voluntary modulation. Over the next two years, Commission proposals and eventual decisions will address some of these variables. Whilst we may not see much increase in the overall EU Pillar 2 budget, there remains considerable scope for increasing the UK's share. The outcome will be significant for countryside policies in Scotland and elsewhere in the UK.

National governments will also need to make decisions about the way in which they will implement the measures set out in the Mid-Term Review. For example, they have the option of determining single farm payments in a variety of different ways, either emphasising a farm's historic receipts from the CAP or adopting a regional approach in which payments per hectare are more consistent between farms. There will be decisions to be made about whether to focus support on key sectors, such as suckler beef production, through the new system of 'national envelopes'. Member States have scope to adopt full or partial decoupling and will have key choices to make in the dairy, beef and sheep sectors, all of environmental significance. It seems likely that individual countries within the UK, including Scotland, could arrive at distinctly different policy measures within the choices available (Scottish Parliament, 2003). Greater use of cross-compliance, now a mandatory element of the CAP, will be required from January 2005. Again, there is scope for national governments to determine specific rules within a European framework.

28.7 Implications for the managed landscape

The overall direction of European policy is now clearer with recent decisions on the CAP and other policies but there are still considerable uncertainties. Some stem from enlargement and contentious decisions about the budget which lie ahead. Others arise from choices which the British and other governments will need to make, particularly over implementation of the CAP.

We can expect less pressure on the countryside from production-linked agricultural subsidies but potentially greater concerns about the lack of incentives for farmers to manage land in an appropriate way – unless required to do so by cross-compliance. Forestry policy will need to adapt to the new agricultural framework, including the new Pillar 2 regulations, which will be agreed in 2005. Whilst there have been considerable strides towards a more integrated approach in the Mid-Term Review proposals and we can expect less conflict between market and agri-environment policies, the scale of funding for rural policies remains in doubt. We should not assume a major transfer of support to rural development in the short term.

Editors' note

There have been some significant developments since this paper was written. Many of the proposals outline in the paper, such as enlargement of the EU, have now been implemented.

The European Common Agricultural Policy reforms of June 2003 began to operate in Scotland in January 2005. The most significant change has been the introduction of the Single Farm Payment (SFP), which is detached from agricultural production. The SFP is subject to meeting compulsory cross-compliance obligations which include environmental conditions. In Scotland, the SFP is calculated on the basis of subsidies received during the period 2000-2002. A national reserve has been created by "top-slicing" 3% of all SFP entitlements. The purpose of the reserve is to address certain transitional issues and to support cases where a farmer is disadvantaged by the move to a decoupled system of support by being, for example, a new entrant to farming or by making investments.

The value of SFP entitlements are also subject to a modulation deduction. By 2007, it will be compulsory for all Member States to implement modulation at a rate of 5%. Furthermore, the UK has decided to apply an optional higher rate of modulation (called national modulation) to help implement rural development plans. After consulting on a number of options, the Scottish Executive has decided to apply an overall rate of modulation that will be at least 10% by 2007. There remains some uncertainty as the EU budget agreement at the end of 2005 has left a shortfall, compared to expected funding, for Pillar 2 of the Common Agricultural Policy.

The Scottish Executive has also consulted on the Land Management Contract menu scheme and this is currently being introduced.

References

European Union (2002). *Final Report on Financing Natura 2000. Working Group on Article 8 of the Habitats Directive.*
http://europa.eu.int/comm/environment/nature/nature_conservation/natura_2000_network/financing_natura_2000/art8_working_group/pdf/final_report_en.pdf.

European Union (2003a). *The Strategy for a Future Chemicals Policy? REACH.*
www.europa.eu.int/comm/enterprise/reach/whitepaper/index.htm

European Union (2003b). Council Regulation (EC) No. 1782/2003 of 29 September 2003 establishing common rules for direct support schemes under the common agricultural policy and establishing certain support schemes for farmers and amending Regulations (EEC) No 2019/93, (EC) No 1452/2001, (EC) No 1453/2001, (EC) No 1454/2001, (EC) 1868/94, (EC) No 1251/1999, (EC) No 1254/1999, (EC) No 1673/2000, (EEC) No 2358/71 and (EC) No 2529/2001. *OJ L270*, 21.10.03, pp. 1-69.

European Union (2004a). *Third Report on Economic and Social Cohesion – COM(2004) 107 of 18 February 2004.* Office for Official Publications of the European Communities, Luxembourg.

European Union (2004b). *Proposals for the New Structural Funds Regulations for the period 2007-2013.*
http://europa.eu.int/comm/regional_policy/sources/docoffic/official/regulation/newregl0713_en.htm.

European Union (2004c). *LIFE – III. The Financial Instrument for the Environment.*
http://europa.eu.int/comm/environment/life/life/index.htm.

Scottish Parliament (2003). Environment and Rural Development Committee Official Report of Wednesday 3 September 2003, Col. 65. www.scottish.parliament.uk/business/committees/historic/environment/or-03/ra03-0302a.htm.

29 A countryside for everyone: towards multi-benefit land use

John Thomson

Summary

1. There is now general acceptance that other rural businesses and the wider public, as well as farmers and other land managers, have a stake in the countryside, its use and the benefits it provides.
2. The opportunities to secure multiple benefits have never been better than they are currently. Objectives for land use and the environment are converging.
3. What is needed is a widely shared vision to guide the use of the many mechanisms now available to achieve a countryside for everyone.
4. To achieve such a shared vision, society must look forward and discuss the options available.
5. Multi-benefit land use must be the aim throughout Scotland and not just in 'special places'.

29.1 Introduction

This chapter looks forward and considers the role of land use in 21st century Scotland. It highlights the fact that in a post-industrial society the public seeks from the countryside a much wider range of products than previously. Similarly, the land and water in rural areas now support a far greater diversity of economic enterprises. Against this background the chapter poses three deceptively simple questions about Scotland's countryside. Whose is it? What is it for? And how as a society do we make sure that we get what we want?

29.2 Whose countryside is it – the stakeholders

The first of these issues has long been a matter of political controversy. In various guises, it preoccupied the new Scottish Parliament for much of its first session (1997-2001). For better or worse, Scottish Natural Heritage often finds itself at the sharp end of differences of view about the precise balance that is struck between individual property rights and the claims of wider society. But despite the sound and fury, it is still possible to detect an underlying – if generally unspoken – consensus that the countryside is something in which everyone has a stake. The argument is over the precise extent of the claim that different groups, from the public at large to individual landowners, should have.

The most obvious statutory expression of the drive to re-balance the scales is the new access legislation, creating a near-universal right of responsible access (Part 1, Land Reform (Scotland) Act 2003). But just as telling is the acceptance in almost all debates about land management issues – whether concerning agricultural and forestry practices, the management of wild deer, or the competing recreational uses of rivers and lochs – that the outcomes are a matter of legitimate public interest. To that extent, at least, we are genuinely looking to create a 'countryside for everyone'.

29.3 What is the countryside for – the objectives

But if everyone is to varying degrees a stakeholder, what is it that we are looking to the countryside to provide? Traditionally, the answer has been predominantly basic and utilitarian: food, timber and other primary products. There has long – perhaps always – been another strand, still related to human needs but to the less tangible requirements for identity, pleasure and spiritual uplift. In his Ford lectures of 1999, the distinguished historian Professor Chris Smout charted what he perceived to be the protracted struggle for supremacy between these two perspectives on the role of the natural world, which he characterised as 'use and delight' (Smout, 2000).

Professor Smout's guardedly optimistic conclusion was that, as the new millennium approached, reconciliation was perhaps in sight – that the dictates of use and delight were at last converging. There is at least a sporting chance that he may have been right. The rest of this chapter seeks to explain why and to describe some of the steps that will have to be taken to achieve this desirable outcome.

The primary ground for hope lies in just how far we are now removed from a subsistence economy. In the short term, at least, Scotland faces no threat from starvation or lack of shelter. It is part of a wider world economy which will ensure that, even in the event of catastrophic harvest failure or an epidemic of livestock disease, it can import all that is required to satisfy the needs of its citizens. This is not an argument for neglecting Scotland's traditional land using industries or for ignoring the impact that its consumption patterns have on the global environment: its so-called 'environmental footprint'. Nor does it mean that the present generation should not be taking steps to safeguard the long-term productivity of the nation's natural resources. But it removes the necessity that our predecessors faced of according overriding priority and devoting almost all of their energies to feeding themselves and their families through all the vagaries of the Scottish climate.

At the same time, people's dependence on agriculture, forestry and other land-managing activities for employment and wealth creation has also hugely diminished. Even in the least diversified rural areas these industries now account for well under a quarter of the income generated (Allbrooke *et al.*, 1998); in most the percentage, even taking account of associated supply and processing activities, is far lower than that. This decline reflects the growing importance of other sectors of the rural economy, notably all the activities conventionally bundled together under the umbrella term of 'services'. Importantly from a land use perspective, these include tourism and leisure-related enterprises, which across the country as a whole now employ over twice as many people as agriculture. Enhanced mobility and the telecommunications revolution are also making it increasingly possible for people to live in the countryside and work in town, or indeed to conduct a wide range of businesses unrelated to the land from a rural base. As rural communities look ever more to these other types of activity to generate their income, the question arises as to what styles of land and water management will equip them best to survive in a highly competitive, post-industrial world.

There is a growing acceptance that in such a world a high quality natural environment is not only valuable in itself, it is also a major economic asset. Attractive and accessible landscapes, rich in wildlife, are a springboard to prosperity. They provide the foundation for many service sector businesses and in a country such as Scotland, on the edge of the continent, act as a counter to the economic pull of an increasingly congested heartland. Even traditional agricultural products are coming to rely on the distinctiveness and quality image that are founded, at least in part, on such attributes. Thus it is less and less a matter of setting environmental goals *alongside* social and economic ones; instead they increasingly *underpin* them.

29.4 Achieving a countryside for everyone – mechanisms

What are the policy implications of this new economic reality? In short, policies and programmes of public financial support must acknowledge the integrated nature of the rural economy and help to promote the multiple outputs and benefits that sensitive land and water use can deliver.

Forestry policy and practice is already well down this road, albeit that the timescales involved in the industry mean that it will be many years before the full reward is reaped. Agricultural policy – long a laggard – is also beginning to head in the same direction. Hitherto, the strongly sectoral nature of the Common Agricultural Policy, and the central place within it of the commodity support regimes, has constituted a major impediment. But the reforms of the year 2000 began to signal a move to broader-based support, with the establishment of the so-called Second Pillar – the Rural Development Regulation – which at least in principle could fund a much wider range of rural economic activities. The reforms agreed in 2003 offer the prospect of largely severing the link between support and production, potentially opening the way for programmes much more geared to delivering social and environmental benefits, and indeed strengthening a diversified rural economy.

A *Forward Strategy for Scottish Agriculture* (Scottish Executive, 2001) and the *Scottish Forestry Strategy* (Scottish Executive, 2000) capture at least some of the notion of a countryside delivering benefits for all – rural communities and townspeople, consumers, producers and visitors alike. What is more, some of the tools that will be needed to deliver it have already been invented or at least are on the drawing board. We already have substantial experience of designing and running agri-environment schemes, albeit on a scale which still falls far short of what most people involved, be they farmers or environmentalists, believe to be necessary. For the areas of the highest nature conservation importance, Scottish Natural Heritage is rolling out its Natural Care programme. With the Scottish Forestry Grants Scheme, there is a system of incentives for woodland establishment and management better adapted than its predecessors to deliver the multiple benefits sought from 21st century forestry.

Also emerging are some of the planning mechanisms that will be needed if public money is to be spent wisely and efficiently in pursuit of this new agenda for the countryside. Once again, things are most advanced on the forestry side, where Indicative Forestry Strategies, forest design plans and, in some places, Forest Framework Plans provide a guide. Indicative Forest Strategies, such as the *Ayrshire Woodland Strategy* (Ayrshire Joint Structure Plan and Transportation Committee, 2003), are encouragingly comprehensive and visionary. Under the Land Reform (Scotland) Act 2003, local authorities are obliged to take stock of access needs and opportunities within their areas and to prepare Core Path Plans. In prospect, too, are National Park Plans for Loch Lomond and The Trossachs and the Cairngorms which, if they live up to their potential, will come nearer than any other mechanism so far to setting out a truly integrated vision for land and water use across substantial tracts of the country. Finally, but perhaps most ambitiously of all, there are the River Basin Plans that will be required to implement the Water Framework Directive. If properly linked to development plans prepared under the town and country planning system, these could provide a basis for guiding land and water use across the whole of Scotland.

At the level of the individual landholding there has also emerged in embryo what could well be a vehicle capable of delivering the new agenda: the Land Management Contract (Scottish Executive, 2001). A firm aspiration in the Partnership Agreement (Scottish Executive, 2003), Land Management Contracts could, if imaginatively deployed, encapsulate and symbolise a new relationship between land managers and society at large – a bargain based on active stewardship designed to provide the public goods so important to the population as a whole, and to other elements of the rural economy, alongside the more traditional agricultural outputs. This would allow farmers both to make decent a living and to provide the public benefits demanded by society.

29.5 A vision for the future

Many of the tools that are needed to fashion a countryside for everyone are thus to hand. But we still lack any clear – let alone agreed – idea of what it might look like in practice. We have no shared vision for the countryside a generation or two hence.

The *Custodians of Change* report (Scottish Executive, 2002) highlighted this omission some time ago. It called for a concerted effort to generate not just a national consensus about the future of the countryside but a suite of much more local visions. The report implied that only with some sort of plan, however broadbrush and flexible, would it be possible to use all the relevant implements of policy effectively to deliver maximum public benefit. Only then would the individual land manager know how he or she could contribute most constructively to the wider endeavour to create an attractive, accessible and productive countryside, rich in wildlife and locally distinctive. This is particularly relevant at a time when so many farmers are going to be reviewing their future business objectives.

Custodians of Change also pointed out that much of the material needed to piece together a more tangible vision for the future of Scotland's countryside is already available. There is a mass of information about what is already there, not only in terms of biodiversity but also of geology, landscape and cultural interests. Much of this is held in local biological record centres and is increasingly accessible through information technology, whilst local authorities were in most cases joint sponsors of their element of the nationwide programme of Landscape Character Assessments. Ambitions for the future, meanwhile, have been articulated through documents such as Local Biodiversity Actions Plans, the various forestry plans mentioned previously and most notably through the *Natural Heritage Futures* publications produced by Scottish Natural Heritage, which include 21 local prospectuses (Scottish Natural Heritage, 2002).

Most of these have already been the subject of extensive consultation. Brought together, they could provide the basis for the environmental component of the sort of widely-shared vision that is needed. But to be credible, and indeed realistic, they still require the injection of a stronger social and economic perspective. They need not only to spell out what the countryside in any given part of Scotland might look like and contain, but also check this against the economic and social realities. For example, would such a countryside generate the level of economic activity needed to support local communities? What sort of skills would be needed to deliver it? How might these be combined to create satisfying and rewarding jobs, attractive to a highly educated workforce, intolerant of the drudgery of much past rural employment? Many of these, one might expect, would themselves reflect the multiple objectives of land management, combining elements of traditional occupations, such as shepherding and keepering, with others such as conservation, visitor management and interpretation.

Increasingly, Scotland possesses the types of fora that are needed to conduct the debate that is required. These include Community Planning Partnerships, Rural Partnerships, Local Economic Fora, 'community futures' exercises, regional initiatives like the Southern Uplands Partnership, agricultural business rings and local access fora. Not all of these may immediately see the relevance to them of the questions posed in this chapter about the future of the countryside. But if they reflect on the economic realities outlined earlier, quite apart from the wider quality of life issues that are at stake, they will surely recognise that these are matters which are indeed deserving of their attention and where there are real choices to be made.

29.6 The challenge for the future

The outcomes that are desirable, let alone feasible, will no doubt vary markedly from one region of the country to another. Indeed, if they did not they would fail to reflect the diversity which is itself one of Scotland's greatest attractions. People must also be prepared to think radically and experiment, recognising that what currently exists – despite the comfort of familiarity – may not be the best that is achievable. The case for such open-mindedness is becoming all the stronger as the world gears up to the growing reality of climate change.

Above all though, society needs to look forward and to discuss the options. That, indeed, was one of the main purposes of this conference. It was also one reason why Scottish Natural Heritage issued to coincide with it a discussion paper entitled *Scotland's Future Landscapes?* (Scottish Natural Heritage, 2003). This was designed both to highlight the quality and diversity of Scotland's landscapes and to stimulate wider debate about how the nation as a whole wished to see them evolve in future.

This discussion paper pointed up some important choices facing Scotland at the start of the 21st century. The questions that it raised were entirely consistent with current ecological thinking, as embodied in the so-called 'ecosystem approach', which stresses both the need to specify objectives for environmental management and to accept that these are a matter of societal choice. One matter over which there is no choice, however, is the multiple functions that the countryside performs. It is not just a factory floor for farmers and foresters, but nor is it simply an amenity for visitors and commuters. It is both these things and many more. Given that fact, multi-benefit land use must be the aim throughout Scotland, not just in the 'special places' on which conservationists have in the past focused so strongly. Unless it is, a true 'countryside for everyone' will remain a pipedream.

References

Allbrooke, R., Ashworth, S., Cook, P., Copus, A. & Gourlay, D. (1998). *Agriculture and its Future in Rural Dumfries & Galloway*. Scottish Agricultural College, Dumfries.

Ayrshire Joint Structure Plan and Transportation Committee (2003). *Ayrshire and Arran Woodland Strategy*. Ayrshire Joint Structure Plan and Transportation Committee, Prestwick.

Land Reform (Scotland) Act 2003. www.opsi.gov.uk/legislation/scotland/act2003/20030002.htm.

Scottish Executive (2000). *Forests for Scotland: The Scottish Forestry Strategy*. Scottish Executive, Edinburgh.

Scottish Executive (2001). *A Forward Strategy for Scottish Agriculture*. The Stationery Office, Edinburgh.

Scottish Executive (2002). *Custodians of Change*. Agriculture and Environment Working Group, Scottish Executive, Edinburgh.

Scottish Executive (2003). *A Partnership for a Better Scotland: Partnership Agreement*. Scottish Executive, Edinburgh.

Scottish Natural Heritage (2002). *Natural Heritage Futures: An Overview*. Scottish Natural Heritage, Perth.

Scottish Natural Heritage (2003) *Scotland's Future Landscapes? Encouraging a Wider Debate*. Scottish Natural Heritage, Perth.

Smout, T.C. (2000). *Nature Contested. Environmental History in Scotland and Northern England since 1600*. Edinburgh University Press, Edinburgh.

30 Achieving integration on the ground: a view from the land

Robert Balfour

Summary

1. Balbirnie Estate is managed as a farm but it is also attractive to the public for recreation. Part of the estate is within the Lomond Hills Regional Park.
2. Unmanaged access can cause some problems but the main problems arise from criminal activities, such as vandalism, poaching and fly tipping.
3. The key need is to integrate land management and access. Paths, information and rangers can all help to achieve this, and adequate funding is essential.

30.1 Introduction

This chapter will focus on achieving integration on the ground – a view from the land – and will describe how our family estate, Balbirnie, is managed; summarise the various forms of land use and the impact of access on these; and consider areas of conflict or potential conflict and how attempts are made to mitigate them. Finally, this chapter will look at how attempts have been made to integrate access with other land uses and the solutions that have emerged.

30.2 The Balbirnie Estate

The Balbirnie Estate is situated on the northern edge of the former new town of Glenrothes which has a population of around 50,000 people. About half of the estate is situated within the Lomond Hills Regional Park, which is not to be confused with the Loch Lomond and Trossachs National Park (Figure 30.1). Within the Regional Park and just to the north of it are a number of small Fife villages, including Falkland, Auchtermuchty and Freuchie.

The Estate extends to about 2,000 ha and, within that area, about 1,000 ha, including some Less Favoured Area (LFA) land, are farmed in-hand. Let farms comprise 440 ha and 300 ha are given over to forestry. There is also a Site of Special Scientific Interest (SSSI) of about 100 ha and an area of heather moorland on East Lomond which extends to about 120 ha.

The farming operation is predominately cereals and vegetables, with the latter being grown for the supermarkets. There is also a large dairy herd and a suckler cow herd where the stock is finished, mainly for one of the major wholesale buyers of Scottish beef. Achieving and then maintaining quality assurance standards is obviously of great importance as far as the farming operation is concerned. The Estate also runs a sporting department and has one full-time gamekeeper.

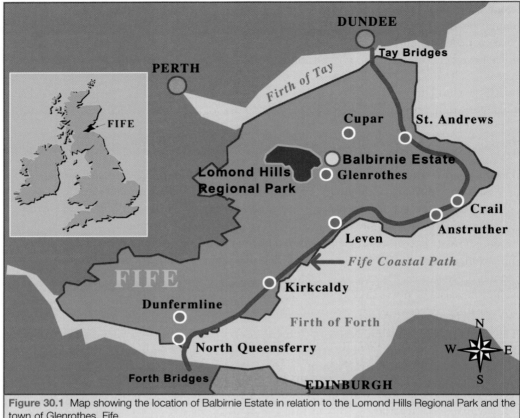

Figure 30.1 Map showing the location of Balbirnie Estate in relation to the Lomond Hills Regional Park and the town of Glenrothes, Fife.

30.3 The Lomond Hills Regional Park

The Lomond Hills have been a magnet for informal recreation for many years and the Balfour family was at the forefront of setting up the original Fife Regional Park some 20 or so years ago. This support was given because it was felt that, even then, access was a fact of life for land managers and that the best way to reduce the impact on the business was to try and manage it. One of the advantages then of being in a Regional Park was that the Local Authority provided a ranger service. The Estate also entered into a footpath agreement to help people to go from the north side of Glenrothes up to the summit of East Lomond. The Regional Park extends to about 5,000 ha and has six reservoirs supplying water to central Fife. In excess of 120,000 people visit the park each year.

30.4 Existing land uses

30.4.1 How are existing land uses affected by access?

The land under forestry is a very useful place for people to take informal recreation as it can absorb large numbers of people without having much effect on either biodiversity or the growing crop. There can, of course, be problems when forestry operations need to be carried out, such as felling and re-planting, or clearing up after wind blow. Within the last five years, the Estate has re-planted quite a large area on East Lomond and has put up deer fencing to keep roe deer (*Capreolus capreolus*) out. However, running through the middle of that plantation is the main path from Glenrothes to the summit. A number of ways have been tried to make the entrance to the forest easy for the public without letting the roe deer in. Stiles have been used, but sometimes the fence has been cut by people wanting even easier access. Self-closing gates have also been used, but unfortunately these do not stop the poachers, who jam the gates open, drive the deer in and then put their dogs in after them. This tends to happen in the spring when the deer are together and has happened three times in the last five years. Unfortunately, because the gate is in the middle of a straight fence run, the only way to get the deer out is to shoot them. We are now trying to replant without deer fencing, although this means that we have to manage the deer carefully. In order to ensure that the deer population remains in balance, we normally try and remove up to 35 head per year, of which about 15 will be bucks. That said, having more people in the forest does in fact help to reduce the poaching problem because individuals are less likely to poach when there are more people about.

Forest roads have also been used by vandals who steal cars and then set them on fire in the wood, as well as by fly tippers who deposit rubbish. One solution to this sort of problem is to put very large stones at the entrance to such roads, as these do not impede the walkers, cyclists or horse riders but do stop any access by vehicle.

A further source of conflict or potential conflict is with livestock. The path referred to above goes through a number of fields which, during the summer, have cows with calves-at-foot in them. In this case, the biggest problem is not actually people with dogs walking through them but people leaving gates open (either deliberately or accidentally). This means that the gates have to be padlocked, as a gate left open can result in 40 cows and calves wandering around with the possibility of them then getting on to a public road. Equally, it is very frustrating for the Estate staff when they have one field which contains cows with male calves and another field which has cows with female calves and they all become mixed. They then have to spend time separating them again.

The arable fields have not proved to be a problem as far as access is concerned and, by and large, people do not like walking in fields which have crops growing in them. The Estate tends not to have fields with vegetables within the Regional Park or, if there are any, they are very near to the public road and most of them have large strips of set-aside round the outside. However, there have been problems on occasion with damage being done to straw bales near to the towns.

All of the land on the Estate is used in some way for country sports and quite a bit of rough shooting takes place. The potential for conflict can arise if people do not keep their dogs under proper control, for example by going through a pheasant pen or disturbing a shoot. On the biodiversity side the use of game strips, beetle banks, hedges and grass strips along burns and ditches have helped to improve the range of habitats for both wildlife and game. There are now more buzzards (*Buteo buteo*) reported, though they appear to be affecting some of the songbirds in particular, and crowds of gulls (*Laridae*) following the plough are now rarely seen.

The Estate is trying to integrate further access with its own objectives, as well as to enable people to enjoy the countryside more. The Estate has been involved in a project with Fife Council and the Regional Park to look at how best to create circular walks and paths within the Regional Park, and a plan of action has been prepared (Donald

McPhillimy Associates, 2003). Interestingly, one of the biggest challenges is to set up a system of paths so that people can walk from within Glenrothes into the park without getting into their car. If people need to get into their car in order to enjoy the countryside, car parks then have to be provided. In the case of private owners, the provision of car parks raises the Estate's rateable value and so increases running costs. A car park, though, is unlikely to generate much income for the owner. In this sustainable world it should be possible to enable people to walk from their front door into the countryside, particularly in a town the size of Glenrothes. It should also be possible for people to walk in the towns. It is a small price to pay to have a path up the edge of a field or along a farm track, provided that it does not go too near a steading, in order to provide people with circular or more long distance walks.

There is some truth in the argument that an estate which has a relatively large acreage can afford to give up a small bit for access. However, the Estate does not want people walking through the main farm steading when up to 300 cattle can be housed there, especially when it is in a position to create paths well away from the steadings and houses. There is both a health and safety issue with people walking through farm steadings, as well as a small risk that animal biosecurity might be compromised. In conclusion, the Regional Park has put together a network of paths, some of which go through Balbirnie and some of which go across other people's land. Some of these farms are very much smaller than Balbirnie, but they have been very willing to accept these proposals.

30.4.2 What can be done to help farmers to manage this access?

It is not always feasible for Scottish Natural Heritage, the local authorities or other bodies to fund either ranger services or advice. Access is another land use and must be integrated alongside other uses such as agriculture, field sports, natural heritage management or forestry, and they must all be integrated with each other. Ranger services can be provided by local people and this has been done as a pilot scheme within the Regional Park. This scheme allows farmers, and particularly shepherds, to do some rangering work and to act as the interface with the general public. This has many advantages: the farmers are on the land every day; they know the land; they know what to look out for; and it helps with integration.

There is also a good case for providing information to people as they make use of the countryside, particularly in a relatively urban area such as Fife. There are information boards at the beginning and end of some of the paths and it might be possible to explain, in very simple terms, what people are seeing in fields at certain times of the year. Standardised signs are, however, required.

There is a real issue about education. Both sides in the discussion need to get together to try and come up with effective approaches to inform people of what is going on because there are a lot of people who do not understand what is happening in the countryside.

There is another issue about the cost of these paths – what is the price to pay for this integration? If the core path network is not set up, particularly in the urban fringe and the intensively-managed lowlands, the access provisions in the Land Reform (Scotland) Act 2003 will ultimately fail because they will not be seen to have been integrated public access with other land uses. Farmers and landowners will see not only their businesses adversely affected but also the environment.

30.4.3 Who will pay for the access provisions?

People will be able to take access at no personal cost. However, providing paths and information and managing access all cost money, and the money has to come from somewhere. There will be some people who will argue that farmers get paid huge sums of money under the Common Agricultural Policy (CAP) and that this should cover the costs associated with access. The payments under the CAP, however, are support payments for growing certain crops (although under the new regulations they might all be de-coupled from production) and have been put in place to compensate for any reduction in income since the ending of guaranteed prices for agricultural outputs. Some of it will go into managing the environment, but this still does not alter the fact that the cost of putting in a core path network will be substantial. The contribution of farmers to this cost will probably be the provision of the land for the actual path.

The Scottish Executive has already allocated some additional funding to local authorities to help them begin planning core path networks, establish local access fora and employ more rangers. This is helpful, but more funds are needed and so there is a need for funding bodies to work closely together.

30.5 Conclusions

Managing land at the estate scale generally allows for more effective integration of the planning and undertaking of farming, forestry and other operations. In addition, this advantage of scale enables the land manager to plan and co-ordinate provisions for public access so that it is best integrated into other land uses.

This does not mean that there are no problems, as practical issues will always arise – especially in urban fringe areas. However, with careful thought and some imagination, many of these can be resolved, especially by using lessons and experience from elsewhere.

Public access routes rarely stop at the farm or even estate boundary, and it is important that neighbouring land managers work together and with others to plan and implement the provision and management of access. This working together is made easier where there is a framework in place, and the Lomond Hills Regional Park has been a good example of this. The establishment of local access fora should help extend this framework for working together to many more parts of Scotland.

Good intentions are a start, but putting access routes in place does cost money and adequate resources must be provided. Farmers, foresters and other land managers have their part to play, but they must be convinced that all the cost is not falling on them – in reality, providing access generates little or no income for individual farms or estates.

References

Donald McPhillimy Associates (2003). Access Action Plan – Lomond Hills from the South. Unpublished report. Fife Council, Glenrothes.

Land Reform (Scotland) Act 2003. www.opsi.gov.uk/legislation/scotland/acts2003/20030002.htm.

30 Farming, Forestry and the Natural Heritage: Towards a More Integrated Future

31 The main priorities for research and advice on farming, forestry and the natural heritage in the next 10 years

Colin A. Galbraith & Richard Davison

Summary

1. This chapter reviews progress towards achieving better integration of farming, forestry and the natural heritage and identifies the main priorities for research and advice in the next 10 years.
2. Key elements of a coherent vision for a multi-benefit countryside exist, mainly through recent national strategies for agriculture, forestry and biodiversity. However, this vision needs to encompass landscape and access.
3. There are gaps in our knowledge of how farming, forestry and the natural heritage interact, most notably in relation to predicting the possible outcomes that can flow from the implementation of different policies and funding regimes.
4. A significant challenge remains that of securing delivery on the ground. Involving people, understanding their motivations and encouraging change through simple and well-targeted advice is very important. Techniques such as demonstration studies and sharing good practice should help raise awareness and enhance overall management. Encouraging co-operative action at a catchment or ecosystem scale, encompassing several farms, is important to securing good outcomes.

31.1 Introduction

Farming and forestry are going through a time of change – at times rapid – which is bringing uncertainty as well as opportunities. Increasing economic pressures, trade liberalisation, policy evolution and growing public interest in the wildlife, landscape and recreational value of the countryside, are all driving the extent and rate of this change. A key opportunity arising from this change is integrating landscape, biodiversity and recreational objectives into land management. It is important that the research and advice underpinning such change is clearly targeted on priorities and supports the development and implementation of policies on farming, forestry and the natural heritage.

The focus of this chapter is to identify the key priorities for research and advice in the next 10 years. The chapter begins by summarising the current position, including the policy context and considers the main issues affecting the development of priorities for research and advice. The chapter then examines in more detail the main priorities for research and advice in the fields of biodiversity, landscape and access. It concludes by outlining the particular research and advisory challenges presented by a more integrated approach to farming, forestry and the natural heritage.

31.2 The need for research and advice

Achieving better integration of farming, forestry and the natural heritage is a major challenge, but is one that must be met to satisfy increasing public pressure for a productive and accessible countryside that maintains biodiversity. Research and advice play an essential role in helping to deliver better integration, particularly so by underpinning the following.

- Developing a vision: if better integration is to be achieved, there must be a clear vision about what we are trying to deliver on the ground, and this should include securing the viability of farming and forestry businesses, protecting and enhancing biodiversity and landscapes, and encouraging responsible access.
- Improving our knowledge: work towards better integration will stand a much better chance of success if it is informed by sound knowledge of the relationships between farming, forestry and the natural heritage, and of the impacts of different policies and funding regimes. Much of this will require new levels of integration and joint working between existing agencies and research bodies.
- Developing plans and strategies: a range of well-focused plans and strategies are needed to ensure that all agencies and land managers work together, at national and local levels, and these need to be underpinned by sound research and advice.
- Supporting local delivery: providing advice that is fit-for-purpose is an essential part of supporting delivery on the ground. Advice needs to be supported by sound research, good awareness, demonstration and the sharing of good practice.

31.3 Farming, forestry and the natural heritage: an overview

Although the subject area covered by this conference was very wide, the chapters in this volume illustrate that our knowledge of the main changes occurring in farming and forestry priorities, and of the main ways in which farming and forestry affect the natural heritage, particularly on biodiversity, is generally good. Gill *et al.* (this volume), Boatman (this volume) and Humphrey *et al.* (this volume) provide valuable reviews of current knowledge of the drivers of change and the impacts of farming and forestry on the environment.

It is clear from these reviews that, for many decades, policies for farming, forestry, biodiversity, landscape and access were developed largely in isolation from each other and objectives did not overlap. However, Gill *et al.* (this volume) highlight the changes to economic support for food production, the development of the Common Agricultural Policy and the shift towards meeting environmental objectives. The last two decades have seen farming and environmental objectives coming closer together, indicating a major change in outlook. Some problems do remain however so, for example, although the use of chemicals in food production has decreased since the 1980s, diffuse pollution risk is still a major concern and was one of three major issues highlighted in the *Custodians of Change* report (Scottish Executive, 2002). A combination of funding support, changes to farming practices and changes to climate appears to have resulted in an increasingly sharp distinction in the use of chemicals between grassland-dominated farming in west Scotland and arable-dominated farming in east Scotland, with consequent impacts on wildlife.

The coming together of land use and environmental objectives is even more the case in forestry. Gill *et al.* (this volume) note that creating a strategic reserve of timber is no longer a central theme of forestry policy. Policy is now very much driven by the need to secure sustainable forest management and to deliver social, economic and environmental outputs (Scottish Executive, 2000). There is also an increasing emphasis on replanting native species and the active restructuring of forests to meet these wider aims.

Although our knowledge of the main drivers for change and the broad impacts of farming and forestry on the environment is generally good, it is clear from Boatman (this volume) and Humphrey *et al.* (this volume) that our detailed knowledge of some of the particular relationships between land use and the natural heritage is less good. In particular, our ability to forecast future impacts of farming and forestry on aspects of the natural heritage, and hence to develop policies that help to deliver positive environmental outcomes, is limited.

Recent years have seen the policy objectives for farming and forestry and those for the environment starting to overlap. A key step has been the development of major new strategies, such as *The Scottish Forestry Strategy* (Scottish Executive, 2000), *A Forward Strategy for Agriculture* (Scottish Executive, 2001) and the *Scottish Biodiversity Strategy* (Scottish Executive, 2004). Considerable progress has been made in relation to objectives for the protection and enhancement of biodiversity, while there is increasing recognition that landscape is an integral part of the environment and is clearly a key area of interest to the tourist industry and others.

It is unclear how the public will judge the development of these strategies, though public funding remains very important, particularly to farming. Sankey (this volume) provides a timely reminder that the views of the public are important. Our knowledge of the views of the general public appears good. As Sankey (this volume) reports, the public know what they would like to see at a very general level and are content for public funding to be provided to farmers, provided that it delivers what they want, such as effective access to land and good conservation of biodiversity. A key challenge, though, remains that of collecting more detailed information on what the public believes should be delivered through such funding.

31.4 Future priorities: biodiversity

The impacts of farming and forestry on biodiversity are generally well-known, and the conservation and enhancement of biodiversity is now more firmly enshrined in Government policy in Scotland. They do identify several priorities for future research and advice:

- the need to agree key indicators for change and to monitor changes to farming, forestry and biodiversity over a period of time;
- the need for modelling of impacts and of the possible effects of policy changes and financial incentives on biodiversity, particularly at the catchment scale; and
- the need to understand better the possible impacts of global influences, such as climate change, on farming, forestry and biodiversity.

The human dimension in all of this is very important – not only in terms of what the public wants, but also in terms of the decisions made by farmers, foresters and other land managers. Several speakers at the conference highlighted the need to understand better the decision-making process employed by farmers and other land managers in deciding which land management practices to adopt. Having a good policy in place, with funding incentives to help secure its implementation, is part of the solution but does not necessarily result in individual farmers or foresters making a particular decision.

The theme of understanding the human dimension is picked up in some detail by Gotts (this volume). Essentially, there is a greater chance of success if farmers, foresters and land managers actually want to change their practices to help conserve and enhance biodiversity, and have a personal commitment towards doing so. Gotts (this volume) emphasises that knowing what the outcomes will be from a change in practice is very important in convincing farmers and foresters. The Targeted Inputs for a Better Rural Environment (TIBRE) initiative is one example of illustrating the outcomes of particular practices (Scottish Natural Heritage, 2004). Some of the case studies in this volume (for example, Macgregor & Warren; Audsley; Jenkins) highlight a few of the challenges that can be encountered in encouraging farmers and other land managers to change course.

Face-to-face advice is considered significant in encouraging people to adopt new practices. Organisations like Farming & Wildlife Advisory Group (Scotland) can help to deliver this. There remains, though, a significant challenge to improve the provision of advice to farmers and crofters, as identified in *A Forward Strategy for Agriculture* (Scottish Executive, 2001) and the *Custodians of Change* report (Scottish Executive, 2002).

31.5 Future priorities: landscape

Although landscapes are important to people, they have only recently been recognised as a key issue and an integral part of the 'environment'. The *Custodians of Change* report (Scottish Executive, 2002) stated that landscape was one of three key issues for the next ten years (the others being biodiversity and diffuse pollution). Swanwick & Martin (this volume), Bennett (this volume) and Macinnes (this volume) provide an excellent picture of our current knowledge and understanding of landscapes.

Landscape is a product of the physical and natural environment, and of social and cultural factors. Assessment tools, such as 'landscape character assessment', exist to help measure the variation in the character of the landscape and hence explain local distinctiveness and sense of place. Our knowledge of the historic environment is less good (Macinnes, this volume) but a programme of work is well in hand to improve this. Landscape character assessment is now seen as a valuable tool in planning, designing and managing landscapes, particularly in environmental impact assessment. The challenge, as suggested by Swanwick & Martin (this volume), is to apply landscape character assessment in a proactive way. Bennett (this volume) reports on how Dumfries & Galloway Council has utilised landscape character assessment proactively, such as planning and design guidance, local forestry frameworks and management strategies for National Scenic Areas.

The proactive application of tools such as landscape character assessment is much more advanced in forestry than in farming. There seems little evidence currently that landscape character assessment is being used in Scotland to help target agri-environment funding, though Farmer (this volume) puts forward some examples from England. The proposal to introduce land management contracts in Scotland (Scottish Executive, 2001) provides an opportunity to put landscape, and access, on an equal footing to biodiversity in deciding the public benefits that should flow from public funding.

The lack of a national policy on landscape is a constraint on the fuller consideration of landscape in land management and its funding. This issue, as explained by Brooks (this volume), is being addressed by Scottish Natural Heritage. For example, the decline of natural features and the reduction in the active management of dykes, hedges and hedgerow trees have resulted in changes to the lowland agricultural landscape which can be addressed through funding for land management. The broader scale changes in agriculture (Boatman, this volume) reinforce these changes.

Unlike biodiversity and access, there is little advice available to farmers on how their work can be changed to improve the landscape. Farmer (this volume) reviews the concerns of land managers about current advice and calls for better communication, education, advice, commitment, partnership and innovation. As with biodiversity, understanding the human dimension and convincing farmers and other land managers about the need for, and benefits of, changes to how they work is important.

31.6 Future priorities: access

Public access to the countryside has been an emotive issue, for both land managers and the public, for many decades. The Land Reform (Scotland) Act 2003 came into force in early February 2005. This law means that everyone has the right to be on most land and water for recreation, education and for going from place to place providing they act responsibly. Access rights and responsibilities are explained in a new Scottish Outdoor Access Code (Scottish Natural Heritage, 2005).

Silvester (this volume) provides a timely reminder of some of the issues involved in the access debate and places the legislative changes in the context of wider change on a number of fronts that farmers and other land managers are having to deal with. Those farming in the urban fringe face a wide range of issues, such as the spillover of social problems from towns and cities, including fly tipping, stray dogs, vandalism and drug use, and do not necessarily receive much support or advice targeted at managing these issues.

Farmers do receive funding support for agri-environment work, and there are grants available from public agencies to support the planting and management of trees, the management of access and recreation, and other initiatives. Hunt (this volume) seeks to answer the tricky question of whether or not these funding schemes are co-ordinated – pulling together in the same direction – or in conflict. Funding schemes are undoubtedly becoming more co-ordinated, but there remains a need to agree a common vision and to co-ordinate action across a wide range of public agencies to deliver this. Hunt concludes that land management contracts do provide a possible way forward but emphasises the need for these contracts to cover biodiversity, landscape and access issues. If they do not, then an opportunity to move strongly towards an integrated approach will be lost.

As highlighted by a number of authors in this volume, the co-ordination of action across a number of farms would deliver better improvements and a more integrated approach – path networks work much better if they provide access to all countryside around a town or village rather than just a few patches. Equally, landscapes do not fit neatly with the boundaries of ownership and nor does wildlife respect such boundaries.

A key challenge, therefore, is to encourage more co-ordinated action. Taylor *et al.* (this volume) report on a review of the ways in which access can be managed positively. Better path networks is one such way, as are good signage and information, and taking opportunities to raise the awareness of the general public, not just their awareness of access rights and responsibilities but also of the needs of the farming and forestry community. Taylor *et al.* emphasise the need for a strategic and planned approach to managing access and that sharing good practice is an important part of the solution. Emphasising the economic benefits that can flow from investment in access provision might be one way of encouraging land managers to integrate access with land management through the provision of paths. Bartlett (this volume) illustrates the economic potential of developing cycle routes in forests.

31.7 Moving towards better integration

Many of the building blocks for achieving better integration of farming, forestry and the natural heritage exist, most notably in terms of our knowledge of the impacts of farming and forestry on the natural heritage and the publication of new strategies for agriculture, forestry and biodiversity. Land management contracts provide the best opportunity so far of achieving better integration. This volume provides a number of case studies showing that there is progress being made towards better integration on the ground.

Moving from action at an individual farm level to co-operative action across a number of farms at a landscape-scale is not without challenges. Angelstam *et al.* (this volume) review the challenges of a landscape-scale approach and of combining the natural and social sciences. One way of meeting these challenges is to use modelling approaches to help plan patterns of land use that incorporate multiple objectives. Involving farmers and other land managers in achieving the balance of land use and how best to integrate farming and forestry with the natural heritage and its enjoyment is essential, as suggested by Anderson (this volume) and Morgan-Davies *et al.* (this volume). It is equally important, though, for other members of local communities to be involved.

Jenkins (this volume) shows what is possible if a group of farmers is committed to improving the local natural heritage. Faced with rising costs and falling prices, a group of ten farmers in North Wales decided to make their businesses more viable by reducing the use of feed, fertiliser and straw, reducing the number of livestock and creating hedgerows and shelter woodlands. The change seems to be remarkable and is an excellent example of what can be achieved when a group of farmers come together to manage whole landscapes.

The case studies presented at the conference looked at two particular issues: diffuse pollution and the encouragement of wild birds through changes in land management. Macgregor & Warren (this volume) look at the motivations and practices of farmers in relation to diffuse pollution. Interestingly, many farmers believed that they were not responsible for the pollution in the Eden catchment in Fife, but were surprised when advised about the economic implications of the nutrients being lost through diffuse pollution. A key challenge is to overcome the belief that change is not needed, and Macgregor & Warren conclude that more proactive promotion of why change is important might be required. On the other hand, Audsley (this volume), reporting on a 4 Point plan to reduce diffuse pollution, notes that if advice is kept simple and is based on simple actions which can reduce the likelihood of problems getting bigger – and thereby save the farmer time and money – then acceptance amongst farmers is good. It is clear from Tharme (this volume) and Grant *et al.* (this volume) that the abundance of wild birds is related to the structure and composition of vegetation. Using predictive models to find the right balance of vegetation restoration to encourage wild birds and monitoring long-term trends is seen as vital.

Most people would agree that land management contracts are an important way forward. Newcombe (this volume) looks at some of this issues that might arise in implementing the land management contract approach. Agreeing the outcome – the vision – is an important part of the process, as is co-ordinating action across a number of contracts. The UK Land Use Policy Group provides a mechanism for examining the effectiveness of European Union rural policies. Blaney & de la Torre (this volume) report on the main findings of a review of the effectiveness of the Rural Development Regulation. In all of the plans reviewed, there were very few objectives relating to environmental outputs and there seems to be little co-ordination across a range of Government policies in the European Union member states. The research is now helping to identify environmental priorities for the future Rural Development Plans (2007-13) in Scotland and other GB countries.

Achieving better integration of farming, forestry and the natural heritage requires a clear vision, accepted by all interests – but what are we trying to achieve? This question is posed by Thomson (this volume) and his conclusion is that we do not yet have a shared vision for the countryside a generation or two hence. Recent strategies (see Section 31.3) will help to move the debate forward, but the challenge remains that of moving from national strategies down to more local visions and individual land management contracts. Balfour (this volume) reinforces the need for this to involve closely those who work and look after the land. The human dimension, ultimately, remains critical in achieving better integration on the ground. Steps that will help to achieve better integration include:

- bringing together advice on biodiversity, landscape and access, and encouraging key representative and advisory groups to promote this advice;
- going beyond the strategies and making sure that these work on the ground to the benefit of farmers, foresters and other land managers, the public and the natural heritage;
- thinking more at a landscape scale – working at the individual farm level is good but so much more would be achieved by working at a larger scale, including better prospects for maintaining viable wildlife populations, for co-ordinating access provision and management, and for conserving and improving the distinctive landscapes of an area;
- better modelling to help people to see what the outcomes could be like and to help determine how different policy and funding options affect the natural heritage as a whole;
- involving people in the implementation of strategies and raising awareness and understanding – implementation fora, such as biodiversity implementation groups, the national and local access fora, the Scottish Forestry Forum, the Moorland Forum and the proposed Landscape Forum, can help to build ownership of the issues by discussing problems and finding solutions; and
- demonstration remains a very good way of seeing what is possible, of showing other land managers what the results are, of learning about the issues involved and what works and what does not – but demonstrations need time and resources, and action needs to be taken now; monitoring of progress in a number of places and feeding results more quickly into the development of policy and advice is essential.

31.8 Conclusions

The subject area covered by this conference has been a challenging one, but better integration on the ground remains something that all interests must strive to achieve. Progress towards better integration is mixed but it is moving in the right direction. Returning to the four strands identified in section 31.2, progress can be summarised as follows.

- Developing a vision: some of the building blocks of a good vision are there, but we need to make sure that the strategies that do exist are co-ordinated and add value to each other.
- Improving our knowledge: our knowledge is quite good overall but there are still gaps, particularly in terms of predicting the possible outcomes that can flow from the implementation of different policies and funding regimes.
- Developing plans and strategies: there are several strategies dealing with key land uses. Further work is needed to incorporate landscape and access issues more clearly into these strategies, though the preparation of new core path plans by local authorities and national park authorities by 2007 will help to set out a clearer strategy for the provision of paths throughout Scotland.
- Supporting local delivery: this remains, perhaps, the biggest challenge and more progress is needed. The key to this, as suggested by many of the authors in this volume, is to involve people. Understanding their motivations and encouraging them to change their practices through simple, sound and well-targeted advice, better awareness, good demonstration, the sharing of good practice and through effective funding of such programmes is essential.

References

Scottish Natural Heritage (2005). Scottish Outdoor Access Code. Scottish Natural Heritage, Perth.

Land Reform (Scotland) Act 2003. www.opsi..gov.uk/legislation/scotland/acts2003/20030002.htm.

Scottish Executive (2000). *Forests for Scotland: The Scottish Forestry Strategy*. The Stationery Office, Edinburgh.

Scottish Executive (2001). *A Forward Strategy for Scottish Agriculture*. The Stationery Office, Edinburgh.

Scottish Executive (2002). *Custodians of Change. Report of Agriculture and Environment Working Group*. Scottish Executive, Edinburgh.

Scottish Executive (2004). *Scotland's Biodiversity – It's In Your Hands. A Strategy for the Conservation and Enhancement of Biodiversity in Scotland*. Scottish Executive, Edinburgh.

Scottish Natural Heritage (2003). *Scotland's Future Landscapes? Encouraging a wider debate*. Scottish Natural Heritage, Perth.

Scottish Natural Heritage (2004). *TIBRE Arable Handbook – 2nd edition*. Scottish Natural Heritage, Perth.

Index

Note: Most entries refer to Scotland, unless otherwise indicated. References to Scottish Executive have generally been omitted as mentions are ubiquitous.

Abbey, E. 230
Aberdeen Environmental Education Centre 222
Aberdeen University 13, 243
  Department of Agriculture and Forestry 174
Aberdeenshire 119, 262
Aberfeldy 245, 247
Abernethy 86
access 31, 97-128, 301-2
  Access Management Appraisal 122
  commercial 103, 104
  to farmland 223
  lack of 29
  see also agri-environment, forestry and access; Land Reform (Scotland) Act; public access; recreation/tourism; Scottish Outdoor Access Code
accidents, risk of 103
acid rain 29
Acrocephalus schoenobaenus 47
Action Plans 155
  see also Biodiversity Action Plans; Local Biodiversity Action Plans
Adams, A. 231-2
'Adaptive Management Experiment Teams' 203-4
ADAS Consulting Ltd 183, 187, 240, 242, 243
added value of farming 112-13
adders 64
advisory and planning tools 37, **85-95**, 155
  future 92-3
  motivation for action 87-9
  practical 89-92
  sources of knowledge 86-7
Ae biking trail 227
Aebischer, N.J. 44, 51
afforestation see forestry
AgCensus 11, 13
Agricultural Holdings (Scotland) Act 2003 102
agriculture see farming
Agriculture Acts 1947 and 1957 8
Agriculture Council (EU) 278
Agriculture Development and Advisory Service 258
Agriculture and Environment Working Group 51-2, 174, 185
Agriculture Working Group 79
  see also Rural Land Use Working Group
agri-environment and farming 50, 51-2
  see also Environmentally Sensitive Areas; Rural Stewardship
agri-environment, forestry and access schemes 99, **109-16**, 263
  added value of farming 112-13
  future 113-14
  policies 109-10, 114-15
  scheme adopted 112

Wales see Pontbren
Aitken, M. 218
Alauda arvensis 44, 45, 47-8, 241, 242
alder 258
Alexander, N. 231-2
Allbrooke, R. 286
allotments 159, 163, 165
amphibians 42, 201-2
an Eilean, Loch 129
Anatidae 48
Anderson, A.R. 60
Anderson, R. xi
  on Breadalbane Initiative 191, **245-8**, 302
Andersson, F.O. 194, 196
Angelstam, P. xi
  on holistic understanding of ecological systems 191, **193-309**, 302
angling 121
Angus 176, 262
animals see fauna
Anthus pratensis 45, 241, 242, 243
Antonine Wall 167
AONBs (Areas of Outstanding Natural Beauty) 181
aphids, cereal 41
applications of Landscape Character Assessment
  conservation and management 139-43
  forestry 141-2
  local and national 140
  practical 138-9
arable see crops
Araneae 41, 47
Areas of Outstanding Natural Beauty 181
Areas of Regional Scenic Significance 151
Argyll 175
Armstrong, H.M. 63
ARSS (Areas of Regional Scenic Significance) 151
arthropods 42, 50
ASH Consulting Group 153
ashwoods, upland 61, 62, 63
Asken Ltd 118
assart 163
Assured Combinable Crops Scheme 47
Auchenorrhyncha 45
Audsley, R. xi
  on 4 Point Plan 86, 191, **217-20**, 300, 302
Australia 181, 214
Austria: Land Use Policy Group study 270-1
Avery, M.I. 60
Aviemore: hill farming and environmental objectives 249, 250, 252
awareness-raising 122, 124
  Breadalbane 247
  events 90-1

Ayrshire
  forestry 167
  Land Management Contracts 262
  Landscape Character Assessment 142
  multi-benefit land use 287
  pollution 218

Baines, D. 48, 49
Balbirnie estate 274, **291-5**
  cost 295
  existing land uses 292-4
  farmers 294
  map 292
  see also Lomond Hills Regional Park
Baldock, D. xi
  on European policy and funding horizon 269, 274, **277-84**
Balfour, R. xi
  on Balbirnie estate 274, **291-5**, 303
Baltic States 19, 278
  forests and rivers 196-7, 200, 201
Bangor: University of Wales 258
BAP see Biodiversity Action Plans
barley 43, 47
barn owl 49, 258
Barnes, R.F.W. 46
Barr, C.J. 44-5
Bartlett, K. xi
  on 7stanes Project 191, **225-32**, 302
bats 51
Bayfield, N.G. 265
BBS see Breeding Bird Survey
beaches 150
  see also coastal areas
bear 59
beaver 59, 202
bees 42
BEETLE (Biological and Environmental Evaluation Tools for Landscape Ecology) 69
beetles 47
  beetlebanks 30, 86, 293
  carabid 48
  ground, leaf and rove 41
Beintema, A.J. 4
benchmarks and thresholds 93, 199-200
Bengtsson, J. 200
Bennett, S. xi
  on supporting landscape design 131, **147-56**, 181, 300
Benton, T.G. 40, 41-2
Bergmanis, R. 196
Berkes, F. 195, 253
Betula
  B. pendula 64, 258
  B. tortuosa 199
BIFF see Breadalbane Initiative
biking see cycling
Biodiversity Action Plans 154, 183, 281
  Ancient and/or Species-rich Hedgerows 45
  farming 40, 42, 45, 51

305

forestry 60, 61-5
Local 82
priority species 235
priority woodland habitats 61-5
see also Local Biodiversity Action Plans
biodiversity and farming 37, **39-57**
   Action Plans 40, 42, 45, 51
   conservation and restoration 50-1
   evidence of changes in biodiversity 41-2
   importance of agriculture 40
   policy and practice, changes in 40, 51-2
   practices, changes in 43-50
biodiversity and forestry 37, **59-75**
   Biodiversity Action Plan on priority woodland habitats 61-5
   landscape scale of management 68-9
   plantations managed for biodiversity effects 66-8
   priority areas for future research 69-72
   priority species 65-6
Biodiversity Habitat Networks 183
biodiversity lost, remedies and 9, 26, 37-95
   changes, evidence of 41-2
   degradation by visitors 250
   future lowland landscapes 176
   holistic understanding of ecological systems 193, 194-5
   landscape management guidelines 182-3
   policies 41-2
   public and countryside 30-1
   research and advice priorities 299-300
   Royal Highland Education Trust 223
   see also advisory and planning tools; holistic understanding of ecological systems; Local Biodiversity Action Plans; Rural Land Use Implementation Plan
Biodiversity Strategy (EU) 281
biological diversity see biodiversity
Biological Diversity Convention 77, 194
Biological and Environmental Evaluation Tools for Landscape Ecology 69
Biomathematics and Statistics Scotland 13
biosecurity 104, 250
BioSS 243
birch 61, 62, 66
   specific species (*Betula*) 64, 199, 258
Bird, S.B. 258
birds 13
   advisory and planning tools 86, 87
   Balbirnie estate 293
   biodiversity and farming 40-1, 42, 45, 46, 47, 49, 50-1
   biodiversity and forestry 60, 65-6, 68, 71
   broadleaf woodland 31, 49, 65-6, 68, 86, 201

Directive (EU) 281
extinctions 44
hole nesting 49, 64, 201, 258
invertebrates as prey 217-18
Land Management Contracts 262, 265
moorland 45, 60, 154, 231-44
Pontbren 258
public and countryside 30
research and advice priorities 302-3
wildfowl/wetland 48, 212, 256
see also Breeding Bird Survey; Common Birds Census; game birds; moorland birds; passerines; raptors; RSPB; Scottish Borders Local Biodiversity; waders
Birnie, R. 10, 40
Bissonette, J.A. 194, 195
blaeberry 66
Blaney, R. xi
   on Land Use Policy Group and EU policies 192, **269-72**, 303
Blaxter, K. 6, 7-8
Boatman, N.D. xi 86
   on biodiversity and farming 37, **39-57**, 298, 299, 301
Boloria euphrosyne 65
Bombidae 42
Borders/Scottish Borders 175, 222, 262
   biodiversity 69, 70
   Council 234, 235, 236, 237
   Forest Trust 21, 233
   forestry 29, 69, 70
   Rural Partnership 234, 237
   see also Scottish Borders Local Biodiversity
boreal forest see holistic understanding of ecological systems
boundaries 120, 158
   bird-friendly 65
   deer fencing 292-3
   dry stone (drystane) dykes 142, 150, 173, 174
   fences damaged 120
   maps showing 159-60, 162, 163, 165
   margins 30, 45
Boutin, S. 203
Boyle, S. 167
bracken 241
Brassica napus 47
Braunisch, V. 201
Breadalbane Initiative 191, **245-8**
   map 246
   policy development 247-8
   raising awareness, building capacity, bridging skill gaps 247
   as special case 246
   timber utilisation projects 247
Breeding Bird Survey (*earlier* Common Birds Census) 41, 42, 235
Brickle, N.W. 46, 47
British Trust for Ornithology 41, 233
broadleaved woodland 152, 199
   biodiversity 61, 62, 63, 64
   decline 173
   new plantations 60

Pontbren 256, 257
see also oakwoods
Brooks, S. xi
   on future lowland landscapes 131, **171-8**, 301
   on invertebrates 46
Broome, A.C. 64, 65
Brown, A.F. 240
brown trout 201, 218
Browne, S.J. 67
Bryman, A. 199, 202
bryophytes 64, 66, 68
BSE (Bovine Spongiform Encephalopathy) 26, 256
BTO (British Trust for Ornithology) 41, 233
Buccleuch 31
Buchan, N. 135
Buchan 176
Buchanan, G. xi
   on moorland birds and grazing **231-44**
built-up areas see urban/built-up
bumblebees 42
buntings 44, 46, 47, 51
Burhinus oedicnemus 51
burning 49
Burton, R. 13
Buteo buteo 293
Bütler, R. 201
butterflies 65
buzzard 293

CAG Consultants 168
Cairngorms 6
   biking trails 231-2
   forestry 12, 22
   grass and grazing 11-12
   hill farming and environmental objectives 249, 250, 252
   Land Management Contracts see CNPA
   Landscape Character Assessment 129, 141
   Local Biodiversity Action Plan 266
   Moorlands Project 266
   multi-benefit land use 288
   Rural Land Use Implementation Plan 81
Cairnsmore of Fleet National Nature Reserve 86
Caithness 9
Caledonian Partnership 21, 91
Calluna vulgaris 86, 151, 241, 242
Campbell, E.H. 46
Canada 47, 204
Canis lupus 59
CAP see Common Agriculture Policy
capacity studies 140
capercaillie 31, 65-6, 68, 86, 201
Capreolus capreolus 293
car parks 30, 230, 294
Carabidae 41
carbon cycle equation, global 18
Carduelis cannabina 47
Careers Scotland 221, 223
carrion crow 235
Carter, C. 60

Index

Carter, S.P. 157
Carterocephalus palaemon 65
case studies 189-272
  habitat planning 202-4
  holistic understanding of ecological systems 199, 200
  Land Use Policy Group and EU policies 192, **269-72**, 301
  natural heritage, contribution to 11-12
  need for international network of 204
  Nitrate Vulnerable Zones 191, **211-15**
  see also Breadalbane; 4 Point Plan; hill farming; holistic; Land Management Contracts; moorland birds; Pontbren; Royal Highland Education Trust; Scottish Borders Local Biodiversity; 7stanes
castles and forts 150
Castor fiber 59, 202
catchment area plans 114
cattle *see* livestock
Cattle Industry Act 1934 8
Causeway Coast (Northern Ireland) case study 181-2
CBC *see* Common Birds Census
CCOW (Countryside Classroom on Wheels) 222-3
CCW (Countryside Council for Wales) 240, 243, 258
CEH (Centre for Ecology and Hydrology) 42, 243
Census 7
  see also Common Birds Census
Central Scotland Forest Initiative 31, 141, 174
Centre for Ecology and Hydrology 13, 42, 243
cereals 8, 41, 46, 47, 291
Cervus elaphus 8, 60, 63, 91, 193, 239, 251
Chamberlain, D.E. 40, 43, 44, 51
Charadriiformes 48, 60
charlock 47
chemicals (fertilisers and pesticides) 10
  biodiversity and farming 43, 45, 46
  European policy and funding horizon 278, 279
  public and countryside 26-7, 28, 29
  see also Nitrate Vulnerable Zones
children *see* education/schools
Chiroptera 51
chough 49, 182
Chrysanthemum segetum 47
CI (Countryside Initiative) 222-3
Ciconia ciconia 201
Cinclus cinclus 218
Circus cyaneus 154
cirl bunting 51
Clark, T.W. 194, 195, 202
climate change 65, 157
Clyde Valley 263
CNPA (Cairngorms National Park Authority) and Land Management

Contracts 261, 263, 264-7
coastal areas 78, 149, 150, 154, 214
Coccinellidae 41
Coed Cymru 256
Coed Y Brenin biking trails 226
Cohesion Report, Third 279
Coleoptera 41, 47
  see also beetles
Coles, B. 158
Columba
  C. livia 235
  C. palumbus 47, 235
commercial access to land 103, 104
Common Agriculture Policy (and reform of) 8, 287, 295
  access 102, 113
  advisory and planning tools 86, 88
  biodiversity and farming 39, 40, 50, 51, 52
  European policy and funding horizon 277, 278, 281-2, 283
  forestry, key changes in 17, 19-20
  future landscape research 180, 184
  future lowland landscapes 174
  Land Management Contracts 262, 263, 264, 266, 267
  Nitrate Vulnerable Zones 211, 214
  public and countryside 25, 30, 31, 32
  research and advice priorities 298
  Royal Highland Education Trust 223
  Scottish Borders Local Biodiversity Action Plan 248
Common Birds Census 40, 42, 47
  see also Breeding Bird Survey
community involvement 236
Community Planning Partnerships 289
computers 89-90, 186
concepts, multiple definitions of 194
Concordat on Access to Scotland's Hills and Mountains 102
conflicts
  inter-use 121
  management (walkers and bikers) 228-30
coniferous plantations and forest 152, 245, 257
  bioversity 60, 63, 64, 66-8
  holistic understanding of ecology 199, 200, 201
  increase 173, 175
  managed 66-8, 158, 159, 165, 168
  maps 159, 160, 165
  see also pinewoods
Connelly, J. 194
Conroy, J.W.H. 265
conservation 166, 281
  areas 29, 198
  biodiversity and farming 50-1
  conflict with forestry 9
  planning using gap analysis 193, 196-7, 200
  public access 105
  water 258
  see also Game Conservancy; hill farming; strategic conservation

contracts *see* Land Management Contracts
Convention on Biological Diversity 77, 194
coordination needed 301-2
Copus, A.K. 6
Corbet, G.B. 59-60
Core Path Networks 106
corn bunting 44, 46, 47
corn marigold 47
corncrake 49, 51, 86
Corvus: *C. corone*, *C. frugilegus* and *C. monedula* 235
Cosgrove, P.J. 265
costs *see* finance
Countryside Agency 122, 142
  future landscape research 181, 182, 187
  supporting landscape design 148, 155
Countryside Character Area 142
Countryside Classroom on Wheels 222-3
Countryside Commission 142, 148
  for Scotland 148
Countryside Council for Wales 240, 243, 258
Countryside Initiative 222-3
Countryside Premium Scheme 81, 142
Countryside Stewardship Schemes 142
Countryside Survey 42, 45, 71
cranefly 49, 64
Creag Meagaidh National Nature Reserve 91
Cree Valley (Galloway) 152
Crex crex 49, 51, 86
Crianlarich 91
crime *see* illegal activities
crofting 25, 43, 142
  Crofters' Commission 81
  hill farming and environmental objectives 249, 250, 252
  Land Management Contracts 263, 265-6
  maps 159, 162
crops
  Arable Aid 11
  Balbirnie Estate 291, 293, 295
  changes and biodiversity 43, 44, 46-7, 51
  continuous cropping 10
  crop rotation 10, 43, 46
  damaged 120
  decline 265
  genetically modified 29
  Nitrate Vulnerable Zones 213, 214
  spraying 120
  unharvested 236
crossbill, Scottish 65
cross-compliance 50
crow 235
CS (Countryside Survey) 42, 45, 71
cultivation changes and biodiversity 47-8
'culturalness' vision 197, 200
Culture, Media and Sport, Department of 227
Cunningham, H.M. 48
Curculionidae 41

307

curlew 241, 242
  stone 51
Custodians of Change v 92, 143, 176, 180, 288
  agri-environment, forestry and access schemes 32, 111, 113
cycling (mainly off-road) 29, 105, 106, 302
  see also 7stanes Project

dairy farms 211, 213-14
Dalbeattie biking trail 227
Dalry 154
Dampney, P. 89
dangerous activities 120, 124
Davidson, D.A. 157
Davison, R. vi, xi
  on research and advice priorities 274, **297-304**
deadwood habitats 63, 64, 66, 67, 68, 202
deciduous trees *see* broadleaved woodland
Decker, D.J. 249
Dee valley 154
deer 8, 91, 239, 251, 293
  biodiversity and forestry 60, 63
  browsing 193
  Deer Commission for Scotland 81
  Deer Management Groups 81
  stalking 102
  see also Cervus elaphus
definitions, multiple 194
Defra (Department of Environment, Food and Rural Affairs) 26-42, 89, 142, 240, 243
Degerman, E. xi
  on holistic understanding of ecological systems **193-209**
degradation, land 173
Déjeant-Pons, M. 167
Delgado, C. 6
demographics 6-7
demonstration sites 86
  see also events
Dendrocopos medius 201
Denmark 271
Dennis, P. 10, 13
Departments, government
  Environment, Food and Rural Affairs *see* Defra
  Environment, Transport and Regions 42
  see also SEERAD
derelict land 173
designed landscapes 158, 184, 187
  future lowland 172, 173, 174-5, 176
  guidance and Landscape Character Assessment 140
  Landscape Design Guidance 155
  see also supporting landscape design
Development Plans 140
  see also Rural Development Plans
dipper *Cinclus cinclus* 218
Diptera 41
diseases, livestock 26, 102, 185, 226,

240, 256
diversification 264
diversity
  loss of 173
  see also biodiversity
Dixon, P. 158
dogs 120, 126, 293
Donald, P.F. 40, 47
Dönz-Breuss, M. 204
Douglas fir 63
Dreweatt Neate 180, 186
driving off-road, illegal 119
drumlins 149
Duelli, P. 195
Duffey, E. 48
Dugan, D. 86
Dumfries and Galloway 222, 281, 300
  biking trails 227
  Land Management Contracts 262
  Landscape Assessment 147-51
  Landscape Character Assessment 142
  see also supporting landscape design
Dumfries and Galloway Council 148, 153, 154, 155
  Structure Plan 147, 151, 152
dumping rubbish 120
Dunbartonshire 119
dwarf shrubs and birds 242
Dwyer, J. 270
Dyck, B. 19
dykes, drystane *see under* boundaries
Dyson Bruce, L. 158

earth science 176
earthworms 45, 47
East Lothian 176, 223
  Local Biodiversity Action Plan 233
East Stewartry coast 154
Easter Ross 66
ecology
  Ecological Site Classification Decision Support System 64
  Ecology and Hydrology, Centre for 13, 42, 243, 258
ecosystems, holistic understanding of 193, 194-5
  research 86
  resilience 195
Eden River 302
  as Nitrate Vulnerable Zone 211-15 *passim*
Edinburgh University 223
education/schools 294
  access 106-7
  birds 236
  countryside, lack of 250
  environment management courses 247
  training and awareness 90-1
  see also Royal Highland Education Trust
Edward-Jones, G. 112
Edwards, C.A. 10, 47
Edwards, K.J. 157
Edwards, T. 19, 180, 184
Egglishaw, H.J. 201

EIA (environmental impact assessment) 139
El Titi, A. 47
ELS (Entry Level Scheme) 142
EMA (Environmental Management for Agriculture) 184-5
Emberiza
  E. cirlus 51
  E. citrinella 46
  E. schoeniclus 47
Enfys 256, 259
England 300
  English Nature 183, 240
  Farm Environment Plans 142-3
  Forestry Strategy 9
  hill farming 253
  landscape research case studies 180, 183, 184, 186
  see also moorland birds and grazing
Entec UK Ltd 182, 187
Enterprise in Education through Business Links 221
Entry Level Scheme (England) 142
Environment and Heritage Service 187
Environment Food and Rural Affairs *see* Defra
environment/environmental 271
  damaged 120-1
  'footprint' 286
  forestry, key changes in 18
  impact assessment 139
  land management *see* hill farming
  legislation 219
  public and countryside 28
Environmental Action Programme 280
Environmental Change Network 212
Environmental Council 281
Environmental Management for Agriculture 184-5
Environmental NGOs 31
  see also National Trust; RSPB; Scottish Wildlife Trust
Environmental Resource Management 142, 151, 152
Environmental Technology Action Plan 280
Environmentally Sensitive Areas (ESAs) 52, 142, 181-2
  see also Breadalbane
erosion 9
ESC-DSS 64
Estate Management Plans 140, 141
Estonia: forests 196-7, 201
European policy and funding horizon **277-84**
  CAP reforms 277, 278, 281-2, 283
  implications for managed landscape 282-3
  LIFE fund 280-1
  nature conservation 281
  Structural Funds 277, 278-9
  see also finance, funding and subsidies
European Union 13, 26, 303
  advisory and planning tools 91
  biodiversity and farming 40

Index

European Action Programme, Sixth 278
European Landscape Convention 143-4, 167
forestry, key changes in 17, 19-20
Land Use Policy Group and EU policies 192, **269-72**
Nitrate Vulnerable Zones and Nitrates Directive 211-12, 215
public and countryside 25
Research and Technical Development (VISULANDS) 71
Single Farm Payment 88, 91, 92
Water Framework Directive 217
see also Common Agriculture Policy; European policy and funding horizon; France; Germany; Single Farm Payment
Europe's Rural Futures study *see* Land Use Policy Group
eutrophication 211
see also Nitrate Vulnerable Zones
Evans, A.D. 51
events
  awareness 90-1
  public access 105
  see also demonstration sites
extinct species 59

face-to-face advice 90
Fahrig, L. 196
Fairclough, G. 165, 166
Falco
  F. columbarius 60
  F. tinnunculus 49
FAO (Food and Agriculture Organisation) 6, 18
Farm Business Diversification Scheme 264
Farm Environment Plans (England) 142-3
Farm Management plan 252
Farm Scale Evaluations 46
farm visits 221, 222
  see also Royal Highland Education Trust
Farmer, A. xi
  on future landscape research 131, **179-87**, 300, 301
farming
  added value of 112-13
  economics 183-4
  Farm Environment Plans 142-3
  future landscape research 180-1, 183-4
  future lowland landscapes 173, 174-5
  land use change/impacts 6, 8, 9-11
  Landscape Character Assessment 142
  multi-benefit land use 287
  natural heritage, contribution to 6, 9-11
  Neolithic 157
  public and countryside 32
  research and advice priorities 298, 299, 302, 303
  Rural Land Use Implementation Plan 80, 82
  see also agri-environment and farming; biodiversity and farming; land managers/farmers; livestock; mixed farming; Nitrate Vulnerable Zones; Scottish Borders Local Biodiversity Action Plan
Farming and Wildlife Advisory Group 87, 90, 92, 183, 185, 189, 300
Farming and Wildlife Advisory Plan 233, 236, 237
farmyards and public access 104
fauna, significant *see* birds; deer; fish; invertebrates; livestock; mammals; wildlife
FC *see* Forestry Commission
FCS *see* Forestry Commission Scotland
feedback loop and advisory and planning tools 87
FEPs (Farm Environment Plans) 142-3
Ferris, R. 60
fertilisers 4, 10, 213
  biodiversity and farming 43, 44-5, 46, 50
  'free' 219
  public and countryside 26-7, 28, 29
FHNs (Forest Habitat Networks) 68-9, 248
fields 44-5
  see also boundaries
Fife 262
  Coast Path 126, 127
  Regional Park 292, 293
  see also Nitrate Vulnerable Zones
finance, funding and subsidies 8, 32, 113, 123, 172
  Balbirnie Estate 295
  cutting hidden costs 218
  hill farming and environmental objectives 250, 251, 252
  key changes in forestry 20
  Land Use Policy Group and EU policies 271, 272
  research and advice priorities 299, 301, 302
  7stanes Project (costs and potential income) 225, 230
  see also European policy and funding horizon; Scottish Forestry Grants
Finglas, Glen 243
Finland 200, 201, 271
fir, Douglas 63
Firbank, L.G. 42, 43
fire risk control 125
First Edition Survey Project 167
fish/fishing 121
  eggs 217
  farming 28, 29
  invertebrates as food 217-18
  salmon 201
  trout, brown 201, 218
Fleet valley 154
flies, predatory 41
flora 41
  see also crops; forests and woodlands; grasslands; moorland
Flow Country 9, 21, 30, 149
fly-tipping 119, 120, 293
focal species modelling 69
fodder 43, 213, 258
food
  demand for 6
  Food and Agriculture Organisation 6, 18
  maximising production 40
  quality 26, 32
  see also crops; farming; livestock
foot and mouth disease 26, 102, 185, 226, 240, 256
forecasting *see* future
forest *see* forests and woodlands
Forest Framework Plans 287
Forest Habitat Networks 68-9, 248
Forest Partnership for Action 18
Forest Research 69
forestry 29, 167, 236, 292
  cessation of planting 11
  design and Landscape Character Assessment 140
  early plantations mainly Sitka spruce 60
  future lowland landscapes 173, 175
  increased planting 12, 26
  Indicative Forestry Strategies 151-2, 153
  land use change/impacts 6, 8-9, 11, 12
  Land Use Policy Group study 271
  Landscape Character Assessment 141-2
  Local Forestry Frameworks 147, 153-4, 155
  multi-benefit land use 287
  natural heritage, contribution to 6
  planned design 11
  public and countryside 28, 29-30, 120
  recreation *see* 7stanes Project
  research and advice priorities 298-9, 303
  stream quality 200, 201
  supporting landscape design 147, 151-4
  see also agri-environment, forestry and access schemes; biodiversity and forestry; coniferous plantations; Forestry Commission; forests and woodlands; key changes in forestry; Scottish Forestry; timber
Forestry Act 1919 8
Forestry Authority 152
Forestry Commission v
  advisory and planning tools 90
  biking trails 225, 226, 227, 228, 230
  biodiversity and forestry 60, 61, 62, 65, 68
  forestry, key changes in 21, 22
  future lowland landscapes 175
  Landscape Character Assessment 141
  natural heritage, contribution to 8, 11

309

Pontbren 256
public and countryside 28, 30, 32
Royal Highland Education Trust 222
Scottish Borders Local Biodiversity
    Action Plan 236
supporting landscape design 148, 151, 152, 153
Woodland Data 158
Forestry Commission Scotland (FCS)
    advisory and planning tools 87, 90, 92
    agri-environment and access schemes 110
    biodiversity and forestry 71
    Breadalbane Initiative 245, 246, 247
    Landscape Character Assessment 183, 187
    public access improvement, help for 118
    public and countryside 30, 31
forests and woodlands
    access policy 226
    boreal 197
    Breadalbane Initiative 245, 246, 247, 248
    Central Scotland Forest initiative 31, 141, 174
    coniferous 60, 152, 245 (see also pinewoods)
    deciduous see broadleaved
    future lowland landscapes 173, 175
    grazing 86
    historic environment 157, 158, 159
    landscape design guidance for 147, 148, 149, 150, 152-3
    loss of 18
    maps 158, 159
    multi-functional 31
    plantations 68, 200, 245, 257
    Pontbren 255-7, 258
    protected 196
    regeneration 250, 252
    Scandinavia 200, 201
    semi-natural 60, 64, 66, 68
    shelter bands 302
    supporting landscape design 149
    Sweden and Baltic States 196-7, 201
    upland 60, 61, 62, 63, 66
    wet 61, 62, 257
    see also forestry; holistic understanding; native woodlands
Forres 20
Forsyth, D.J. 47
Forward Strategy for Scottish Agriculture v 13, 261, 287
    agri-environment, forestry and access schemes 110, 113
    biodiversity and farming 51-2
    public and countryside 26
    research and advice priorities 200, 299
Four Point Plan for livestock and pollution reduction 86, 191, **217-20**, 302
diffuse pollution risks from agriculture 217-18
dirty water around steading 218
feedback from farmers 219
guidance to farmers 218
nutrient use, better 218-19
plan described 218
Risk Assessment for Manure and Slurry 219
Fowler, J. 89, 248
France 26, 52, 181, 253
    Land Use Policy Group study 270-1
Franklin, J.E. 61, 66, 195
fritillary butterfly, pearl bordered 65
Froud-Williams, R.J. 48
fuel wood 18
Fuller, R.J. 44, 45, 48, 49, 63, 67, 239
funding see finance
fungi 46, 66
future 273-304
    advisory and planning tools 92-3
    agri-environment, forestry and access schemes 113-14
    forestry, key changes in 21-2
    historic environment 168
    impacts on natural heritage 12-13
    multi-benefit land use 288-9
    public access 105-6
    public and countryside 31-2
    see also Balbirnie Estate; European policy and funding; multi-benefit land use; research and advice priorities
future landscape research 131, **179-87**
    context for change 179-80
    farming economics 183-4
    farming needs 180-1
    landscape management guidelines and biodiversity, case-study of 182-3
    self-help for farmers, case-study of 184-6
    Strategic Management Plans case-study 181-2
future lowland landscapes 131, **171-8**
    agricultural 174-5
    SNH and 171-2, 173-7
    wooded 175
FWAG see Farming and Wildlife Advisory Group

Galbraith, C.A. vi, xi
    on research and advice priorities 44, 274, **297-304**
Gale, M. xi
    on key changes in forestry 3, **17-23**
Gallinago gallinago 45, 241, 242
Galloway 22, 152, 154
    see also Dumfries and Galloway
game birds 86, 201
    biodiversity and farming 31, 44, 45, 46, 47-8, 49
    biodiversity and forestry 65-6, 68
    moorland birds and grazing 239, 241, 242
    public and countryside 26, 31

Scottish Borders Local Biodiversity Action Plan 233-7
    see also capercaillie; grouse; partridge
Game Conservancy Trust 87
    biodiversity and farming 41, 46
    Scottish Borders Local Biodiversity Action Plan 233, 234, 236, 237
Gamhna, Loch 129
gap analysis, strategic conservation planning using 193, 195, 196-7, 198
gardens, historic 158
Gaston, K.J. 194
GCT see Game Conservancy Trust
Geaumannomyces graminis var. *tritici* 46
Gee, A.S. 9
geese 87
Gelan, A. 6
Geltsdale 240
genetically modified crops 29
Geographic Information Group of SNH 138
Geographic Information System see GIS
Germany 201, 270-1, 279
Gibb, L. xi
    on Royal Highland Education Trust 191, **221-4**
Gibbons, D.W. 41, 86
Gill, M. xi, 63
    on natural heritage, contribution to 3, **5-16**, 298
Gillings, S. 45
GIS (Geographic Information System) 69, 133, 137-8, 197
Glendinning, M. 157, 158
Glenrothes 291, 292
Glentress 28
Glentress biking trail 227, 230
globalization 18-20
golden plover 60, 239, 241, 242
golf 28
    courses 120, 159, 165
Gorelick, S. 26
gorse 149, 150
Gotts, D. xii
    on advisory and planning tools 37, **85-95**, 99, 300
Gough, S.J. 49, 239
grain see cereals
Grampians 22
granite uplands 149, 151
Granström, A. 200
Grant, A. 140
Grant, M. xii
    on moorland birds and grazing 191, **231-44**, 302
Grant, S.A. 10
grants see finance; Scottish Forestry Grants Scheme
grasslands 10, 86, 239
    changes and biodiversity 43, 44, 48-9
    cutting 48-9
    decline 11, 43
    drainage and reseeding 48
    height, birds and 262
    intensive 86

Index

mat grass 10
ryegrass 45
strips 293
see also grazing
grazing/rough grazing 13, 148
  biodiversity 40, 49
  decline in area 11-12
  management 250, 251
  maps showing 159, 160, 163, 165
  see also grasslands; livestock;
    moorland birds and grazing
great crested newt 202
Greece 271
green belt studies 140
Green, R.E. 48, 49, 86
green woodpecker 258
greenshank 60
Greenspace for Communities 90
Gregory, R.D. 40, 41
grey partridge 26, 44, 46, 233-7
Grime, J.P. 49
grouse
  black 49
  red 239, 241, 242
GRUB (Grazing and Upland Birds) 243
Guisan, A. 197
gulls 235, 293
Gunderson, L.H. 194, 195
Guruswamy, L. 194

Habitat Action Plans 61, 62, 68, 248
habitat networks, functionality of 81, 195-9
  Biodiversity 183
  Forest Habitat Networks 68-9, 248
  gap analysis, strategic conservation planning using 193, 195, 196-7, 198
  suitability models, tactical planning using 196-7
Habitats and Rare, Priority and Protected Species 72
Hampshire Farm Research case study 184
Hancock, M.H. 44
Hanson, W.S. 158
Hansson, L. 61
HAPs (Habitat Action Plans) 61, 62, 68, 248
hare, brown 44
Harper, D.G.C. 47
HARPPS (Habitats and Rare, Priority and Protected Species) 72
Harris, S. 44
harvesting changes and biodiversity 47
Haughton, A.J. 46
Hawes, C. 46
Haysom, K.A. 48
health
  hazard and public access 120, 124
  Health and Safety on Educational Excursions 221, 224
  Health and Safety Executive 126, 222
Heard, M.S. 46
heather (*Calluna vulgaris*) 86, 151, 241, 242

see also moorland
Heberlein, T.A. 194
Hebrides 40, 43
hedgerows 30
  BAP for 45
  changes and biodiversity 44-5
  decline 26, 42, 43, 44-5, 173, 175
  Pontbren 256, 257, 258
hen harrier 154
Henderson, I.G. 233
Henderson, J. xii
  on Rural Land Use Implementation Plan 37, **77-83**
Henderson, R. 247
Henrickson, L. xii
  on holistic understanding of ecological systems **193-209**
Hepburn, L.V. 45
Herbert, R. 60
herbicides *see* pesticides and herbicides
Heritage Lottery Fund 155
Hierarchical Monitoring Methods 215
Highlands
  advisory and planning tools 91
  biodiversity and forestry 64
  future landscapes 175
  hill farming and environmental objectives 249, 250
  Land Management Contracts 262
  Landscape Character Assessment 142
  see also Breadalbane; Cairngorms
hikers 106, 229
hill farming and environmental objectives and suggested solutions 191, **249-53**
  access 250
  benefits of discussions 252-3
  livestock ratio 250, 251
  woodland regeneration 250, 252
Hill Sheep and Native Woodland project 91
Hillphones scheme 102
HIMOM (Hierarchical Monitoring Methods) 215
Hingley, R. 158
historic environment 131, **157-70**, 196
  future 168
  management 166-8
Historic Land Use Assessment 158-68
  additional benefits 167-8
  key strengths 166-7
  Landscape Character Assessment 157, 164, 165-6, 167
  maps 159-65
Historic Scotland 158, 166, 168
  maps 159-65
Hocknell, S. 78
Hole, D.G. 47
holidaymakers *see* recreation/tourism
holistic understanding of ecological systems 191, **193-209**
  sustainability, biodiversity and ecosystem 193, 194-5
  see also habitat networks; research

and development needs
Holland, J.M. 50
Homoptera 41
horse-riding 106
housing *see* urban/built-up areas
hoverflies 67
HSI modelling 198
Hughes, R. 135
Humphrey, J.W. xii
  on biodiversity and forestry 37, **59-75**, 92, 298, 299
Hungary 19, 270-1, 278
Hunt, I.S. 114
Hunt, S. xii
  on agri-environment, forestry and access schemes 99, **101-16**, 301
Huntly 13
Hutchings, M.R. 44
Hymenoptera 41, 42, 44

Iason, G.R. 13
IEEP (Institute for European and Environmental Policy) 269-70
IFS (Indicative Forestry Strategies) 151-2, 153
illegal activities 119, 120, 121, 293
IMBA (International Mountain Biking Association) 231
incentive schemes 90
  see also finance
income
  loss 26
  7stanes Project 225, 230
  see also finance
Indicative Forestry Strategies 151-2, 153
industry and commerce 159, 172
  globalization of 18-19
  increased 173
  maps showing 162, 163, 165
  mining 150, 162, 163, 199, 200
  tourism as biggest 27
information
  biking trails 229
  geographic *see* GIS
  land management 185-6
  see also signs and notices
infrastructure 172
  built-up *see* urban
  damage 120
  see also transport system
Inglis, I.R. 47
insects and biodiversity
  farming 41, 42, 46, 49, 51
  forestry 64, 65
  see also beetles
Institute for European and Environmental Policy 269-70
Institute of Grassland and Environmental Research 243
insurance costs and public access 120
integration 50, 181, 302-3
International Organisation for Biological Control 50
inter-use conflicts 121

311

Inventory of Historic Gardens and
  Designed Landscapes 158
Inverness: forestry 28, 29
invertebrates 13
  biodiversity and farming 41-2, 45,
    46, 47, 49
  biodiversity and forestry 64, 65, 66,
    67
  dung-dwelling 50
  as food for fish and birds 217-18
  see also beetles
IOBC (International Organisation for
  Biological Control) 50
Ireland 271, 281
islands 142, 262
  hill farming and environmental
    objectives 249, 250, 252
  see also Northern Isles; Western
    Isles
Islay 49
Italy 271, 278

jackdaw 235
Jakobsen, C.H. 202
Jasinski, K. 204
Jenkins, D. xii
  on Pontbren Farmers' Group 191,
    255-9, 300, 302
Johannesburg Summit (2003) 18
John Muir Trust 31
Jokimäki, J. 197
Jones, A.T. 61
Jones (James) & Sons 20
Jougda, L. xii
  on holistic understanding of
    ecological systems **193-209**
Jukes, M.R. 67
Juniperus communis 64, 65

Kaennel, M. 194
Karr, J.R. 194
Kendall, D.A. 47
Kenny, I. 88
Kent Lifescape programme case study 183
kestrel 49
key changes in forestry 3, **17-23**
  global context 18-20
  Scotland and sustainability 20-2
  see also biodiversity and forestry
Kinross 223, 245, 262
Kirby, K.J. 60, 63, 195, 197
Kirkham, F.W. 45
Kirkton (near Crianlarich) 91
Kirroughtree biking trail 227
Kleijn, D. 51
knowledge
  creating demand for 88
  sources of 86-7
Kola peninsula 202
Kortland, K. 66
Kristensen, E.S. 6
Kupfer, A. 202
Kurki, S. 200
Kurlavicius, P. 197, 201
Kuuluvainen, T. 197, 200

Kyle of Lochalsh 249, 250, 252

labour 13, 27
ladybird 41
Lagopus lagopus scoticus 239, 241, 242
Lahemaa National Park (Estonia) 201
Lambeck, R.J. 69, 197
LaMIS (Land Management Information
  System) 185-6
Lanarkshire 119
land cover 172-3
  see also forests and woodlands;
    grasslands; moorland
Land Management Contracts 191-2, **261-8**,
    283, 288
  biodiversity and farming 52
  development 261-3
  forestry, key changes in 20
  future landscape research 181, 184
  Landscape Character Assessment 143
  modelling exercise 263-4
  public and countryside 32
  research and advice priorities 300,
    301, 303
  Scottish Borders 237
  Steering Group 253
  see also CNPA; land managers
Land Management Information System
  case study 185-6
land managers/farmers 301
  advisory and planning tools 92
  agri-environment, forestry and
    access schemes 111-12, 114, 115
  biking trails 228
  environmental land management *see*
    hill farming
  future lowland landscapes 176
  help for *see* public access
    improvement
  Land Management Contracts 263, 265,
    266
  pollution dangers 219
  public and countryside 31
  Scottish Outdoor Access Code 103,
    106-8
  supporting landscape design 155
  time available 7
  training 223
  see also Land Management Contracts
Land Reform (Scotland) Act 2003 v 31,
    99, 294
  biking trails 229, 231
  multi-benefit land use 285, 287
  public access improvement 117, 118,
    119, 120, 126
  research and advice priorities 301
  Scottish Outdoor Access Code, draft
    101, 102, 103, 104
land use change and natural heritage 157
  drivers of 6-9
  impacts 9-11
Land Use Consultants
  historic environment 158, 167, 168
  Landscape Character Assessment 134,
    136

supporting landscape design 148, 155
Land Use Policy Group and EU policies
  192, **269-72**, 301
Land Use Research *see* Macaulay
  Institute
landscape 129-87, 300-1
  see also designed landscapes; future
    landscape research; future
    lowland landscapes; historic
    environment; Landscape Character
    Assessment; supporting landscape
    design
Landscape Character Assessment 131,
  133-45, 175, 182, 288
  Dumfries and Galloway 147, 148,
    151-2, 154, 155
  Historic Land Use Assessment 157,
    164, 165-6, 167
  key stages 134-5
  Landscape Character Areas 135, 138
  meaning and importance of landscape
    133-4
  nature of 134-5
  Scottish programme of 135-7
  supporting landscape design 147, 148,
    152, 154, 155
  see also applications of Landscape
    Character Assessment
Landscape Character Types 135, 138, 141
landscape design *see* designed
  landscapes
Landscape Design Associates 184, 187
Landscape Forum 303
landscape management guidelines,
  biodiversity 182-3
Langholm 153
Langton, T. 202
Lantra 223
lapwing 44, 45, 47, 265
larch 64, 152
Laridae 235, 293
Larsson, T.-B. 194
Latvia 19, 278
  forests 196-7, 200, 201
Lawers, Ben (Tayside) 167
Lazdinis, M. xii
  on holistic understanding of
    ecological systems **193-209**
LBAPs *see* Local Biodiversity Action
  Plans
LCAs (Landscape Character Areas) 135,
  138
LCTs (Landscape Character Types) 135,
  138, 141
lead mining 150
LEADER (Liaison Entre Actions de
  Development de l'Economie Rurale)
  256
LEAF (Linking Environment and Farming)
  91, 222
Leake, A.R. 50, 51
Lee, K.N. 203
legal liability of occupier 120, 121
legislation 5, 7, 8
  Agricultural Holdings Act 2003 102

Index

environmental 219
Water Environment and Water Services Act 2003 217
see also Land Reform (Scotland) Act; Natural Heritage (Scotland) Act
leisure *see* recreation/tourism
Lelong, O. 158
Lepus capensis 44
LERAPS (Local Environment Risk Assessment for Pesticides) 214
Less Favoured Area 291
Support Scheme 262
LFPs (Local Farming Plans) 114-15
liability (legal) of occupier 120, 121
lichens 64, 66
LIFE projects 91, 280-1
Lifescape Project case study 183
Lindenmayer, D.B. 61, 195
Linder, P. 200
Lindsay, J.M. 60
Linking Environment and Farming 91, 222
Linnaea borealis 65
linnet 47
Linyphiidae 47
Lisbon agenda 279
litter 120
livestock (increased sheep and decreased cattle numbers) 11-12, 63, 163, 239, 302
  advisory and planning tools 87
  Balbirnie Estate 291, 294
  biodiversity and farming 49
  Breadalbane Initiative 247
  diseases 26, 102, 185, 226, 240, 256
  future lowland landscapes 173
  hill farming and environmental objectives 250, 251
  Land Management Contracts 265
  meat labelling 27
  Nitrate Vulnerable Zones 211, 213-14
  parasites 50
  Pontbren Farmers' Group 255, 256, 258
  public access improvement, help for 120, 126
  stocking rates changes and biodiversity 49
  see also 4 Point Plan; grazing
Llanfair Caereinion *see* Pontbren
LMCs *see* Land Management Contracts
Local Authorities 31, 81, 114, 123, 262
Local Biodiversity Action Plans 82, 114, 168, 186, 265, 288
  Cairngorms 266
  LBAP Network and Partnership 81
  see also Scottish Borders Local Biodiversity
Local Economic Fora 289
Local Enterprise Companies 148, 262
Local Environment Risk Assessment for Pesticides 214
Local Farming Plans 114-15
Local Forestry Frameworks 147, 153-4
Local Plans (Dumfries and Galloway) 151
Lochaber 28, 29
Lockerbie 153, 227

Lofty, J.R. 47
Lolium perenne 45
Lomond Hills Regional Park 291, 292, 295
Lomond, Loch 288
  forestry 22
  hill farming and environmental objectives 249, 250-1
  Landscape Character Assessment 137, 141
  National Park Plan for 167, 168
Loobu, River (Estonia) 201
Lothians 176, 223, 233, 262
Lohmus, A. 196
Lottery Fund, Heritage 155
lowland landscapes *see* future lowland landscapes
Loxia scotica 65
Luxembourg 271

Mabie biking trail 227
Macaulay Institute 10, 11-12, 174
  Land Cover maps 158
  moorland birds and grazing 240, 243
McEachern, C. 249
Macey, E.C. 173
Macgregor, C.J. xii
  on Nitrate Vulnerable Zones 191, **211-15**, 300, 302
machair 40, 43
Macinnes, L. xii
  on historic environment management 131, **157-70**, 300
McIntosh, R. xii, 60
  natural heritage, contribution to 3, **5-16**
Mackechnie, A. 157, 158
MacKenzie, N. 61-2
Mackey, E.C. 26
McNeely, J. 194
McPhillimy Associates 293-4
Malmgren, J.C. xii
  on holistic understanding of ecological systems **193-209**
mammals 51, 59, 202
  biodiversity and farming 42, 44, 46, 49, 51
  see also deer; dogs; livestock; squirrel
management
  historic environment 166-8
  see also land managers
manure and slurry 213, 218, 219
Mar Lodge Estate 141
marigold, corn 47
marine landscapes *see* coastal areas
Marine Working Group 78, 82
markets, proximity to 21
Markland Report 281
Marrs, R.H. 45, 240, 241
Marshall, G.R. 249
Martin, J. xii, 175
  on landscape character assessment 131, **133-45**, 300
Martin, P. 18, 47

Mason, W.L. 61
mat grass 10
Mayer, P. 194
Mayle, B. 63
meadow pipit 45, 241, 242, 243
mechanisation 44, 174
media 230
Medieval land use 163, 164
Menu Scheme 267
merlin 60
middle-spotted woodpecker 201
Mid-Term Review (EU) 282
Miliaria calandra 44, 46, 47
mining 150, 162, 163
  Sweden 199, 200
mires *see* wetlands
Mitchell, P.L. 60
mixed farming 111
  changes and biodiversity 44, 51
  decline 43, 173
  Nitrate Vulnerable Zones 212, 213
Moffat 227
'monitor' farms 93
Montgomeryshire Wildlife Trust 258
  see also Pontbren
Moorcroft, D. 47
moorland 86, 151, 241, 242, 291
  birds *see* moorland birds
  Cairngorms Moorlands Project 266
  decline and loss 26, 173
  historic environment 168
  map of 160
  restoration 242
  supporting landscape design 149, 150
  see also heather
moorland birds and grazing 45, 60, 154, 191, **239-44**
  bird abundance and vegetation 240-2
  moorland restoration 242
  spatial models and determining mechanisms, verifying 242-3
Moorland Forum 303
Moray 262
Moreby, S.J. 50
Morgan-Davies, C. xii
  on hill farming and environmental land management 191, **249-53**, 302
Morris, A.J. 46
Morris, M.G. 48
moss 149
Moss, R. 201
Motacilla 48
  M. flava 45
moths, clear-wing 64
motivation for action 87-9
Motray Water 212, 213
mountain bikes *see* 7stanes Project
Mountaineering Council of Scotland 123
Mountford, J.O. 45
Mull of Galloway 150
multi-benefit land use in 21st century 274, **285-90**
  future 288-9
  mechanisms 287-8
  objectives 286-7
  stakeholders 285-6

313

multiple case studies and habitat planning 202-4
Murray, R.D. 234
Myhrman, L. xii
  on holistic understanding of ecological systems **193-209**

Napier University, Centre for Timber Engineering 20
Nardus stricta 10
Näslund, I. 201
National Assessment of Scotland's Landscapes (SNH) 141
National Biodiversity Network 72
National Farmers' Union *see* NFU
National Forest Inventory for Scotland 60
National Goose Forum 87
National Inventory of Woodlands and Trees 71
National Landscape Character Assessment 143
National Monuments Record of Scotland 158
National Nature Reserves 86, 91
National Parks 163, 201, 288
  Land Management Contracts 262-3, 266
  Landscape Character Assessment 140
  Plan for Loch Lomond 167, 168
  see also Cairngorms; Trossachs
National Scenic Areas 176, 181, 300
  historic environment 163, 166, 167, 168
  Landscape Character Assessment 140, 141
  Management Strategies 147, 152, 154-5
National Trust for Scotland 31, 118, 141
native woodlands, restoration of 11, 21
  see also broad-leaved
Natura 2000 279-80
Natural care programme 287
natural heritage 3, **5-16**, 299
  case studies 11-12
  forecasting future impacts 12-13
  land use change, drivers of 6-9
  land use, impacts of 9-11
  Natural Heritage Areas 141
  Natural Heritage Futures (SNH) v 139, 140
Natural Heritage (Scotland) Act 1991 5
'naturalness' vision 197, 200
Nature Conservancy Council 6, 8-9
nature conservation *see* conservation
nature reserves 86
NBs (nutrient budgets) 211, 213, 214
Nedwell, D.B. 211
Neolithic farming 157
nestboxes 236
Netherlands 271
New Castleton biking trail 227
new and potential member states of EU 278, 279
new settlements 140
Newcastleton (Roxburghshire): maps of land use 159-65
Newcombe, F. xii
  on Land Management Contracts 184, 191-2, **261-8**, 303
newt, great crested 202
Newton, I. 46
NFU Scotland 26, 223
  public access improvement 118, 123
  Scottish Borders Local Biodiversity Action Plan 233, 237
NGOs (non-governmental organisations) 27
  advisory and planning tools 87, 92
  see also Environmental NGOs
night-time access to land 105
Niklasson, M. 200
Nilsson, S.G. 195
Ninnes, R.B. 241
Nith estuary 154
Nitrate Vulnerable Zones 191, **211-15**
Nixon, C.J. 60
non-governmental organisations *see* NGOs
Northern Ireland 181-2
Northern Isles 167
  biodiversity and farming 40, 43
  forestry 22
  Orkney 40, 43, 167
Norton, B.G. 194, 201, 202
Noss, R.F. 194, 195
NSAs *see* National Scenic Areas
nuclear waste disposal 28
Numenius arquata 241, 242
nutrient budgets 211, 213, 214
NVZs *see* Nitrate Vulnerable Zones

oakwoods 66
  Atlantic 60
  upland 61, 62, 63
OAS 52
O'Brien, M. xii
  on moorland birds and grazing **231-44**
Obrist, M.K. 195
obstruction by visitors 120
Oenanthe oenanthe 241
off-road
  biking *see* cycling
  driving, illegal 119
oilseed rape 47
Olson, D.M. 196
operational management and ecological systems 194, 195-6
Ordnance Survey maps of Historic Land Use 159-65
organic farming 50
Orkney 40, 43, 167
Ovis aries 49, 163, 239, 250, 251, 258

Pain, D.J. 41, 49
Pakeman, R.J. 241
Parasitica 41, 50
Parish, D.M.B. 233
Parker, V.T. 194
Parks *see* National Parks; Regional Parks
Parr, T.W. 44-5
Partnership Agreement 288
partridge, grey 26, 44, 46, 233-7
Passer
  P. domesticus 41
  P. montanus 233-7
passerines 41, 44-8 *passim*, 51, 233-7, 241, 242, 243
  winter monitoring 235-6
pasture 148
paths 102, 103, 288, 293, 301
  for bikes *see* 7stanes Project
  Core 288
  creation 124
  long-distance 154
  Paths for All Partnership 90, 106, 123
  vehicle tracks 173
Patterson, G.S. 60
Peace, A.J. 67
Pearce-Higgins, J. xii
  on moorland birds and grazing **231-44**
peat/peatland 60
  decline 9, 173
  working 162, 163
Pembrokeshire Coastal path 125-6
Peninsula landscape type 149, 150, 151
Penn, D.J. 194, 195
Pennanen, J. 200
Pennines 242
Pentland Hills Regional Park 124
PEPFAA Code *see* Prevention of Environmental Pollution from Agricultural Activities
Perdix perdix 26, 44, 46, 233-7
Perth 223, 262
Perthshire *see* Breadalbane
pesticides and herbicides 214, 278
  biodiversity and farming 43, 45, 50
  public and countryside 26-7, 28, 29
Peterken, G.F. 61, 64, 68, 195
Petersen, D.L. 194
Petit, S. 45
Phasianidae 47-8, 235
  Perdix 26, 44, 46, 233-7
  Phasianus colchicus 235
pheasant 235
phosphorus pollution 211
Picea
  P. abies 67, 199
  P. sitchensis 60, 63, 66, 67, 201
Picus viridus 258
pigeon
  feral 235
  wood 47, 235
Pilgrim, A. 214
Pillar 1 and Pillar 2 concepts 20, 31
pinewoods 12-13, 61, 62, 63, 64, 65, 66, 67
  old 68
Pinus sylvestris (Scots pine) 12-13, 64, 66, 67
planned village, maps of 159, 160, 165
planning
  conservation, gap analysis used in 193, 196-7, 198

Landscape Character Assessment 140
Planning Advice Notes 140
policy and Landscape Character
   Assessment 151-2
tactical habitat, suitability models
   used in 196-7
see also advisory and planning tools
plant hoppers 45
plantations see coniferous
   plantations; forestry
plants see flora; vegetation
plateau, moorland 149
plover, golden 60, 239, 241, 242
Pluvialis apricaria 60, 239, 241, 242
poaching 293
Poff, N.L. 200
Poland 19, 270-1, 278
policies
   agri-environment, forestry and
      access schemes 109-10, 114-15
   biodiversity 41-2, 51-2
   farming 40, 51-2, 253
   food 253
   land use change 7-9
   public and countryside 30-2
Pollock, M.L. xii
   on hill farming and environmental
      land management **249-53**
pollution 10, 14, 218, 302
   noise and light 173
   reduction see 4 Point Plan
   see also chemicals; Nitrate
      Vulnerable Zones *and under* water
ponds 42, 256, 257, 258
Pontbren Farmers' Group 191, **255-9**
   maps 255, 256, 257
   work of 256, 258
population and land use change 6-7
Porter, K. 183
Portugal 271
Potts, G.R. 46, 233
practical advisory tools 89-92
prehistoric land use 161, 163, 164
Preston, C.D. 41
Preston, S. 183
Pretty, J.N. 9
Prevention of Environmental Pollution
   from Agricultural Activities, Code
   of Practice for 218
Principal Components Analysis 42
Pringle, D. 8
printed advisory tools 89-90
priority areas for future research
   into biodiversity and forestry
   69-72
priority species 235
   woodland habitats 60, 61-6
   see also Scottish Borders Local
      Biodiversity Action Plan
Pritchard, L., Jr 195
privacy and access to land 104-5, 120
Proctor, R. 86
Project Steering Group, public access
   improvement 118, 119
protected forests 196

Pseudotsuga menziesii 63
Pteridium aquilinum 241
public access improvement, help for 99,
   117-28, 182
   appraisal form 123
   background to research 117-18
   good practice, issue-specific
      examples of 124-6
   land managers, issues of concern to
      119-21
   methodology 118-19
   Project Steering Group 118, 119
   recreation/tourism 105-6, 120
   solutions 122-6
   see also access; recreation/tourism
public awareness 113-14
public and countryside 3, **25-33**
   environment 28
   major issues 29
   public liability 103, 104
   see also public access; tourism
public inquiries and Landscape
   Character Assessment 140
Public Sector Agreement 51
Public Service Agreement 42
Pyrrhocorax pyrrhocorax 49, 182

Quality of Life 171
   Capital 168
   indicators 42
Quality Meat Scotland 26
Quine, C. xiii
   on biodiversity and forestry 37,
      59-75

Rabeni, C.F. 197
radio masts 150
Rae, R. 47
railways 163
rainfall, acid 29
Ramblers' Association 123
Rametsteiner, E. 194
RAMS (Risk Assessment for Manure and
   Slurry) 219
Ranke, W. 201
raptors 49, 60, 64, 182, 235, 258
Rare Breeds Survival Trust 10
Ratcliffe, D.A. 60
Raven, M.J. 41
Ray D. 64
RCAHMS see Royal Commission on Ancient
   and Historical Monuments of
   Scotland
RDR see Rural Development Regulation
REACH (Registration, Evaluation and
   Assessment for Chemicals) 279
Read, H.J. 63
recreation/tourism 27-8, 176, 182, 200,
   286
   Dumfries and Galloway 150, 153
   hill farming and environmental
      objectives 250
   maps of 159, 165
   public access 105-6, 120, 121
   see also access; public access;

7stanes Project; sports; tourism
recycling 278
red deer see *Cervus elaphus*; deer
red grouse 239, 241, 242
red squirrel 65, 69, 70
redshank 45, 265
reed bunting 47
Reforesting Scotland 90
regeneration, forests and woodlands
   250, 252
Regional Advisory Committee 153
Regional Parks 124
   Balbirnie estate 291, 292, 293, 294,
      295
Regional Scenic Area 154
regionalisation see Land Management
   Contracts
relict land use, maps of 162, 167
relief and land use 162-8
renewable energy 280
renewal 195
reptiles 42, 64
research and advice priorities 274,
   297-304
   access 301-2
   biodiversity 299-300
   farming 298
   forestry 298-9
   integration 302-3
   landscape 300-1
   natural heritage 299
   need for 298
research and development needs for
   habitat planning 199-204
   benchmarks and thresholds 199-200
   holistic perspective encouraged 200-2
   multiple case studies, towards 202-4
research, ecological 86
resilience, ecological 195
responsibility for landscape, shared 176
RHET see Royal Highland Education Trust
Ribbens, J. 200, 201
Riley, J. 71
Rio de Janeiro conference (1992) 61
risk assessment
   for Manure and Slurry 219
   nitrates 214
   public access 103-4
risk management 125-6
rivers and valleys 202, 213
   buffer strips along 214
   coordinated management 91
   Dumfries and Galloway types 149
   fluvial dynamics 193
   quality and afforestation 200, 201
   River Basin Plans 82, 288
   see also Eden River; water; Ythan
RLUWG (Rural Land Use Working Group)
   79-80
RNCI (Royal Northern Countryside
   Initiative) 222-3
Roberge, J.-M. 197
Roberts, A.J. 60
Roberts, J. 88
Robertson, N. 6, 7-8

315

Robins, M. 48
Robinson, R.A. 40, 41, 42, 44, 45, 46
roe deer 293
Rollinson, T. 64
rook 235
rotation, crop 10, 43, 46, 51
Rothamsted Insect Survey 41-2
Rothiemurchus 31
rough grazing see grazing
Roxburghshire 227
Royal Commission on Ancient and Historical Monuments of Scotland 158, 166
  maps 159-65
Royal Highland Education Trust 191, **221-4**
  biodiversity 223
  Countryside Classroom on Wheels 222-3
  Countryside Initiative 222-3
  resources 222
  Royal Highland Show 221, 223
  training and placements 221-2
  working with others 223
Royal Highland Show 221, 223
Royal Northern Countryside Initiative 222-3
Royal Society for Protection of Birds see RSPB
RSPB (Royal Society for Protection of Birds) 13, 239-243
  advisory and planning tools 86
  future landscape research 182
  Land Management Contracts 265
  moorland birds and grazing 240, 243
  public and countryside 31
  Scottish Borders Local Biodiversity Action Plan 233-4, 235, 236, 237
RSS see Rural Stewardship Scheme
rubbish dumping 120
Rural Care Scheme (Scottish Power) 256
Rural Development Plans (EU) 26, 32, 266, 271, 272, 287, 303
Rural Development Regulation 52, 262, 264, 270-2, 303
Rural Land Use Implementation Plan 37, **77-83**
  developing 79-82
  draft Biodiversity Strategy 77-9
  future 82
Rural Land Use Working Group 78, 79-80, 82, 89
Rural Partnerships 289
Rural Stewardship Scheme
  agri-environment, forestry and access schemes 110
  biodiversity and farming 51, 52
  Breadalbane Initiative 246, 247
  4 Point Plan 218
  future lowland landscapes 175
  Landscape Character Assessment 142, 143
  Rural Land Use Implementation Plan 80, 81, 82
  Scottish Borders Local Biodiversity Action Plan 236

Russia 200, 202
Rydén, L. 195
ryegrass 45

SAC see Scottish Agricultural College
safety
  public access improvement 120, 126
  school trips 221, 222, 224
Salaca, River (Latvia) 201
Salmo
  S. salar 201
  S. trutta 201, 218
salmon 201, 218
Sankey, S. xiii, 78
  public and countryside 3, **25-33**, 298
SAPs (Species Action Plans) 61
Savills, FPD 174
sawfly larvae 41
Saxicola
  S. rubetra 241, 242
  S. torquata 239, 241, 242
SBC (Scottish Borders Council) 234, 235, 236, 237
SBF see Scottish Biodiversity Forum
Scandinavia 200, 201, 271
  see also Sweden
scenery 151, 152, 154
  see also National Scenic Areas
Schabetsberger, R. 202
Schneider, R.L. 204
schools see education
Schusler, T.M. 249
Sciurus vulgaris 65, 69, 70
Scotland's Future Landscape (SNH) 172, 289
Scots pine 12-13, 64, 66, 67
Scott, J.M. 196, 197
Scott, M. 78
Scott, P., xiii
  on public access improvement, help for 99, **117-28**
Scott, Peter 118
Scottish Agricultural College 13, 234, 243, 245
  advisory and planning tools 87, 90
  Hill Sheep and Native Woodland project 91
  see also 4 Point Plan
Scottish Biodiversity Forum 77-8, 79, 82, 88
  see also Rural Land Use Working Group
Scottish Biodiversity Strategy v 31, 77-9, 93, 168, 275, 299
  see also Rural Land Use Implementation Plan
Scottish Borders Local Biodiversity Action Plan for birds 191, **233-7**
  breeding birds surveys 235
  community involvement 236
  future development 236
  links to other schemes 236
  methods 233-4
  monitoring 234-6
  winter passerine monitoring 235-6
Scottish Burgh Survey 158

Scottish Crofting Foundation 81
Scottish crossbill 65
Scottish Environment LINK 81, 267
Scottish Environment Protection Agency 92, 213, 217
Scottish Executive Environment and Rural Affairs Department see SEERAD
Scottish Forestry Forum 22
Scottish Forestry Grants Scheme 287
  agri-environment, forestry and access schemes 110
  biodiversity and forestry 68, 71
  Breadalbane Initiative 246, 248
  forestry, key changes in 19
  future lowland landscapes 175
  Land Management Contracts 264
Scottish Forestry Strategy v 13, 22, 31, 68, 92, 110, 141, 153, 175, 275, 299
Scottish Landowners' Federation (now SRPBA) 102
Scottish Mountaineering Club 102
Scottish Native Woods 90, 91
Scottish Natural Heritage v 5-6, 31
  advisory and planning tools 86, 87, 91, 92
  agri-environment, forestry and access schemes 109, 114
  biodiversity and forestry 68
  Concordat on Access 102
  future landscape research 182, 183, 187
  future lowland landscapes 171-2, 173-7
  Landscape Character Assessment 11, 134, 135-6, 137, 138, 141, 143, 144, 148
  multi-benefit land use 285, 287, 288, 289
  Natural Heritage Futures 139
  public access improvement 117, 118, 122, 123, 124
  research and advice priorities 300, 301
  Royal Highland Education Trust 222
  Rural Land Use Implementation Plan 81
  Scottish Borders LBAP 234
  Scottish Native Woods 90, 91
  7stanes Project 232
  Sharing Good Practice 91
  supporting landscape design 147, 148, 151-5
  see also Landscape Character Assessment; Scottish Outdoor Access Code; *Sustainable Development*; Targeted Inputs
Scottish Nature Woods 245
Scottish Office Development Department 171
Scottish Ornithologists' Club 234
Scottish Outdoor Access Code, draft v 99, **101-8**, 110, 301
  approved (2005) 108
  background 101-3
  concerns 103-6
  future 105-6
  land managers and 103, 106-8

Index

public access improvement 117, 118, 119, 124, 128
Scottish Power, Rural Care Scheme 256
Scottish Rural Property and Business Association 223
  public access improvement, help for 119, 123
Scottish Borders Local Biodiversity Action Plan 234, 237
Scottish Outdoor Access Code, draft 102-3, 105, 106, 107
Scottish Wildlife Trust 31, 32
sedge warbler 47
SEERAD (Scottish Executive Environment and Rural Affairs Department) 236
  advisory and planning tools 81, 87, 90, 92-3
  Breadalbane Initiative 245, 246, 247
  Land Management Contracts 262
  moorland birds and grazing 240, 243
  see also Rural Stewardship Scheme
Self, M.J. 41, 42
Sellars, H. 71
service industries 286
  see also recreation/tourism
set-aside 52
setting scene see key changes in forestry; natural heritage, contribution to; public and countryside
7stanes Project case study 191, **225-32**
  conflict management 228-30
  costs and potential income sources 225, 230
  forest access policy 226
  impact 230-1
  map of trails 227
  planning 225, 228
sewage disposal 28
SFGS see Scottish Forestry Grants Scheme
SFM (Sustainable Forest Management) 194, 299
SFP (Single Farm Payment) 88, 91, 92, 283
Sharing Good Practice (SNH) 91
Sharrock, J.T.R. 41
Sheaves, J. 213
sheep see livestock
Shepherd, K.B. 240
Shetland 43
shielings 163
Shrubb, M.J. 47, 49
signs and notices, public access and 121, 122, 123, 125, 127, 301
silage replaces hay 48
Silvester, M. xiii
  on Scottish Outdoor Access Code, draft 99, **101-8**, 301
Sim, I.M.W. 239
Sinapsis arvensis 47
Single Farm Payment 88, 91, 92, 283
single-species protection 194
Siriwadena, G.M. 233
Sites of Community Importance 281

Sites of Special Scientific Interest 40, 42, 291
skill gaps, bridging (Breadalbane) 247
skipper butterfly, chequered 65
Skye 119
skylark 44, 45, 47-8, 241, 242
Slater, D. xiii
  on natural heritage, contribution to 3, **5-16**
Slee, W. 9
slurry see manure and slurry
smallholdings on maps 163
Smart, S.M. 42, 45
Smith, G. 194
Smith, K.W. 240
Smout, T.C. 59, 109, 157, 158, 286
Smyth, K. xiii
  on hill farming and environmental land management **249-53**
SNH see Scottish Natural Heritage
snipe 45, 241, 242
social sciences 194
soil 9, 157, 278
  Soil Association 50
  testing 214
Solway Coast 166-7, 168
Sotherton, N.W. 41, 42, 233
sources, of knowledge 86-7
Southern Uplands
  landscape type 148, 149, 150, 151
  Partnership 289
  Way 154
Sowa, S.P. 197
Spain: Land Use Policy Group study 270-1
sparrow
  house 41
  tree 233-7
spatially explicit suitability modelling 198
Special Accession Programme 270-2
special habitats 21
species
  loss 41
  single-species protection 194
  Species Action Plans and forestry 61
Spey Catchment Management Plan 265
spiders 41, 47
sponsorship 230
sports and leisure pursuits 293
  see also cycling; golf; recreation/tourism; walking
spruce
  Norway 67, 199
  Sitka 60, 63, 65, 66, 67, 201
Sprung, G. 229
squirrel
  grey 65
  red 65, 69, *70*
SRPBA see Scottish Rural Property and Business Association
SSSIs see Sites of Special Scientific Interest
stakeholders, multi-benefit land use 285-6

stalking 121
Staphylinidae 41
starling 41, 45
Statistical Accounts 158
Stay Safe 222
Stevenson, J.B. 158
Stirling University 243
Stirlingshire 243
Stiven, R. 89, 248
Stoate, C. 40, 43
stone curlew 51
stonechat 239, 241, 242
Stoner, J.H. 9
Storch, I. 194, 195
Store, R. 197
stork, white 201
Stowe, T.J. 49
strategic access management 122
strategic conservation planning using gap analysis 193, 195-7, 198
Strathmore see Nitrate Vulnerable Zones
Strathspey, forestry 66
Strazds, M. 201
Stroud, D.A. 9
Structural Funds 277, 278-9
Structure Plan, Dumfries and Galloway 147, 151, 152
Sturnus vulgaris 41, 45
subsidies see finance
success, deonstrating 91
Suchant, R. 201
suitability models, tactical habitat planning using 196-7
Summers, R.W. 86, 200
Sunart 91
supporting landscape design 131, **147-56**
  forests and woodlands, landscape design guidance for 147, 152-3
  Landscape Character Assessment 147, 148, 152, 154, 155
  Local Forestry Frameworks 147, 153-4
  map 149
  National Scenic Area Management Strategies 147, 152, 154-5
  statutory planning policy and Landscape Character Assessment 151-2
sustainability 13
  biking trails 227
  forestry 9, 20-2, 227
  holistic understanding of ecological systems 193, 194-5
  Sustainable Development and Natural Heritage (SNH) 5
  Sustainable Forest Management (SFM) 194, 299
Sutherland 9
Sutherland, W.J. 40, 41, 42, 45, 46, 51
Svensson, J. 199, 204
Swanson, F.J. 204
Swanwick, C. xiii, 175
  on landscape character assessment 131, **133-45**, 155, 158, 300
Sweden 271
  capercaillie 201

317

case studies 199, 200
holistic understanding of ecological systems 196-7, 201, 204
Land Use Policy Group study 270-1
swimming 218
Symon, J.A. 7
Symphyta 41, 44
Syrphidae 67

tactical habitat planning 195, 196-7
Tapper, S.C. 46, 234
Targeted Inputs for a Better Rural Environment 89-90, 300
Taylor, K. xiii, 43
 on public access improvement 99, **117-28**, 301-2
Taylor, S. 86
Tayside 167
Tay River Purification Board 213
teacher training 223
technological innovation 172
Tetrao
 T. tetrix 49
 T. urogallus 31, 65-6, 68, 86, 201
Thansborg, S.M. 6
Tharme, A. xiii
 on Scottish Borders Local Biodiversity Action Plan for birds 191, **233-7**, 302
theft 120
theme parks 28
Thiesmeier, R. 202
Thirgood, S.J. 240
Thomas, D.R. 213
Thompson, D.B.A. 60, 239
Thompson, G.K. 253
Thompson, R.N. 60, 63, 64
Thomson, J. xiii
 on multi-benefit land use 274, **288-90**, 303
thresholds and benchmarks 199-200
TIBRE (Targeted Inputs for a Better Rural Environment) 89-90, 300
TIDE (Tidal Inlets Dynamics and Environments) 214
tillage, non-inversion 47
Tilzey, M. 40
timber and wood processing 6
 Breadalbane Initiative 247
 future 17, 21, 173
 habitat quality affected 60
 markets 21
 Pontbren 256
 strategic reserve 8-9
 Sweden 199, 200
 see also forestry
Tipula 49, 64
Tir Gofal Scheme (Wales) 142
Törnblom, J. xiii
 on holistic understanding of ecological systems **193-209**
Torre, M de la xiii
 on Land Use Policy Group and EU policies 192, **269-72**, 303
tourism *see* public access;

recreation/tourism
towns
 townscape analyses and Landscape Character Assessment 140
 see also urban/built-up areas
training
 inadequate 180
 practical advisory tools 90-1
trampling 120
transport system 42, 173
 maps 162, 163
 planning and appraisal 140
tree sparrow 233-7
Trefaldwyn *see* Pontbren
Tringa
 T. nebularia 60
 T. totanus 45, 265
Triturus cristatus 202
Trossachs
 forestry 22
 historic environment 167, 168
 Landscape Character Assessment 137, 141
 multi-benefit land use 288
 public access improvement 119
trout, brown 201, 218
Tucker, G. 49
turf stripping 163
Turnbull Jeffrey Partnership 140, 141
Tweed Valley
 biking trail 227
 Forest Park (Glentress) 28
twinflower 65
Tyldesley, D. 137-8, 140, 165, 167
Tyto alba 49, 258

UK Biodiversity Action Plan priority species 234, 235
Uliczka, H. 199
United Nations Conference on Environment and Development (Rio de Janeiro, 1992) 61
universities 223, 243, 258
 Aberdeen 13, 174, 243
upland glens and fringe 149, 151
urban/built-up areas
 future lowland landscapes 172, 173, 174
 maps 158, 160
 urban fringe 140, 182, 301
 Urban Working Group 78, 82
 see also transport system
Ursus arctos 59
Usher, M.B. 158, 200, 201

Vaccinium myrtillus 66
valleys *see* rivers and valleys
vandalism 120
Vanellus vanellus 44, 45, 47, 265
Västerbotten County (Sweden) 194
Vaughan, N. 44
vegetables 291
vegetation 302-3
 bird abundance *see* moorland birds and grazing

 see also flora; forestry; forests and woodlands; grasslands; hedgerows; moorland
Vickery, J.A. 43, 45, 47, 48, 49, 50
Vidzeme Biosphere Reserve (Latvia) 201
visitors *see* recreation/tourism
Visitor's Guide 124-5
VisitScotland 27-8
VISULANDS 71
Vos, C.C. 69

waders 212, 240, 265
 and biodiversity 45, 48, 51, 60
 see also lapwing
wagtails 45, 48
Wales 49, 226, 302
 Countryside Council for 240, 243, 258
 Forestry Strategy 9
 Tir Gofal Scheme 142
 University of 258
 Welsh Assembly 258
 Welsh Council for Voluntary Action (WCVA) 256, 259
 Welsh Development Agency 258
 see also moorland birds and grazing; Pontbren
walking 22, 27, 28, 29, 92, 106, 229
walls *see under* boundaries
warbler, sedge 47
Warren, C.R. xiii
 on Nitrate Vulnerable Zones 191, **211-15**, 300, 302
wasps, parasitoid 41
waste disposal problem 28, 120, 162, 278
water
 conservation 258
 freshwater biodiversity 78
 margins 219
 pollution 9, 29, 218 (*see also* 4 Point Plan)
 quality and swimming 218
 reservoirs 292
 Water Environment and Water Services (Scotland) Act 2003 217
 Water Framework catchment management 81
 Water Framework Directive (EU) 217, 280, 288
 watersheds 193
 see also coastal areas; rivers; wetlands
Water of Ken 154
Waterhouse, A. xiii
 on hill farming and environmental land management **249-53**
Watkins, C. 195, 197
Watson, A. 47
Watson, R.T. 18
Watts, K. xiii
 on biodiversity and forestry 37, **59-75**
WBC (Wild Bird Cover) 233, 234, 235, 236
Welch, D. 240
Welsh *see under* Wales

Index

Welshpool 256
Wester Ross 154, 166, 168
Western Isles 22, 41, 119
  Hebrides 40, 43
  Islay 49
wet 26, 60
wetlands and mires 26, 60, 242, 256
wheat 46, 47
  Wheat Act 1932 8
wheatear 241
Wheeler-Bennett, G. 180, 186
whinchat 241, 242
Whittington, G. 157
Whole Farm Plans 236, 266
Wickramasinghe, L.P. 51
Wiens, J.A. 194, 204
Wild Bird Cover 233, 234, 235, 236
wildfowl 48, 212
wildlife 27, 154
  protection 29, 61
  see also birds; deer; fish;
    invertebrates; mammals
Wilkins, R.J. 10
Williams, G. 86
Wilson, A. 77, 78
Wilson, E.O. 194
Wilson, J.D. 10, 51
Wilson, P.J. 44, 45, 46, 47
wind farm development and potential
  147, 150, 151
winter passerine monitoring 235-6
wolf 59
wood
  woodchips, new uses for 247, 256-7,
    258
  see also forestry; timber
wood pigeon 47, 235
Wood-Gee, V. xiii
  on public access improvement, help
    for 99, **117-28**
woodland *see* forests and woodlands
Woodland Trust 31, 125
woodlark 64
woodpecker, middle-spotted 201
Worcester, R.M. 27
Work Programme (EU) 279
Working Countryside Box 222, 224
Working Group, Land Management
  Contracts 261-2
World Heritage Sites 167
World Summit on Sustainable
  Development 18
World Trade Organisation 40, 282
World Wildlife Fund Europe 269
Worldwide Fund for Nature Scotland 245
worming agents 50
Worrell, R. 61-2
Wright, I. xiii
  on natural heritage, contribution to
    3, **5-16**
WTO (World Trade Organisation) 40, 282

Yalden, D.W. 60
yellowhammer 46
Ythan river and catchment 6, 11-12

Zimmermann, N.E. 197
Zuckerman, S. 9

Farming, Forestry and the Natural Heritage: Towards a More Integrated Future